Please note that some diagrams referred to in the text have been moved to the end of the book.

Courtesy Western Electric Co.

WESTERN ELECTRIC ATR RADIO APPARATUS FOR THE DOUGLAS DC4, FOUR ENGINE TRANSPORT

The 250 watt, ten frequency transmitter is mechanically connected to the ten frequency receiver below it. Beside the receiver and to the left is the auxiliary radio receiver. Below the auxiliary is the range receiver. Control and switch panels are shown. The interphone amplifier is on top of the rack. Apparatus operates on 110 volts, 800 cycles.

AIRCRAFT RADIO
AND
ELECTRICAL EQUIPMENT

by

HOWARD K. MORGAN, B.S. IN E.E.
SUPERINTENDENT OF COMMUNICATIONS
TRANSCONTINENTAL & WESTERN AIR, INC.

Originally published in 1939

ISBN: 978-1-940001-43-2

The Aviation Collection
by
Sportsman's Vintage Press
2015

PREFACE

There are many people connected with aviation who would like to know more about electricity and radio. This book has been written especially for them and deals with electricity from the very beginning with only one prerequisite: that they are mechanically minded. Most of the classic explanations of electricity have been abandoned and instead, hydraulic analogies have been used to a greater extent than ever before in explanation of electrical and radio phenomena. Every bit of the material is of practical importance in everyday use.

Certain chapters will appeal to one group while other chapters will appeal to another group.

It is suggested that chapters be read in the order in which they appear with the possible omission of the chapters on Commercial Aircraft Transmitters and Receivers. These chapters may be studied in greater detail after reading through the book.

The air transport of 1938 was equipped with one transmitter and three receivers, while the air transport of 1945 will be equipped with at least two transmitters and with six to ten receivers. The number of men maintaining and using aircraft radio equipment will increase tremendously with the equipment complement increase.

While radio equipment is increasing there will be still greater strides in the use of electrical power aboard aircraft. Transports today have generators capable of supplying 600 watts while new airplanes, such as the Douglas DC-4, have generation capability of 15,000 watts.

This evolution will have its effect on the itinerant flyer to say nothing of the maintenance mechanic, the flight mechanic, the airline inspector and the transport pilot. It is for these men that this book is prepared.

Questions appear at the end of each chapter which are of material assistance to those desiring to master the fundamentals of radio and electricity by studying this book alone. Answers appear in an appendix at the end of the book.

Acknowledgment and thanks are extended to manufacturers and airlines who have supplied much material. They are: American Airlines, Bendix Radio, Collins, Lear Radio, Radio Corporation of America, Transcontinental & Western Air, United Air Lines and Western Electric.

The following gentlemen of Transcontinental & Western Air, Inc. have contributed many suggestions and have been of invaluable assistance: Mr. R. C. Ayres, Aircraft Radio Engineer; Mr. Keith Bell, Navigation Department; Mr. Joseph D. Mountain, Link Trainer Instructor; Mr. Peter H. Redpath, Navigation Engineer. Mr. Ayres has given special assistance in that he has been over the entire text and has made numerous and very valuable corrections.

In addition, I wish to thank my mother and my brother-in-law, Douglas M. Whitaker, for their help in going over the original manuscript.

HOWARD K. MORGAN

CONTENTS

CHAPTER I

HYDRAULIC ANALOGIES

Direct electric current is supplied on present transport aircraft for the operation of lights, motors, instruments and radio equipment. The hydraulic analogy will be used to illustrate all of the electrical terms to follow.

Voltage (E)—Voltage is analogous to the pressure exerted at the bottom of a tank by water. As the water is in a state of rest there will be no flow of energy.

The pressure is measured in pounds per square inch so that, regardless of the diameter of the tank, the pressure will always be the same at a given depth (Fig. 1).

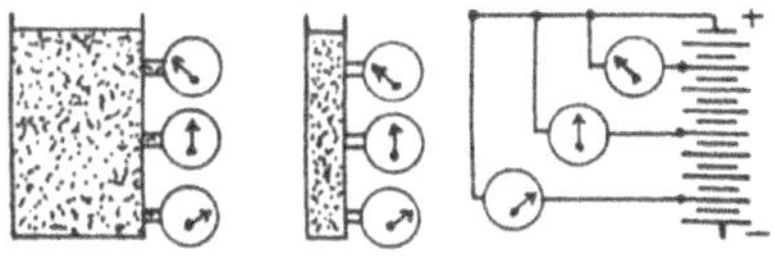

FIG. 1—PRESSURE AND VOLTAGE

In electrical terms this pressure is expressed in volts, the symbol for which is "E." The number of cells in a battery determines its voltage. Dry cells have an average voltage of 1.5 volts each. If there are two cells in series the total voltage will be 3 volts. Wet batteries, such as the storage batteries on aircraft, have an average cell voltage of 2 volts. Aircraft usually have storage batteries of 12 volts, six cells connected in series. More recently there has been a trend from 12 to 24 volt batteries for reasons which will be explained later.

Radio equipment on most aircraft requires not only 12 volts for operation of relays and filaments but also a high voltage supply of about 200 volts for receivers and about 1000 volts for transmitters. These voltages are obtained from dynamotors and vibrator power units.

Current (I)—Current is analogous to the rate of flow of water. The ampere flow of electricity in a wire is similar to the gallon per minute flow of water through a pipe. The symbol for current is "I" and is usually expressed in amperes (Fig. 2).

There can be no flow of water through a pipe unless there is pressure to cause it. In the same manner there can be no flow of electricity through a wire unless there is voltage to cause this flow.

FIG. 2—FLOW AND CURRENT

Starters on large airplanes require currents of 200 to 500 amperes, depending upon the temperature and viscosity of the oil in the motor. Many landing lights require thirty-five amperes at twelve volts each. Small lights require from one-quarter ampere to three amperes, depending upon their size.

Power (W)—Power, the product of voltage and current, is ex-

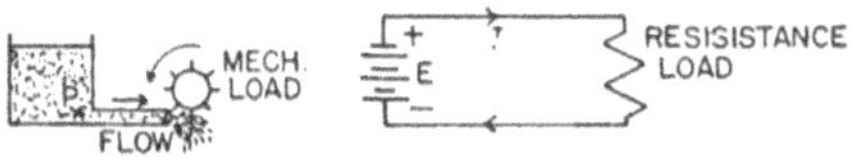

FIG. 3—LOAD

pressed in watts. The symbol for power, or wattage, is "W" (Fig. 3). The formula for power is:

$$E \times I = W$$

Thus, a current of ten amperes at ten volts gives 100 watts. There are 746 watts to one horsepower. One horsepower is equal to the work done in lifting 33,000 pounds vertically one foot in one minute, or one horsepower is equal to 33,000 foot pounds per minute.

It is very important to realize that power is merely a rate of doing work and does not give the amount of work done unless the time is given. Thus, one horsepower will accomplish as much in ten minutes as ten horsepower will in one minute. Electrically the amount of work is expressed in watt hours, or the number of watts operating for one hour.

Ampere Hours (AH)—When considering storage batteries it is often convenient to use "ampere hours" in calculating the charge and discharge of a battery (Fig. 4). The voltage of most battery systems is nearly constant and in this case the ampere hour product is proportional to the volt ampere hour (watt hour) product. When a battery is being charged, a certain number of ampere hours will be added to the battery. When discharge is taking place a certain number of ampere hours will leave the battery. With both taking

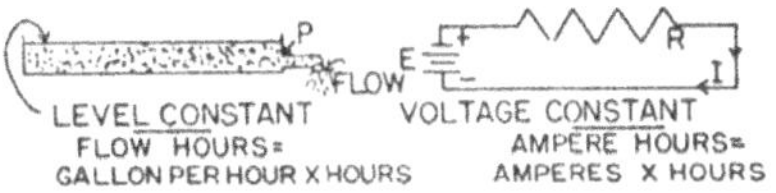

FIG. 4—FLOW HOURS AND AMPERE HOURS

place the difference will indicate whether the battery is being charged or discharged. If the work in watt hours done, rather than the battery condition, needs to be determined the ampere hours must be multiplied by the voltage.

The usefulness of ampere hour calculations will become evident when it is realized that a part of an electrical system, such as a generator, may fail and that it will be necessary to reduce the load on the battery to a minimum. Large current loads which operate over a short period are usually not as important as smaller current drains which may be continuous in flight.

Example—Use of Ampere Hours—An airplane was equipped with an eighty-five ampere hour battery. This battery supplies 8.5 amperes for ten hours or seventeen amperes for five hours and so forth. At extremely high currents, which are well over ten per cent of the ampere hour capacity (as expressed in amperes), the battery would deliver less total energy due to excessive heating in the battery. This same airplane was also equipped with a generator capable of supplying thirty amperes continuously. It must be kept in mind that if the load were thirty amperes the generator would supply the entire load but would not be charging the battery. If the load were only fifteen amperes there would be available another fifteen amperes for charging the battery. Control boxes, which will be treated later, prevent the generator from overcharging the battery. The loads which might be encountered on a flight of five hours' duration and the time of use of each load, are given in the table below (see Fig. 5).

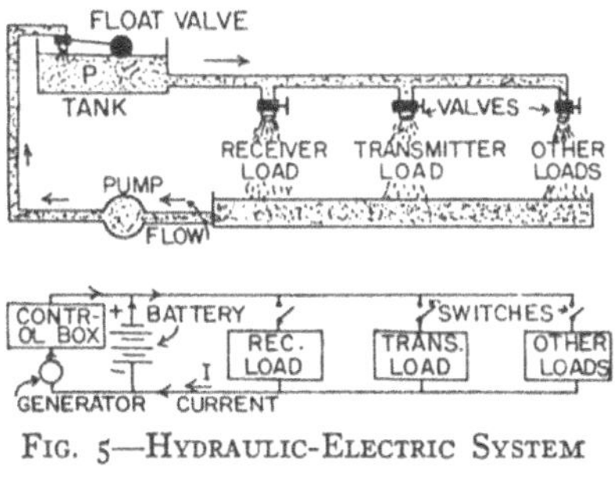

FIG. 5—HYDRAULIC-ELECTRIC SYSTEM

TYPE EQUIPMENT	CURRENT	TIME USED
Radio receivers........	10 amperes	5 hours
Radio transmitter......	80 amperes	30 minutes
Landing lights.........	70 amperes	30 minutes
Cabin lights (average)...	10 amperes	5 hours
Starters...............	200 amperes	3 minutes
Miscellaneous..........	10 amperes	2 hours

The question is to determine the charge of the battery after landing at the airport to which the plane was destined.

The first step is to calculate the number of ampere hours consumed by each load, remembering to convert minutes into hours before multiplication.

TYPE EQUIPMENT	AMPERE HOURS
Radio receivers..........................	50
Radio transmitter........................	40
Landing lights...........................	35
Cabin lights.............................	50
Starters.................................	10
Miscellaneous............................	20
Total ampere hours (AH) consumed in flight..	205

Some of the loads with very high currents were not serious in battery drains as their time of use was short. Actually, the starters will probably cause a larger drain as the battery cannot supply very heavy currents efficiently. However, the total ampere hours will not be substantially higher than shown.

The generator was supplying thirty amperes during the entire flight so that it could return 150 ampere hours to the battery. From the above information the following calculation can now be made:

Total AH consumed in flight...................	205
Total AH supplied by the generator............	150
Remaining AH supplied by battery.............	55

As the battery was fully charged to eighty-five ampere hours before the trip the battery at the end of the trip will have $85-55=30$ ampere hours still available.

The length of the trip and the total number of landings and take-offs determine the generator capacities required. In the example above, such a plane would actually be equipped with a fifty ampere generator and the plane would land after a five hour trip with the battery fully charged. The main function of any battery should be to supply the high current, short time peak loads. The plane battery is often detached from the electrical system while at a station and a battery cart plugged in to the plane to supply loads while the plane's generator is not running, as well as for engine starting. The generators on large planes are driven by one or more of the main engines. Wind driven generators are sometimes used on small planes.

Resistance (R)—Voltage, current and power have been defined. The next important unit is resistance. The hydraulic analogy of resistance is mechanical friction, as found on the inner walls of a pipe. A pipe attached to the bottom of a tank of water will empty the tank in a time largely determined by the size and length of the pipe (Fig. 6). Thus a quarter-inch pipe fifty feet long attached to a tank might empty it in one hour. If the length of the pipe were doubled (twice the resistance) it would take about twice as long to empty the tank. A pipe one inch in diameter and fifty feet long would empty it in a few minutes, due to its low resistance, because of its larger diameter.

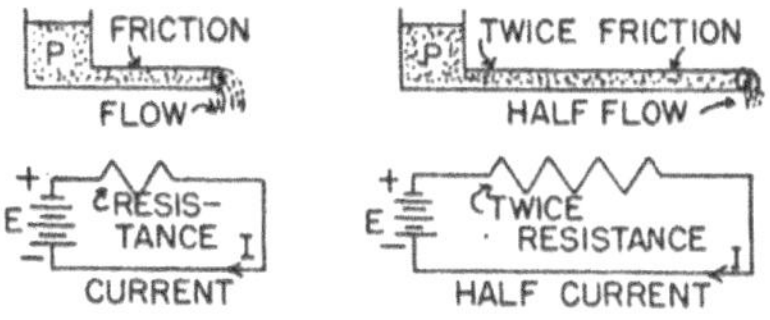

Fig. 6—Friction and Resistance

The unit of electrical resistance is the ohm and is usually designated as "R." An ohm is that amount of electrical resistance which, when placed across a one volt source, will allow one ampere to flow. The second important formula to remember is:

$$E = I \times R$$

Thus, the current times the resistance will give the voltage existing across a resistor. When current flows through a resistor power is dissipated in heat.

If the starter wire on a plane had a 0.01 ohm resistance the voltage, when 200 amperes flowed, would drop from twelve volts at the battery to ten volts at the starter. This is found by multiplying 200 amperes by 0.01 ohms, giving a drop of two volts. Therefore the starter will have but ten volts to operate it.

The starter in the case given required 200 amperes at twelve volts or 2400 watts. If the battery system had been a twenty-four volt system the current required would have been but 100 amperes for the same power. Thus the wire size (cross section area) could have been halved, which would almost halve its weight, and the voltage drop would have been the same, two volts. The new resistance would be twice the old resistance as resistance is inversely proportional to the wire cross section area. The new voltage drop would be 0.02 × 100 = 2 volts. The drop in voltage would be two in twenty-four or half of the previous percentage drop. Thus the wire size could be halved once more so that the new starter wire would weigh about one quarter of the old weight and the percentage drop would be the same as before, four volts in twenty-four.

ALLOWABLE CARRYING CAPACITIES OF AMERICAN WIRE GAUGE (AWG) COPPER WIRE

No. AWG	Dia. Mils*	Area Circ. Mils†	Lbs. per 1000′	Ohms per 1000′	Amperes Rubber Insul.	Amperes Other Insul.
0000	460.	211,600	641.	0.0499	225	325
000	410.	167,800	508.	0.0630	175	275
00	364.8	133,100	403.	0.0794	150	225
0	324.9	105,500	319.5	0.1003	125	200
1	289.3	83,700	253.3	0.1262	100	150
2	257.6	66,400	200.9	0.1591	90	125
3	229.4	52,600	159.3	0.2008	80	105
4	204.3	41,700	126.4	0.2533	70	90
5	181.9	33,100	100.2	0.3193	60	80
6	162.0	26,250	79.5	0.403	50	70
8	128.5	16,510	50.0	0.640	35	50
10	101.9	10,380	31.43	1.017	25	30
12	80.8	6,530	19.77	1.618	20	25
14	64.1	4,110	12.43	2.575	15	20
16	50.8	2,583	7.82	4.09	6	10
18	40.3	1,624	4.92	6.51	3	5
20	31.96	1,022	3.092	10.34		
22	25.35	642	1.945	16.44		
24	20.10	404	1.223	26.15		
26	15.94	254.1	0.769	41.6		
28	12.64	159.8	0.484	66.01		

* A mil is one thousandth of an inch.
† A circular mil is the area of a circle whose diameter is one mil.

Starter wire weights on present large planes with twelve volt systems may total seventy-five pounds or more. Reducing the size of small wires which carry low currents is not a large saving as the insulation forms a substantial portion of the actual wire weight. This is true because the resistance of a wire varies inversely as the square of the diameter and the insulation weight varies approximately as the first power of the diameter. Another important factor is wire heating. Heat production varies directly with the resistance and as the square of the current. If the wire is operated at too high a temperature it will burn its insulation. Energy used in heating the wires is wasted so that wire sizes are determined on the basis of electrical efficiency as well as weight.

Resistors—One of the most useful elements in electrical and radio equipment is the resistor. Resistors are commonly made by winding wire on a fireproof form and coating it with fireproof, vitreous insulating material. Small resistors are also made of carbon in the form of short rods. In either case the resistance of the material is purposely high.

The resistance of most conductors changes somewhat with temperature variations. When resistance must be constant, as in gas analyzers and temperature indicating bridge systems, the resistor is made of special material, such as nichrome wire, which does not change resistance with temperature.

Series Connection—Two resistors may be connected, either in series or in parallel, as shown in Fig. 7.

If the resistors are connected in series, one end of one resistor is joined to one end of the other and connections to the circuit are made at the two remaining ends. The resistance of such a series combination is the sum of the resistance of the units. If a ten ohm resistor is connected in series with a five ohm resistor the total resistance will be fifteen ohms. In like manner, if two pipes are connected end to end, the resistance to water flow would be the sum of the resistances of the two pipes.

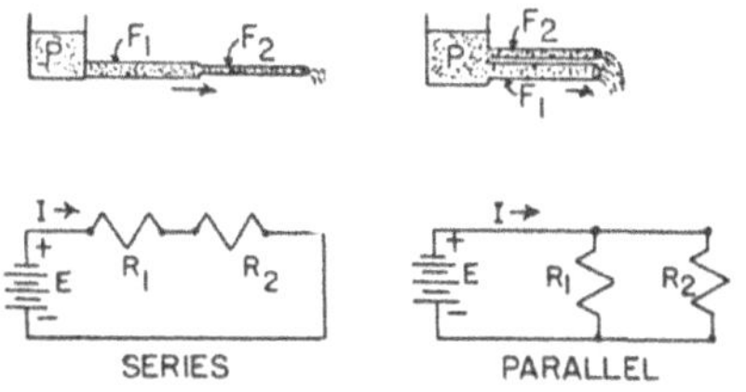

FIG. 7—SERIES AND PARALLEL CONNECTIONS

Current is common to the two resistors in series, as it flows first through one and then through the other. If fifteen volts were placed across the fifteen ohm resistor combination a current of one ampere would flow. The voltage drop across each resistor would be found by multiplying the current times the resistance of each. There would be a drop of five volts across the five ohm resistor and a drop of ten volts across the ten ohm resistor.

Parallel Connection—Two resistors are connected in parallel by joining the top ends together and the bottom ends together

(Fig. 7). This is equivalent, in the hydraulic analogy, to laying two pipes side by side and connecting the intake ends to the water pressure source. The two pipes together will carry more water, i.e., show less resistance, than either pipe alone.

The total electrical resistance of two resistors connected in parallel is found by dividing the product of the resistances by their sum. Thus, a five and a ten ohm resistor in parallel will have an equivalent resistance of fifty divided by fifteen, or three and one-third ohms.

It is simple to find the current through each of the resistors as each of the resistors is effectively across the voltage source. The two resistors, if attached to a fifteen volt battery, would draw currents of three amperes for the five ohm resistor and one and one-half amperes for the ten ohm resistor. Note that the sum of these currents is four and one-half amperes. If the current were computed by using the equivalent resistance of three and one-third ohms the current would be four and one-half amperes as before.

The voltage is common in the parallel combination and the current is common in the series combination.

It is relatively easy to compute the equivalent resistance of three resistors which are connected in parallel. The simplest method is to find the equivalent parallel resistance of two together and then find the result of this equivalent resistance in parallel with the remaining resistor. Suppose that a two, a four and a two ohm resistor are placed in parallel. The two and two ohm resistors together give an equivalent resistance of one ohm. This in parallel with the four ohm resistor, gives a total equivalent of four fifths of an ohm. The formula for any number of parallel resistors is:

$$\frac{1}{R} = \frac{1}{r_1} + \frac{1}{r_2} + \frac{1}{r_3} + \frac{1}{r_4} + \text{Etc.}$$

Where R is the equivalent resistance and the small r's are values of each resistor.

Alternating Current and Voltage—Thus far the discussion has been concerned with direct currents in which the polarity of the voltage remains the same at all times. Alternating currents and voltages vary in magnitude with time and they vary in polarity as well. Perhaps the best manner in which to describe alternating voltages is to visualize an alternating voltage at a very slow rate.

Let two ten volt storage batteries be connected together in series so that the positive terminal of one is connected to the negative terminal of the other. This connection will be called the junction point. (See Fig. 8.) If one of the terminals of a voltmeter is connected to the junction point and the other terminal to battery A, the voltmeter would read ten volts, which is the voltage of that battery. Further, the terminal connected to the end of the battery would indicate on the meter that the battery terminal was positive with

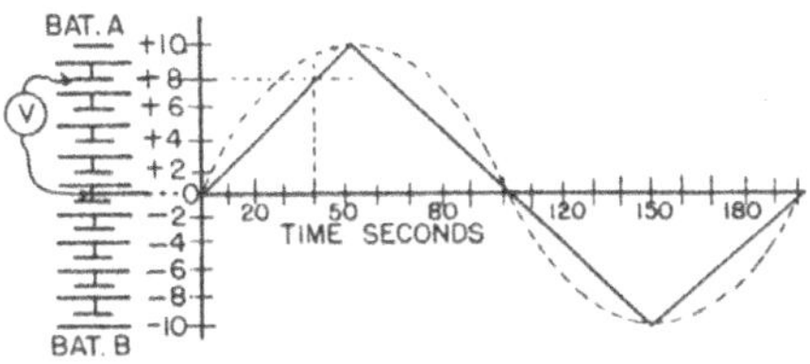

FIG. 8—ALTERNATING VOLTAGE

respect to the junction. If the meter terminal connected to the open end of battery A were moved to the open end of battery B it would indicate ten volts but it would be negative with respect to the junction point.

As each cell of the battery (two volts per cell) is touched with the voltmeter lead, leaving one meter terminal on the junction point, the meter will read positive and negative voltages, in two volt steps.

Start with the movable voltmeter wire at the junction point, the meter wire is moved toward the positive battery ten volt terminal at a rate of a two volt cell per ten seconds and the readings will be recorded as contact is made. The terminal will be moved downward toward the junction after reaching the maximum positive potential in steps of two volts to negative ten volts and then back to the junction in steps. The plot of voltages in this operation is shown by the crosses on the graph, Fig. 8.

In 200 seconds the voltage will read zero volts again and an alternating cycle will be completed. The frequency of this alternating voltage is one two-hundredths of a cycle per second. Alternating current supplies to most cities are at the rate of sixty cycles per second, or 12,000 times as fast. The alternation of voltage is actually quite smooth and is produced by rotating machinery.

If a load is connected to the alternating voltage source, the current varies at the same rate as the voltage. The question arises: What happens to an electric light if the voltage actually reaches zero twice each cycle? The answer is that the light would go out were it not that the frequency is so high that the thermal lag in cooling allows the filament temperature to decrease and increase by only a small amount. The flickering of the light cannot be seen for another reason. The eye cannot respond easily to intensity changes at rates of over about ten to fifteen cycles per second. Neon lights have no appreciable thermal lag, even at sixty cycles per second, so these lights actually go out 120 times a second on a sixty cycle supply.

In order to make calculations of power when using alternating current, some definite current and voltage values must be chosen. The value of alternating current or voltage is the peak current or voltage times 0.707 for pure sine waves. This is called the root mean square (RMS) value and is always used. The most common house supply voltage is 110 volts. The peak value of the voltage reaches about 154 volts but is of little interest to the consumer as he is interested in only the RMS, or effective value of 110 volts.

Frequency—The number of cycles of voltage change per second gives the frequency, usually in cycles per second. The following table gives the frequencies commonly used. A kilocycle is one thousand cycles per second and a megacycle is one million cycles per second.

USE	FREQUENCY (per second)
Power and lighting	50 or 60 cycles
Voice frequencies	100 to 8,000 cycles
Musical frequencies	40 to 15,000 cycles
Radio range frequencies	200 to 400 kilocycles (KC)
Broadcast frequencies	540 to 1650 kilocycles
Aircraft "day" frequencies	4900 to 6600 kilocycles
Aircraft "night" frequencies	2900 to 3600 kilocycles
Ultra high frequencies	50 to 500 megacycles (MC)

Radio transmitters will use but one carrier frequency at a time for transmission. Thus the "day" carrier frequency of an airplane might be 4967 KC. The voice frequencies which are transmitted on the carrier may be within the voice range shown above and may vary rapidly in frequency and amplitude as the words are spoken. The operation of carrier modulation with voice is explained later.

Power Factor (PF)—Alternating current may not be treated as easily as direct current. The reason for this is that the current peak may not occur at the same time as the voltage peak. There will be exactly as many current cycles as there are voltage cycles, but they may start at different times. The current and the voltage are said to be out of phase with one another. Means of creating such a phase difference will be treated in the next chapter.

The term "power factor" is used to calculate the effect of the current and the voltage being out of step in time (out of phase). If the current and the voltage reach their positive peaks at the same time, the power factor is said to be unity. However, if the voltage reaches its peak (positive or negative) at the same time that the current reaches zero there would be no power represented at any time and the power factor would be said to be zero. There are intermediate values between these two extremes which are usually represented by a decimal such as: 1.00, .80, .50, etc.; although they are often shown as a per cent power factor such as: 100%, 80%, 50%, etc., respectively.

Power in alternating current work is always found by multiplying the voltage (RMS) by the current (RMS) and then by the power factor (PF):

$$W = E \times I \times PF$$

The product of current times voltage is given as volt amperes (VA) and does not represent power in watts at all. The watts divided by the volt amperes give power factor.

An example of the use of such a formula is found in the operation of remote indicating instruments of the autosyn (or Selsyn) type. These indicators are small motors whose power factor is approximately 30% (.30 PF). On a certain plane it was found that the total load from such machines was twelve amperes at thirty-two volts, sixty cycles. The volt ampere load was then about 380 VA. Actually, the watt load represented is but $380 \times .30 = 114$ watts. The current flowing in the circuit is about three times as high as necessary to supply the watt load at unity power factor. Voltage drops are calculated as before by the formula:

$$E = IR$$

One large transport (Douglas DC-4) is equipped with an alternating current supply which has advantages over a direct current supply. Present transports are equipped to generate not more than about 1200 watts. The larger planes will require powers of 15,000 watts (15 kilowatts) or more which will be used for raising and lowering the landing gear, supplying power to operate many electric motor-driven pumps, cooking in the galley, etc. Lighting loads alone on large transports will increase greatly and the use of more radio receivers will increase power requirements.

QUESTIONS ON CHAPTER I

It is suggested that reference be made to Chapter XIV before answering these questions as a short explanation is given which should be read.

1. Voltage is analogous to pressure, therefore:

   ( ) a. The larger the tank diameter the greater the pressure. ( ) b. Dry batteries have less voltage than wet batteries. ( ) c. Tank depth and battery voltage are inversely proportional. (✓) d. Tank depth and battery voltage are directional proportional.

2. Most transport aircraft have a battery voltage of:

   ( ) a. 24 volts. ( ) b. 16 volts. (✓) c. 12 volts. ( ) d. 6 volts.

3. Receiver and transmitter plate supply voltages often encountered are respectively:

   ( ) a. 100 and 500. (✓) b. 200 and 1000. ( ) c. 1000 and 300. ( ) d. 12 and 24.

4. The following expression gives wattage:

   ( ) a. W × E. ( ) b. E × E. ( ) c. I × W. (✓) d. E × I.

5. Power is:

   ( ) a. The amount of work done. (✓) b. The horsepower required to produce one hundred watts. (✓) c. The rate of doing work. ( ) d. The rate of current flow.

6. Columbus sailed in 1492. Half of this number is:

   ( ) a. The date of the discovery of electricity. (✓) b. The number of watts in one horsepower. ( ) c. The number of horsepower in a watt. ( ) d. The plate voltage of usual transmitters.

7. The ampere hour is used in battery calculation rather than wattage because:

   ( ) a. Voltage means very little in most electrical calculation. ( ) b. Batteries are charged without wattage. ( ) c. Current is all that counts. ( ) d. It makes calculation easier.

8. A generator charges at 50 amperes for 4 hours. The drain on the battery is 40 amperes for 5 hours. At the end of this time how many ampere hours will be lost by the battery?

   ( ) a. 10. ( ) b. 200. ( ) c. none. ( ) d. 10.

9. A generator charges at the rate of 35 amperes while the load attached to the battery and generator is 45 amperes. If the battery had an initial charge of 100 ampere hours about how long will it take for the battery to discharge?

( ) a. 2 hours. ( ) b. 600 minutes. ( ) c. 60 minutes. ( ) d. 4 hours.

10. A load is attached to a battery and draws 30 amperes. The battery was initially charged so that it contained 120 ampere hours. If the battery is completely discharged about how long will it take for it to reach full charge at the same rate as the discharge?

( ) a. Question impossible. ( ) b. Insufficient data. ( ) c. Time must be given. ( ) d. 4 hours.

11. A battery is attached to a load and a generator. The generator is supplying 40 amperes and the load is taking 40 amperes at 14.2 volts. The battery is:

( ) a. Charging at 40 amperes. ( ) b. Continually gaining in charge. ( ) c. Neither charging nor discharging. ( ) d. Fully discharged.

12. The product of resistance and current is equal to:

( ) a. Power. ( ) b. Voltage. ( ) c. Ampere hours. ( ) d. Current.

13. A wire with twice the circular mil area of another wire will safely take:

( ) a. Half the current. ( ) b. Half the wattage. ( ) c. The same current. ( ) d. Twice the current.

14. Two ten ohm resistors in parallel are equivalent to a resistor of the following number of ohms:

( ) a. 30. ( ) b. 20. ( ) c. 5. ( ) d. 2.5.

15. Two ten ohm resistors in series are equivalent to a resistor of the following number of ohms:

( ) a. 30. ( ) b. 20. ( ) c. 5. ( ) d. 2.5.

16. A ten ohm, a five ohm and a ten ohm resistor are connected in parallel. The equivalent resistance in ohms is:

( ) a. 30. ( ) b. 20. ( ) c. 5. ( ) d. 2.5.

17. The same three resistors as in Question 16 are connected in series. The equivalent resistance in ohm is:

( ) a. 15. ( ) b. 20. ( ) c. 25. ( ) d. 30.

18. Each alternating current cycle goes through zero current the following number of times per cycle:

(a) a. Once. ( ) b. Twice. ( ) c. Three times. ( ) d. Only the voltage does this.

19. Electric lights do not go out completely on the zero voltage portions of each cycle because:

( ) a. Otherwise there would be no light. ( ) b. The light is faster than the eye. ( ) c. Lights are brighter on alternating current. ( ) d. Thermal lag delays cooling.

20. The following frequency group is composed only of voice frequencies:

( ) a. 100, 400, 1000, 5000 and 7000 kilocycles. ( ) b. 40 and 50 megacycles and 1000 megacycles. ( ) c. 5 kilocycles, 400, 300 and 1000 cycles. ( ) d. 10, 200, 400, 1000 and 800 cycles.

21. The following frequency is an ultra high frequency:

( ) a. 1,000,000 cycles. ( ) b. 100 megacycles. ( ) c. 75 kilocycles. ( ) d. 28,000,000 cycles.

22. Power factor times voltages times current equals:

( ) a. Ampere hours. ( ) b. Volt amperes. ( ) c. Kilowatts. ( ) d. Watts.

23. Ampere hours times voltage give:

( ) a. Kilowatt hours. ( ) b. Wattage. ( ) c. Watt hours. ( ) d. Current.

24. Airplane starters at 12 volts may draw:

( ) a. 10 volts. ( ) b. 200 amperes. ( ) c. 200 watts. ( ) d. 300 to 400 milliamperes.

25. An airplane was operated with the following loads:

200 amperes for 6 minutes
20 amperes for one hour
40 amperes for thirty minutes
100 amperes for 0.4 hour

The generator was charging at 50 amperes for two hours. The 120 ampere hour battery was half charged at the start of the two hour trip. What was the final charge of the battery?

( ) a. Full. ( ) b. Almost discharged. ( ) c. 100 ampere hours. ( ) d. Half charged.

CHAPTER II

POWER CONVERSION

Power conversion is accomplished in a number of ways. Motors are used to convert electrical to mechanical power, generators are used for converting mechanical to electrical power and electricity may be converted from one form to another, such as alternating current to direct current, by motor generators.

Electrical circuits were dealt with in Chapter I and this chapter will be confined principally to magnetic circuits.

Magnetic Fields—Magnetic fields are found in all electrical machinery as well as in space, due to the Earth's magnetism. Certain materials may be magnetized and these are usually made of iron or one of its combinations with other elements.

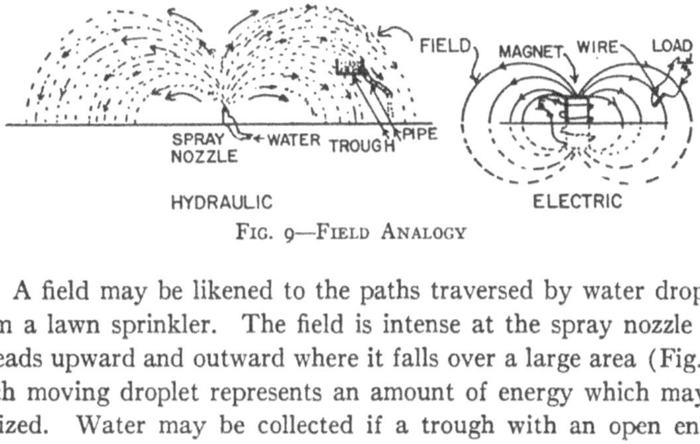

FIG. 9—FIELD ANALOGY

A field may be likened to the paths traversed by water droplets from a lawn sprinkler. The field is intense at the spray nozzle and spreads upward and outward where it falls over a large area (Fig. 9). Each moving droplet represents an amount of energy which may be utilized. Water may be collected if a trough with an open end is placed in the area of the falling spray. If the ends of the trough are closed the water will rise in the trough and added pressure will exist on the bottom of the trough. If a pipe is led from the trough water will flow through it.

This is exactly analogous to the behavior of a wire in a magnetic field (Fig. 9). If the field is moving, as in the case above, a voltage will appear on the wire giving rise to current when the ends of the wire are attached to an electrical load. Iron possesses the property of confining the electrical field, preventing it from being dissipated over a large area. Iron also possesses the property of being magnetized, in which case a steady permanent field will exist around the magnet.

A still photograph of the spray is like the steady magnetic field. There is no movement of the water drops and the only way in which energy can be collected is to move the trough through the fixed spray.

In a like manner, the only way in which voltage may be induced in a wire in a steady magnetic field is to move the wire through the field. Current will flow if the wire is attached to an electrical load.

It makes no difference whether the field is in motion or the wire is in motion, as voltage will be induced in the wire in either event. Voltage may be generated in a rotating machine (generator) with wires on its rotor and a steady field on the stator, or stationary part of the machine. Mechanical energy will thus be converted to electrical energy when the machine is turning as the wires will be passing through the steady field.

Poles—When a bar of iron is permanently magnetized and suspended in space it will turn so that one end points to the North Magnetic Pole of the Earth. This north-seeking end of the magnet is conventionally called the north pole. The other end of the magnet is called the south pole.

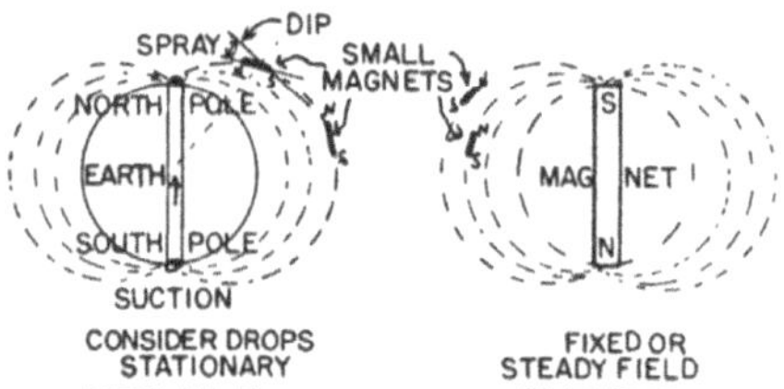

FIG. 10—FIELDS

The Earth has two magnetic poles and lines of force extend from the center of the Earth and enter space at the North Magnetic Pole. Here the magnetic lines curve outward and downward and are again concentrated at the South Magnetic Pole where they enter the Earth once more.

The Earth's field is like the field existing around any magnet (Fig. 10) and hydraulically, this may be compared to a spray at the North Magnetic Pole with water spraying outward and downward past the equator and being sucked into a similar concentrating spray nozzle at the South Magnetic Pole. The two nozzles are connected with a pipe extending through the earth from the South to the North

Pole, completing the circuit. No motion of the water exists in this case as the field is a steady one.

When two permanent magnets have their north poles brought together they will tend to repulse one another (Fig. 11). Hydraulically, this is similar to the forces existing when two spray

FIG. 11—REPULSION

nozzles are forced together, as one spray will repulse the other. When a north and a south pole are brought together (Fig. 12) it is similar to the attraction existing when two nozzles are brought together, one of which is sucking water in (south) and the other is

FIG. 12—ATTRACTION

forcing water out (north). The fixed field equivalent case may be imagined but the analogy is not so clear cut.

Steady fields can do no work but they may create a force. When poles of two fixed magnets are placed together the magnets may move apart when released but the energy represented is a return of part of the energy originally required to bring them together.

Right Hand Rule—When a current flows through a wire it will produce a magnetic field surrounding the wire. The magnetic lines exist circularly around the wire. The direction of the lines of force, or magnetic lines, may be determined by the "right-hand rule."

If a conductor is grasped in the right hand with the thumb extended and pointing in the direction of current flow (toward the negative supply voltage terminal) the fingers will point in the direction of the magnetic field, or flux.

Thus if a wire is placed in a steady magnetic field and current is allowed to flow through the wire it will tend to move out of the field in a direction determined by the direction of current flow in the wire and the direction of the magnetic field.

Generators—If a coil of wire is spun in the Earth's magnetic field a voltage will be generated in the wires. The earth inductor compass operates on this principle but, as the field is very weak, the generation of power is very inefficient.

Electrical generators operate on this same principle. A very strong field is generated by passing current through a field coil which is placed on the iron stator, or frame structure of the machine. A number of turns of wire are wound on the rotor which is revolved within the field created on the stator and voltage is efficiently induced on the wires. If two connections are brought out through slip rings fastened to the rotor and on which brushes fastened to the stator are riding, a current will be generated when the machine is attached to an electrical load.

Fig. 13 shows the elements of such an electrical generator. It will be realized that the only difference between using one wire and a coil of wires is that the individual voltage on each wire in a coil

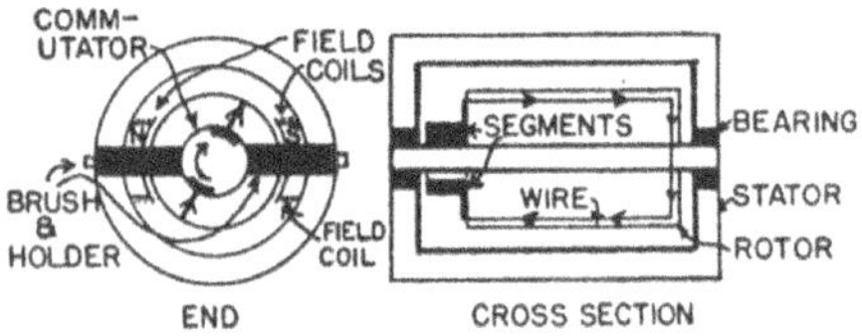

Fig. 13—One Turn Generator

will add, producing a higher total voltage in proportion to the number of turns on the rotor in a given field.

The voltage generated on one wire in the coil is in one direction as the wire passes one pole piece and in the other direction when it rotates past the other pole piece. When the wire is half-way between the two pole pieces the voltage is zero. Thus the voltage generated will be an alternating voltage at one cycle per revolution. Such a generator is called an alternator.

Commutation—Direct current is desired from a generator in many cases. Suppose a turn of wire is placed on the rotor of a generator and each end of the turn is attached to one of two very small metallic sectors on an insulated ring (Fig. 13). Two brushes are held on the stator so that they are opposite one another and resting on the ring with its two sectors. Every time that the sectors pass under the brushes a pulse of voltage will exist and further, the polarity of each brush is always the same, one is negative and the other positive. The side of the turn passing under the north pole might show a positive voltage. Thus when it passed under the south pole it would show a negative voltage but would be touching the negative brush at this moment. Fig. 14 shows the condition.

Such a generator would be very inefficient as voltage pulses would last for a short time. However if many segments are placed on a rotor and a separate turn attached to pairs of segments which are on opposite sides of the rotor, the voltage generated will be fairly steady. Such an arrangement of segments is called a commutator.

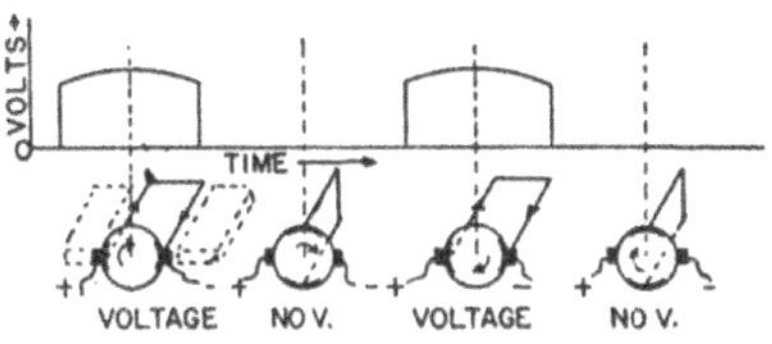

Fig. 14—Single Turn Armature Voltages

Actually the windings on the rotor are continuous, with taps on the windings as shown in Fig. 15. The voltage generated is quite smooth with very little pulsation. The pulsation that does exist is due to the fact that the brush rests on a segment for a short time so that the voltage actually varies slightly between the time that the brush first makes contact to the time that the brush last makes contact. There will also be some pulsation as the brush bridges the insulated gap between two segments. These pulsations are called commutator ripple and may be neglected for many purposes. The

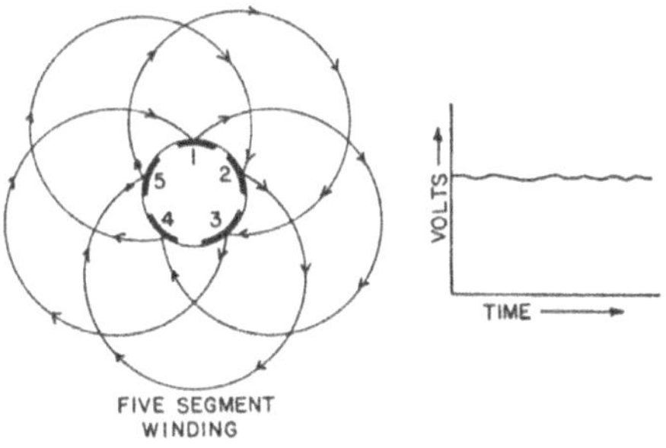

Fig. 15—Armature Winding

small sparking that occurs on the brush faces causes radio interference and this must be eliminated as will be explained in a later chapter.

Field voltage is usually supplied by the same commutator that is supplying the load on the generator. This type of generator is said to be self-excited. The current and output voltage of the generator are proportional to the voltage placed across the field coil which creates the magnetic field in the machine. If a series resistor is placed in one of the field leads, the voltage may be regulated (Fig. 16).

A third brush is utilized which may be moved around the commutator with relation to the other two main brush positions which are opposite one another. If one side of the field is attached to such a brush and the other side of the field to one of the main brushes, the voltage may be regulated by moving the third brush to regulate field voltage. This method is illustrated in Fig. 16 and has been used extensively in automotive equipment.

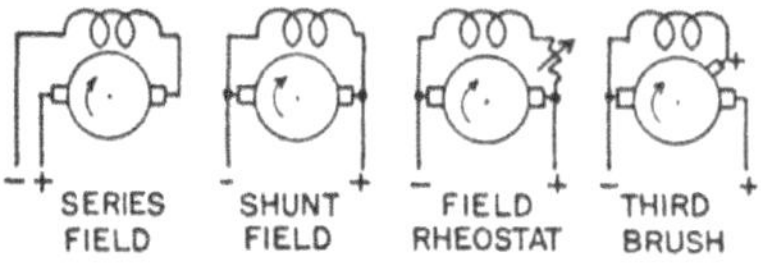

FIG. 16—GENERATOR VOLTAGE CONTROL

The best way to regulate the charge rate and voltage of the generator independently is shown in the later section on generator control boxes, which are widely used on aircraft.

The polarity of a generator can be determined by the polarity of the field wires. It is interesting to note that a self-excited generator cannot theoretically start to produce power unless a field exists. Merely turning the rotor will create no voltage unless a small amount of permanent magnetism remains in the field. Occasionally a generator will not start charging, in which case the generator may be temporarily connected to a battery to create the field.

Motors—Direct current motors operate on the same principle as the direct current generator. Voltage is supplied to the machine, which sets up a field and also sets up a current in the rotor windings through the commutator. Just as it required mechanical power to drive the generator windings through the field, so will the machine rotate when fields and currents bear this same relation. As the rotor tends to reach a position where less torque (twisting force) is produced the voltage supplying brush will have passed to the next commutator segment, the force will continue and the machine will rotate.

Some alternating current motors operate on a somewhat different principle. The rotor is supplied with a number of heavy bars of copper which are attached at the ends of the rotor to a solid copper ring. The alternating current generates voltage in these bars, which, because they are short-circuited, will cause large currents in them. The action of these currents in the field causes rotation in the machine. The machine requires no commutator, slip rings or brushes. With alternating current supplied to the field the polarity of the field will change at a rate determined directly by the frequency. The rotor cannot turn as fast as the rate of alternation for no transfer of power to the rotor would exist. The rotor can turn no faster than the field, which is making a revolution effectively every sixtieth of a second (60 cycle supply). The maximum speed of such a machine is therefore 3600 RPM. If the machine is equipped with four poles (two sets) the maximum speed which the machine may reach is 1800 RPM. On heavy loads these machines will "slip" somewhat and speeds will be some five per cent less than the maximum or synchronous speeds.

The frequencies which will be used in the power supplies being developed for large airplanes are 400 cycles in some cases and 800 cycles in others. The maximum speed of the motors, if of the induction type just referred to, is 24,000 and 12,000 RPM for the two and four pole machines respectively, on 400 cycles. At 800 cycles the speeds are twice as great.

Direct and alternating current motors are also made with the field and rotor connected in series. This type of machine has very high starting torques; when the rotor is not turning and voltage is applied the current becomes very high. After the machine starts it is behaving as a generator to some extent and the voltage it generates tends to reduce the currents to running values. Such machines, if not permanently geared to a load, may run away and destroy themselves because of excessive speed. Almost all universal motors are of the series type and operate on either direct or alternating current. They usually give more power when operated on direct current.

The field structure in all machines must carry a magnetic flow. If the alternating current frequency is low the field structure must be large. This is equivalent to a pipe in which water runs alternately in one direction and then in the other. As the frequency is increased the same amount of work may be done with a smaller field structure as the alternation of field takes place more rapidly. Thus a smaller equivalent amount of water in the analogy is transferred oftener to do the same amount of work as before.

For this reason an 800 cycle motor which can supply about six horsepower can be built to weigh less than forty pounds, whereas the same motor at sixty cycles would weigh somewhere in the neighborhood of 300 pounds.

It would seem that if this were true there would be no reason for stopping at 400 to 800 cycles. Efficiencies do not change radically between sixty and 800 cycles; above that frequency the machines become less efficient and therefore become heavier. The exact frequency at which the lightest machine of good efficiency may be built is at some frequency in the neighborhood of 400 cycles.

Transformers—If two coils are wound on the same iron core and alternating current is supplied to one coil a voltage will be produced in the second coil in the same ratio as the turn ratio of the two coils. This is similar to the action in rotating machines but as the field is alternating in the transformer there is no need to supply any rotating parts.

The main purpose of a transformer is to change voltage, although the power on both sides of the transformer is (almost) equal. Thus, if it is desired to secure 1000 volts from a 100 volt source a transformer with a ten to one turn ratio will be employed. In addition, the transformer must be large enough to transfer the necessary power. If the current through the load were ten amperes at 1000 volts the secondary power would be ten kilowatts. The primary power would be the same so the primary current would be 100 amperes at 100 volts.

Large transformers are ninety-nine per cent efficient. Small transformers on aircraft are about ninety to ninety-five per cent efficient.

The main limiting factor in the size of electrical machinery and transformers is the temperature rise of the equipment. A rise of ninety degrees Fahrenheit (50 degrees Centigrade) is considered about as high a rise as is safe, provided the total temperature is not over 200 degrees Fahrenheit. By supplying forced air cooling, the power handling capabilities can be increased greatly. The frequency has the same effect on weight as it has on motors and generators. The higher the frequency, the lighter the transformer.

Dynamotors—Transformers cannot be used with direct current, as the field in the core would not change or alternate. To generate high voltage from direct current it is necessary to create an alternating field which is done by rotation. A dynamotor is actually a motor and generator in one. Both use the same field and rotor structure but have two separate commutators. Thus, the high voltage windings on the rotor have the same turn relation to the low voltage windings as the voltages. In reality, the machine is a whirling transformer with two commutators.

Small dynamotors have efficiencies of less than fifty per cent and are much less efficient than transformers which, for the same power, are better than ninety per cent efficient.

The same general type of machine is used where it is desired to turn direct currents into alternating currents. The alternating current side of the machine is equipped with slip rings, as commutation is not desired. Such machines also can be used to supply direct currents from alternating currents but this is usually done with a rectifier.

Two-phase Systems—The discussion up to this point has been concerned with single-phase alternating currents. A pair of wires is sufficient to supply single-phase power to any device.

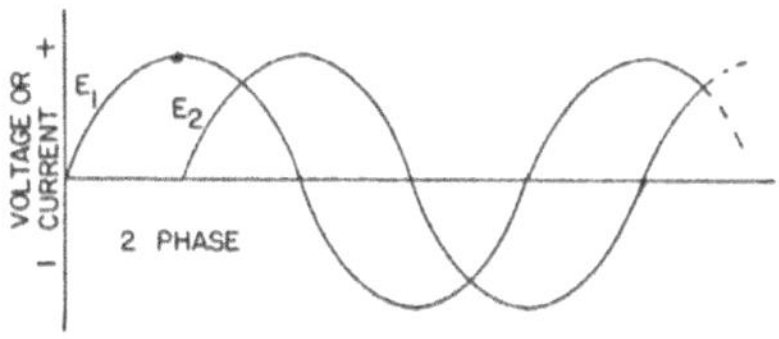

FIG. 17—TWO PHASE

Two-phase voltage supply is also possible when three wires are used (one of which may be ground). Two-phase systems are equivalent to two single-phase systems which use one common wire. The voltages in the two systems are not in phase, although at exactly the same frequency, Fig. 17. One voltage starts a quarter cycle after the other one.

For all heating and lighting purposes, the two-phase system can be considered as two single-phase systems, either of which may be used separately. Induction motors, on the other hand, can be started directly on a two-phase system as the two voltages can be used to produce separate fields which have a time interval between them. Two coils may be placed on the circular field structure at ninety degrees to one another and will supply a starting torque to the rotor. This is required for motors such as the induction type.

If two such fields do not exist, an induction motor will not start as there is equal torque in either direction until the motor starts turning.

Most induction motors are designed for single-phase operation and it is also very desirable that they be self-starting. A starting condenser is used with the motor to make it self-starting and the condenser is actually producing a two-phase supply by causing a lead in the voltage supplied to the coil attached through the condenser. In some cases the condenser is thrown out of operation with a centrifugal switch on the machine after starting. In other cases the condenser is merely left in the circuit.

Two-phase systems, while once quite common, are almost unused today as the three-phase system has numerous advantages not found in two-phase systems. The condenser starting device on single-phase motors is one of the few devices that produces a two-phase supply and it is only useful for starting. The direction of starting can be determined by the proper connection of the condenser and its field coil where it is desirable to have a motor which may be run in either direction.

Three-phase Systems—The three-phase system requires three wires, one of which may be ground but in most cases three metallic conductors are used.

The three voltages are displaced by 120 degrees from one another (Fig. 18). The voltage between wire 1 and wire 2 may reach a given positive maximum at one instant of the alternating cycle. One third of a cycle later (120 degrees) the voltage will reach this same positive maximum between wires 2 and 3. One third of a

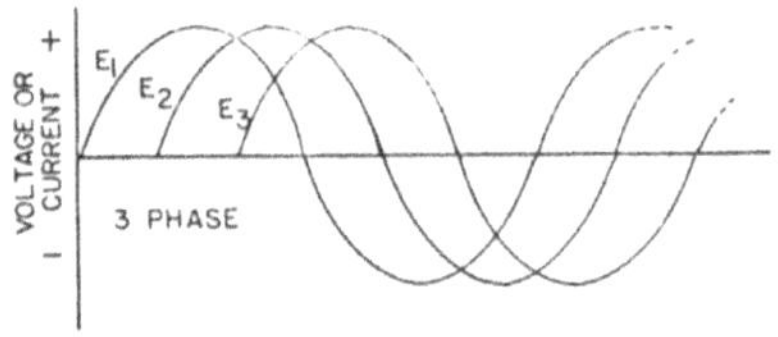

FIG. 18—THREE PHASE

cycle later the voltage between wires 3 and 1 reaches the positive maximum. This completes the circle and the positive voltage makes the complete circle every cycle.

By attaching the three wires to three coils, displaced 120 degrees around the stator, the field will rotate around the stator once each cycle.

Three-phase systems may be attached to a transformer or a rotating machine in two ways. The first is the delta connection (often written Δ) in which each of the windings is connected together and the three-phase wires are attached to the winding junctions. The other method is the "Y" connection. Both are illustrated in Fig. 19.

The center of the "Y" connection is grounded in some cases. The three voltages measured from ground (the Y center) to the three-phase supply wires are 120 degrees apart in phase.

When a three-phase system is used for motors, the field rotates around the field structure and the rotor may follow in exact step. When this is true the motor is said to be a synchronous motor and will stay in exact step with the frequency of the system, even with voltage fluctuations.

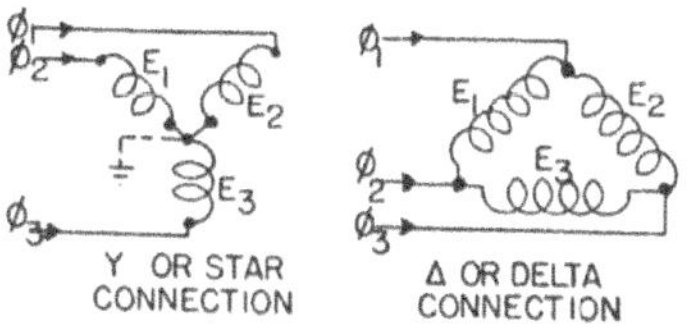

FIG. 19—Y AND DELTA CONNECTIONS

The autosyn or selsyn motors, which are described elsewhere in this book, are a good example of the use of a three-phase system but

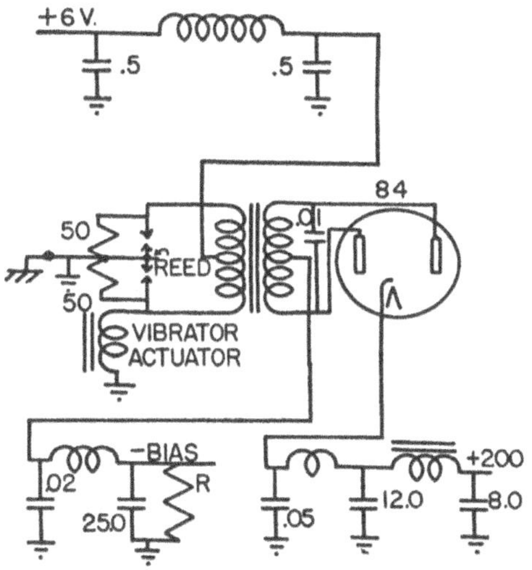

FIG. 20—NON-SYNCHRONOUS VIBRATOR AND FILTERS

in these cases the field, while three-phase, does not rotate except when the machines revolve.

The direction of any motor used on three-phase may be reversed by interchanging any two wires to the motor. This forms a convenient means of changing direction of rotation.

Three-phase systems are valuable in aircraft work and are used extensively for motors of large capacity in industrial installations, where large amounts of mechanical power are to be generated. The three-phase motor also does away with the starting condensers required for single-phase motors. Condensers may become extremely large for big machines and may represent a sizeable loss of power.

Rectifiers—One type of rectifier, which has already been described, is the commutator. Another type of crude commutator is the vibrator. In vibrator systems a direct current is applied to a coil which has a vibrating armature, making electrical connection through the coil. When current flows the coil attracts the armature

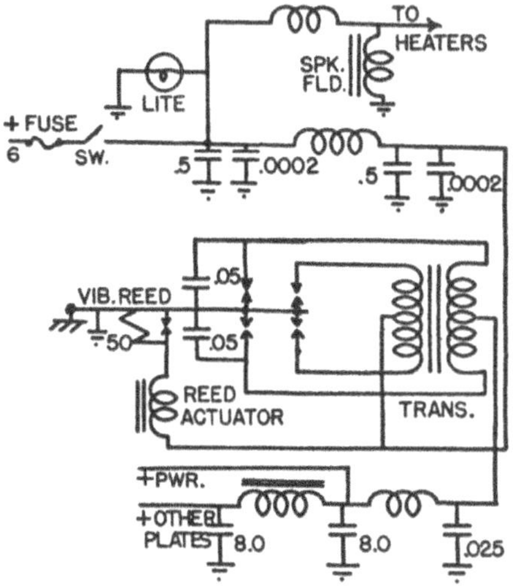

Fig. 21—Synchronous Vibrator and Filters

and breaks its own circuit. The armature thus sets up a vibration in the order of 100 cycles per second.

The battery is connected through another set of contacts to the primary of a transformer, as shown diagrammatically in Fig. 20 and Fig. 21. Thus, pulsating direct current is set up in the transformer primary which is rectified either by another group of contacts, Fig. 21, or by a tube rectifier, Fig. 20. Such vibrator systems are used in many mobile receivers in order to allow the low battery voltage to be transformed into a higher voltage for the receiver. Such rectifying schemes must be well shielded and filtered electrically to prevent interference from being set up, due to sparking, which will disturb receiver operation.

Discs utilizing copper oxide coatings are used for rectification at low voltages. These discs have the peculiarity of passing current only in one direction. When a rectifier of this sort is placed between a transformer and a load there will be only positive (or only negative) peaks of voltage passed on to the load. Fig. 22 shows the shape of

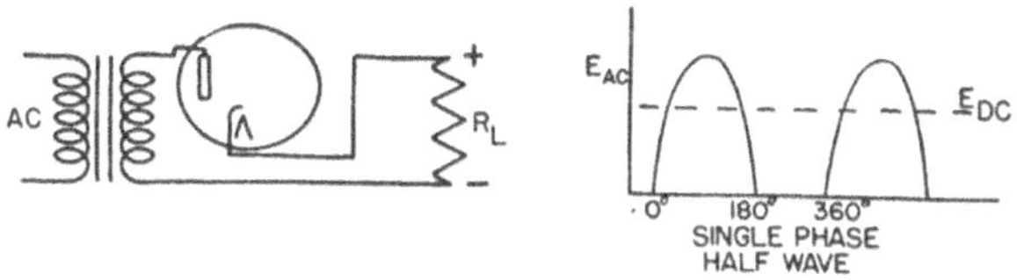

FIG. 22—RECTIFIER WITHOUT FILTER

the voltage curves. If this voltage is properly filtered the result is almost pure direct current.

Vacuum tubes operate as rectifiers in a like manner. They will pass voltage only when the plate is more positive than the filament. Tubes will pass no current in the reverse direction, although copper oxide rectifiers will pass a current of about ten per cent or less in the reverse direction, not being quite as efficient as tubes.

QUESTIONS ON CHAPTER II

1. The Earth is like a large fixed magnet. Its North magnetic pole effectively attracts the north pole of a magnet. Even so, two north poles of a magnet repel one another.

   ( ) a. True. ( ) b. False.

2. If a field and a wire are both moving together in the same direction:

   ( ) a. Voltage will be induced in the wire. ( ) b. No voltage will be induced in the wire. ( ) c. An alternating field will be produced. ( ) d. Current will flow.

3. If the field or stator part of a generator were revolved and the part ordinarily used as a rotor were stationary:

   ( ) a. The machine would not be generating voltage. ( ) b. The field would heat excessively. ( ) c. The operation would theoretically be the same. ( ) d. No moving field would be produced.

4. A generator with many segments on the commutator is better than one with few segments because:

   ( ) a. Brushes would be made narrower. ( ) b. Higher voltage would be produced. ( ) c. The commutator mica separators would be thicker. ( ) d. The ripple voltage would be less pronounced.

5. The synchronous speed of a two pole synchronous motor is proportional to frequency. A two pole synchronous machine at 200 cycles would then turn at:

( ) a. 12,000 RPM. ( ) b. 24,000 RPM. ( ) c. 200 RPM. ( ) d. 1000 RPM.

6. A transformer has a turn ratio of one (primary) to ten (secondary) so that if 500 volts appears on the secondary the primary voltage must be:

( ) a. 100. ( ) b. 50. ( ) c. 5. ( ) d. 25.

7. A transformer has a steady direct current of five amperes flowing through the primary. The turn ratio is one to ten as before. What is the secondary current in amperes?

( ) a. 50. ( ) b. 0.5. ( ) c. 0. ( ) d. 5.

8. A dynamotor is like:

( ) a. Two generators connected together. ( ) b. A storage battery. ( ) c. Machine which has its two fields in opposition. ( ) d. A revolving transformer with commutation.

9. The principal difference between two phase and three phase is:

( ) a. Two phase has two voltages, three phase has three voltages. ( ) b. One employs two wires and the other three wires. ( ) c. One wire may be grounded in two phase only. ( ) d. There is no difference.

10. Simple single tube rectifiers turn alternating into direct current because:

( ) a. They reduce the number of alternations. ( ) b. They eliminate one half of the alternating wave cycle. ( ) c. They pass current in both directions. ( ) d. They cannot turn alternating to direct but only direct to alternating current.

11. Generator third brushes are used on automotive equipment:

( ) a. To regulate the speed of the generator. ( ) b. To reduce commutator ripple. ( ) c. To set the charge for average conditions. ( ) d. To maintain the battery voltage at exactly 7.1 volts.

12. Magnetic north poles attract one another because:

( ) a. Their fields are the same. ( ) b. They have like forces. ( ) c. Like poles attract one another. ( ) d. They actually repel one another.

13. Voltage can be induced in a wire by moving it through a steady field and likewise:

( ) a. Voltage can be induced in a stationary wire in a moving field. ( ) b. Current will flow even though the wire is not attached to an electrical load. ( ) c. Transformers employ steady fields. ( ) d. Transformers can transfer energy due to their steady field.

14. The earth inductor compass works because:

( ) a. The earth rotates. ( ) b. Wires are rotated in the earth's field. ( ) c. The earth induces voltage in all wires. ( ) d. It uses a fixed magnet.

15. The polarity of a direct current generator may be reversed by:

( ) a. Reversing the connections to the load. ( ) b. Reducing the number of turns on the field winding. ( ) c. Reversing the polarity of the field leads. ( ) d. By stopping the generator and starting it again.

Chapter III

CONDENSERS AND COILS

Condensers and coils exhibit a quality called reactance when used in electrical circuits. This chapter is devoted to a discussion of reactance, which is most important in all alternating current and radio circuits. It accounts for power factor and for behavior of tuned circuits found in all receivers and transmitters. Hydraulic analogies will be used to explain the behavior of electrical circuits with condensers and coils.

There are two conditions met in all electrical circuits. One is the transient condition at the instant of applying or removing a voltage. The other is the steady state condition when steady voltage exists on the circuit.

The first mechanical analogy explains the behavior of a resistor in a circuit. Fig. 23 shows the electrical and hydraulically equivalent

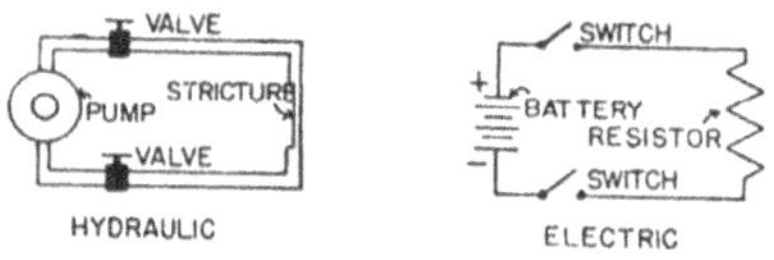

Fig. 23—Resistance

circuit. The battery is likened to a centrifugal pump which applies pressure to the circuit. The pump is equivalent to any generator, but the battery is used for simplicity. It must be remembered that when the wire is removed from a battery or the other element it is equivalent to closing valves in the hydraulic circuit and separating the pipes.

The resistor is equivalent to a stricture in a pipe returning to the pump. Therefore, it will restrict the water flow and heat will be generated by friction of the liquid on the walls of the pipe. Naturally, some friction will exist on the walls of the larger section pipe but this resistance will be small in proportion.

Electrically, the battery exerts a certain voltage (pressure) which forces current through the resistor and produces heating of the resistor.

The transient and steady state conditions are alike, as the current will flow instantly, at a constant rate, as the switch is closed. This will not be true with condensers and coils, as will be shown.

Condensers—A description of the construction of a condenser and its behavior in electrical circuits is necessary before discussing the various uses of condensers in circuits.

Any two sheets of metal which are insulated from and are adjacent to one another form a condenser. The insulating medium (dielectric) between the plates may be air, mica, sulphur, oil or paper. The capacity is proportional to the number and area of adjacent pairs of metallic surfaces and the type of dielectric. Capacity is inversely proportional to the thickness of the insulating dielectric. The more surfaces and the closer the spacing, the higher will be the condenser capacity. The unit of capacity of a condenser is the farad. Usually the value will be given in microfarads (one millionth of a

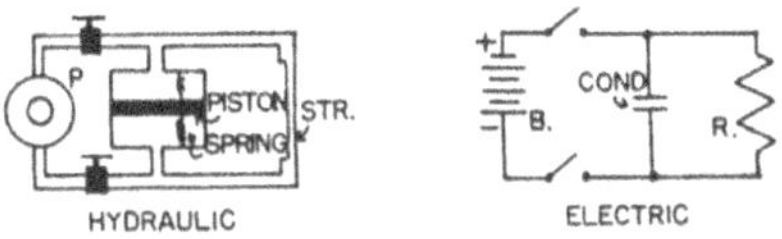

Fig. 24—Parallel Capacitance and Resistance

farad) or micromicrofarads (one millionth of a microfarad). The dielectric will withstand a certain voltage above which the voltage will break down the dielectric, puncture and spoil the condenser (unless air is used). Fig. 24 shows the hydraulic and electric circuits of a condenser attached to a battery.

Mechanically, the condenser is analogous to a piston in a cylinder. The piston is held at center with two springs. When connected to a pump the piston will be displaced against spring pressure and will remain at a new position. When the cylinder is removed from the circuit, if a valve at its outlet has first been closed, the piston will remain in position with pressure caused by the spring tension. No liquid may flow through the cylinder at any time due to the piston. The hydraulic capacity of a condenser is determined by its volume displacement.

Electrically, a condenser attached to the battery will gain a potential (voltage) equal to the battery voltage. If the condenser is removed from the circuit it will remain charged. In the steady state there is no way of determining that the condenser has been placed across the circuit, as no direct current will flow through the condenser. Current will pass through a resistor which is in parallel with the condenser but the condenser will have no effect upon this current.

The transient condition is quite different. In the hydraulic analogy, there will be a high flow of liquid into the cylinder when it is first attached to the pump. The piston will come to rest quickly as the liquid compresses the spring, so that its pressure is equal to that of the pump.

Electrically, this same condition is encountered when a condenser is attached to a battery. A high instantaneous current will flow, after which the condenser will reach its full electrical charge.

If the hydraulically analogous condenser and strictured pipe are removed together from the pump the spring will cause liquid to be

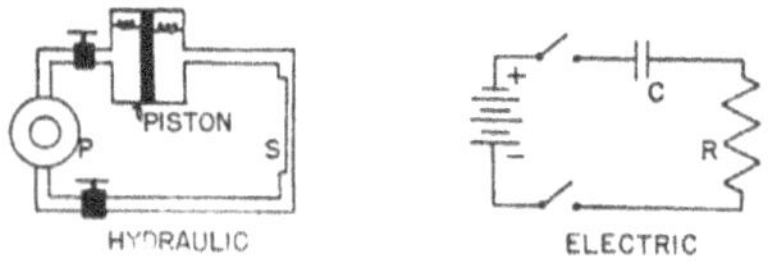

FIG. 25—SERIES CAPACITANCE AND RESISTANCE

forced through the stricture until the piston reaches a point of equilibrium at center.

If the condenser and resistor are removed from the battery together (and remain connected) the condenser will discharge through the resistor until the condenser has lost its (voltage) charge.

It will be noticed that no energy is lost during a complete on-off cycle, as the amount of electricity stored in the condenser while charging is restored to the circuit upon discharge.

With the behavior of a condenser in mind, the case in Fig. 25 will be clear. In the steady state condition, the current cannot flow through the condenser and resistor. When the voltage is first applied, the condenser charges.

Current flow is limited by the resistor in charging so that the charging current can never be greater than that of the current through the resistor alone with the same voltage applied. When the battery is removed the condenser will retain its charge. If the open end of the condenser is attached to the open end of the resistor there will be a discharge of the condenser. Therefore, a condenser connected in series with a direct current supply will allow no flow of current, except a transient current when first connected if the condenser is in a discharged condition at the time.

Coils—Coils are variously referred to as chokes, choke coils, reactors, retard coils, solenoids and inductors. Perhaps it is best to refer to them as inductors.

A description of their construction and operation will be given before describing their many uses. Inductors are composed of a number of turns of wire wound on a form. The form may be of some non-magnetic material or an insulator but if it is of iron or one of its alloys it will exhibit far more inductance for the same number of wire turns (at low frequencies). Inductance is expressed in henries, millihenries (thousandth of a henry) and microhenries (millionths of a henry).

The behavior of an inductor in parallel with a resistor and also in series with a resistor will be considered. All inductors will have a certain amount of resistance due to the resistance of the wire, but this will be disregarded in the following explanation as it is not serious in its effect, as a rule. The resistance may be considered as a certain amount of added resistance in series with the coil.

The hydraulic analogy of an inductor is a paddle wheel water motor in the pipe circuit which has a flywheel attached to it. The size of the flywheel is analogous to the amount of inductance of the coil. The larger the flywheel the greater the equivalent inductance. Naturally, the motor will exert a certain amount of back pressure as it comes up to speed and will tend to coast as pressure is removed from its rotor blades.

Fig. 26 shows the hydraulic equivalent of the inductor and resistor in series. The water motor will run after pressure has been applied by the pump and will require little energy to operate, as the water motor is considered to be frictionless. Power from the pump will be required to force water through the restriction in the pipe.

In this same steady state condition, the electrical circuit will show a flow of current through the circuit limited by the resistor alone and a steady field will be maintained in the core of the inductor.

The transient condition presents a different picture. In the hydraulic analogy in Fig. 26 consider that pressure is applied to the system. The water motor will tend to prevent water flow until the motor wheel has reached its equilibrium speed, at which time the steady state condition will exist. When the water motor is disconnected from the circuit by closing valves on both sides of it, there will be a high pressure built up due to the paddle wheel tending to force water against the closed valve. This pressure will diminish as the motor slows down, due to frictional churning of the water. Finally the motor will stop.

Electrically, the transient conditions will be represented by the battery being connected to the inductor and resistor in series. The inductor will tend to prevent flow of current in the circuit and some time will be required for the current to reach a steady value which is limited in value by the resistor alone. When the battery is removed a large voltage will build up across the inductor due to the collapse of the field of the inductor. The voltage will diminish near the end of the collapse.

An excellent example of this last phenomenon is exhibited in the usual battery ignition system in an automobile. A transformer is

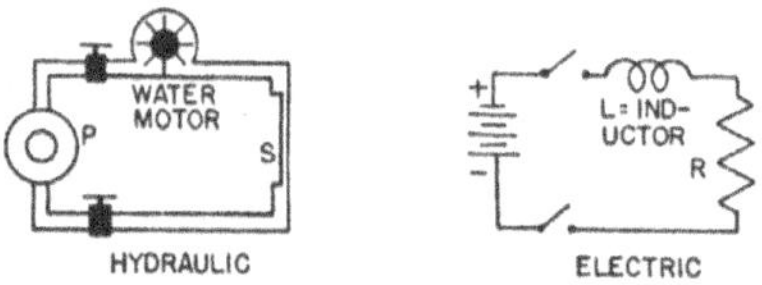

FIG. 26—SERIES INDUCTANCE AND RESISTANCE

connected with its low resistance primary in series with the battery and the breaker points in the distributor. Heavy current then flows in the primary and sets up a field in the iron core. A secondary with a large number of turns is wound on this same core. At the instant that the breaker points open, the field collapses and a high voltage, of the order of ten to twenty thousand volts, is produced across the secondary, which fires the spark plug selected by the distributor.

In servicing relay windings, a rather severe shock may be encountered if a circuit connected to the relay is opened with the fingers across the coil. Actually, the wave shape of the impulse is very steep and the energy represented is small, so that the shock may hurt but is rarely very dangerous. Large inductors may produce a fatal shock under such conditions.

The hydraulic analogy of the inductor in parallel with a resistor is shown in Fig. 27. After pressure is applied from the pump a large flow of water will take place through the water motor. Nothing will limit the flow except the resistance of the connecting pipes and some resistance represented by the water motor.

In the transient condition, the water motor will present resistance to the flow of water but the flow will rapidly build to a very high value. When pressure is removed there will be a very large pressure from the exceedingly high speed of the water motor. This pressure will cause water flow in the pipe with the stricture if it still remains in parallel with the water motor.

Electrically, the steady state condition will represent a very high flow of current in the inductor. Usually this high flow will blow out a fuse, cause damage to the battery or destroy the inductor due to heating. Relay coils which must remain across a voltage are wound with wire whose resistance is adequate to limit the current flow sufficiently. This is equivalent to the case of the resistor and inductor

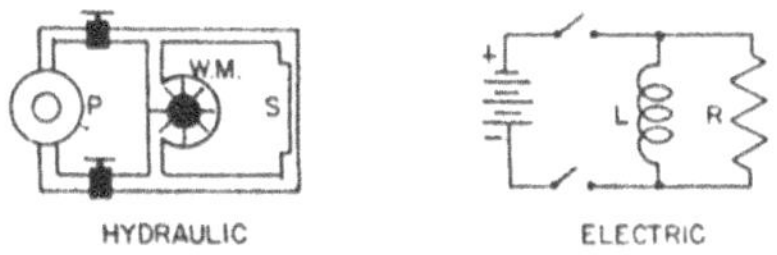

FIG. 27—PARALLEL INDUCTANCE AND RESISTANCE

in series as described before. In the transient condition, the current flow will be limited by the inductor but the current will increase rapidly in value. When the inductor is removed from the circuit a very high voltage will exist across the inductor due to the heavy field collapse. If the resistor is connected in parallel with the coil as shown, the current will be of high value through the resistor until the field finally collapses.

Unlike the condenser, the inductor is not capable of storing energy for an appreciable time and, as soon as it is removed from the circuit, the field will collapse and usually will not return its energy to the circuit. Voltage at the terminals will reach high values on field collapse if they are not connected to a load.

The combination of inductance, resistance and capacitance is shown in Fig. 28. In the steady state condition, the condenser will remain charged and the current will flow through the inductor and be limited in value by the resistor in series with it.

The transient condition is more interesting. The hydraulic analogy will show a diminishing water flow into the condenser as pressure is applied but little flow through the inductor branch of the circuit at first, due to the inertia in starting the water motor. Finally, the water will flow normally through the pipe stricture and water motor and no water will flow to the condenser.

When the pump is removed from the circuit the water motor, aided by the condenser pressure, will tend to sustain flow through the stricture. The piston will move to the neutral position and, due to the inertia in the water motor, it will pull the piston well beyond neutral. This stores pressure in the reverse direction. As the water motor slows down, the pressure exerted by the springs in the piston will cause the direction of flow to reverse and the motor will start revolving in the opposite direction. When this part of the cycle ends, the condenser will again reverse water flow. Each time this happens,

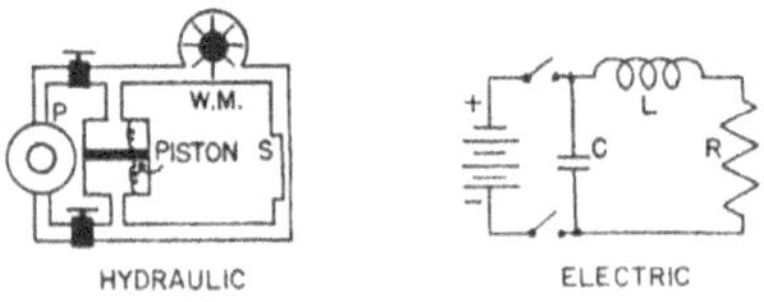

FIG. 28—INDUCTANCE, CAPACITANCE AND RESISTANCE IN RESONANT CIRCUIT

energy will be lost by forcing water through the stricture and producing heat.

The system is actually in an oscillatory condition as water flows first one way and then the other until finally the flow becomes increasingly feeble and stops. In this case the energy stored in the inductor and condenser at current turn-off is finally dissipated in the load resistor.

Electrically, this condition is equivalent to the operation of a tuned circuit. As the voltage is applied, the current to the condenser starts at a high value and decreases until the condenser reaches full charge. Current through the coil increases as time goes on until current flows through the resistor, limited by its resistance alone. When the battery is removed current flows in one direction and then the other and the energy is transferred first to the inductor's field and then to the capacitor until, after repeated oscillation, the current is reduced to zero by the resistive loss in the circuit.

Alternating Current Equivalent Circuits—The transient condition in each of the direct current cases gives a clue to the operation of these same circuits with alternating voltage supplied.

Fig. 29 shows a reciprocating pump which should be substituted mentally for the unidirectional pump in each of the figures. The frequency of alternation is determined by the number of complete pump cycles (or shaft revolutions) per second. Electrically, the battery is replaced by an alternator (alternating current generator).

The condition existing in Fig. 23, when the reciprocating pump replaces the centrifugal pump, is very much the same as before. Instead of the loss in the resistor being due to continuous flow of water

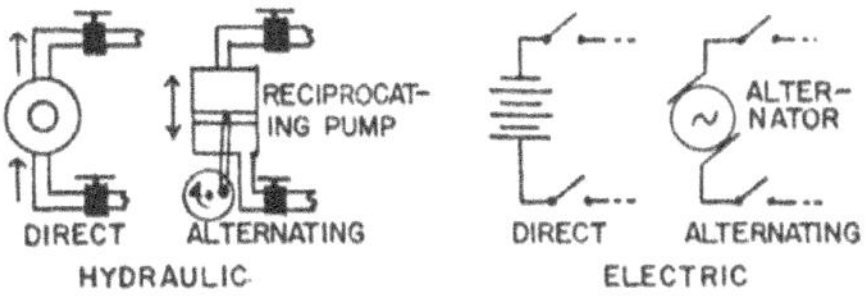

FIG. 29—DIRECT AND ALTERNATING CURRENT GENERATOR ANALOGY

through the stricture the loss will occur as the water passes first one way and then the other through this stricture. The peak flow of water will be greater than the equivalent steady flow. This is necessary as the same amount of lost power must be represented and the pulsing flow must stop twice during each cycle.

The condition above is electrically equivalent to current flow through the resistor which reaches peak values greater than the direct current value for the same heat loss. The root mean square (RMS) value of the alternating current is equal to the direct current value. The RMS value of voltage must likewise be equal to the direct voltage. Highest water flow occurs when highest pressure exists at the pump. This condition represents unity power factor as voltage and current reach a peak at the same time.

The hydraulic condition represented in Fig. 24, for the reciprocating pump, is quite different from the above case. The piston in the cylinder tends to follow the pump with little back pressure. Recalling the transient condition, the piston moved immediately upon application of pressure. As the reciprocating pump reaches its top or bottom position the water flow into the cylinder will be less than when the pump is at midstroke (and moving fastest). Thus the water flow is a maximum at center stroke when the pressure is least (considering the pressure at one end of the stroke as positive and the pressure at the other end of the stroke as negative). The water passing through the stricture may represent but a small flow in comparison with that to the cylinder.

This case is equivalent to the electrical condition. The current flow to the condenser is least when the voltage reaches the peak value and greatest when the alternating voltage is at its most rapid change, i.e., from zero upward to positive or zero downward to negative. The resistor current may be small in comparison to the condenser current. If the resistor has a very low value the current through it might be greater than that through the condenser. In this case the current peak would more closely approach the voltage peak. If the flow through the resistor were far more than to the condenser, the condition would be practically the same as in the case represented by Fig. 23.

The power factor is said to be leading when the current reaches a peak before the voltage peak. The amount of lead is based entirely on the amount of current flowing through the condenser in comparison to that flowing through the resistor. When all current flows through the condenser, the power factor is zero; when all flows through the resistor, the power factor is unity. For intermediate values the power factor is some fraction of unity.

The reciprocating pump case in Fig. 25 is rather simple as the piston in the cylinder follows the pump oscillations. Water flows through the resistor much in the same manner as it would were the cylinder and piston not in the circuit. If the cylinder is small the flow of water may be decreased, for the cylinder and piston will have insufficient capacity for high flow.

Electrically this same condition means that a condenser in series with an electrical load will allow alternating (not direct) current to flow and if the condenser is large enough there will be little voltage drop across it. The current will lead the voltage in this case.

When the reciprocating pump is substituted in Fig. 26 the condition is quite different. The water motor cannot respond to the reciprocating pump motions as it has inertia and is late in its operation. Thus, when the pressure is maximum the water motor is still gaining speed and even after the pressure starts to fall the motor increases speed until the pressure drop across the motor is zero. Shortly thereafter the motor decelerates. The maximum water flow occurs some time after the maximum pressure.

Electrically, the condition is represented by maximum current flow occurring after the maximum voltage. There is a lag dependent upon the inductance of the inductor. The lag is great when the inductance is large. The resistor merely serves to limit the maximum current possible, especially with a very small inductor.

The reciprocating pump case in Fig. 27 hydraulically represents the water motor being placed directly across the reciprocating pump. The stricture in the pipe is quite independent of the action of the motor unless the stricture allows very large flows in comparison to those occurring through the motor. If the inertia of the motor is small the flow through the water motor may be very large.

Electrically, this condition is represented by the current flow through an inductor placed directly across the voltage source. If the inductor has small inductance the current flow may be great enough to cause excessive heating if there is any resistance within the inductor, which is always true from a practical viewpoint. The current will lag the voltage, however the larger the inductor, the less will be the lag as the principal current will pass through the resistor.

The reciprocating pump case in Fig. 28 represents a case where both the water motor and the cylinder with piston are connected to the circuit. It is possible to arrange the inertia of the water motor and the capacity of the cylinder so that they have equal and opposite effects. In this case the flow will neither lead nor lag the pressure alternations. The circuit will be similar then to the case where only a stricture is attached to the pump.

Electrically, this case represents resonance which may occur in any mechanical or electrical system. If the lead and the lag are identical the current will follow the voltage variations. If little resistance is present in the circuit the case is very similar to the transient condition explained previously. As a small amount of resistance must be present the circuit will draw a very small current and the principal flow of current will be in oscillation between the capacitor and the inductor.

Capacity—The capacity of a condenser can be likened to the size of the cylinder in which the spring-returned piston is placed. The strength of the spring corresponds to the dielectric constant of the insulator. A large piston area represents high capacity. A small area with a weaker spring will allow greater piston movement at the same pressure and store an equal amount of energy.

There are two methods of storing an equal amount of energy in a condenser. The first method is to have a large piston area (large capacity) and charge with a given pressure. Thus, the pressure might amount to ten pounds per square inch and the volume for that area at a given displacement might be one cubic foot. The amount of energy available would be given in gallons at a given pressure. The other method of storing the same amount of energy is to reduce the volume (area or displacement) of the cylinder but to increase pressure so that the product is the same as before. Electrically, the charge in a condenser is given as the product of the capacity and the voltage. Although there will be few occasions where the following formula may be used, the quantity of electricity (Q) stored will be:

$$Q = CE$$

The quantity is given in coulombs and a one farad condenser will store one coulomb at one volt.

When the condenser is used on alternating currents the frequency of the current will determine the amount of energy that can be passed. At a very low frequency the analogous piston will slowly move with the alternating pressure and will limit the current. On the other hand, if the alternating current has a high frequency the same amount of water in the same cylinder may allow a large transfer of energy, as it is displaced many times oftener in a given time. Thus a condenser of fixed size will allow alternating current flow in proportion to the increase of frequency.

For example, a one microfarad condenser will pass just as much energy at twenty cycles as a ten microfarad condenser at two cycles per second.

Inductance—Inductance can be compared with the flywheel mass of the water motor. The heavier the flywheel the more time will be required to accelerate it at a given pressure. The energy (Q) that can be stored in the field of an inductor (L) is:

$$Q = LI$$

One henry at one ampere will represent a storage in the field of one coulomb.

The higher the frequency the less time per cycle will be allowed for the flywheel to accelerate. The higher the frequency the less current will be passed in proportion. This is exactly opposite to the case of the capacitor, which passes more current at a given voltage with increase of frequency.

Reactance—Resistance was used to express the current limiting quality of resistors. A similar expression to show the current limiting value of an inductor or capacitor is reactance.

Reactance depends on the capacity or inductance of the unit as well as upon frequency. Reactance can never cause more than a quarter cycle (ninety degree) lead or lag in an electrical circuit. This will be evident in any of the hydraulic analogies.

Reactance can be obtained from the following formulas:

$$X_L = 2\pi f L$$

$$X_C = \frac{1}{2\pi f C}$$

Where C = the capacity in farads (a microfarad, the usual quantity used, is one millionth of a farad).

L = the inductance in henries.

π = 3.1416, a constant (3.14 is close enough for most ordinary calculations).

f = frequency in cycles per second.

X_C and X_L = the capacitive and inductive reactances, respectively.

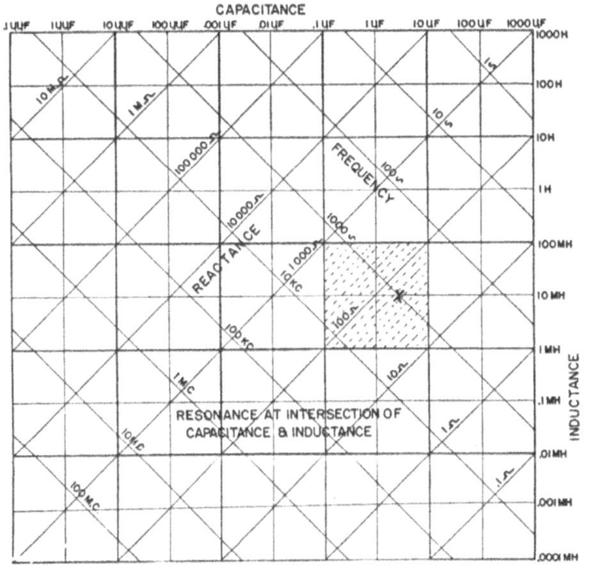

Fig. 30—Reactance Chart

The value of reactance may be calculated but it is usually easier to use the table of Fig. 30. The method of use is to select the value of the inductor or capacitor on the chart. This line is then followed to a point where it intersects the desired frequency. The intersection of these two lines gives the reactance directly.

For example: What is the reactance of a ten millihenry coil at 1000 cycles? Follow the 10 MH line to the left until it intersects the 1000 line at X. The reactance will be roughly seventy ohms, as the intersection is between the ten and one hundred ohm lines. To find the answer more accurately refer to Fig. 31 which is a finer scale and replaces the shaded section of Fig. 30. Follow the ten

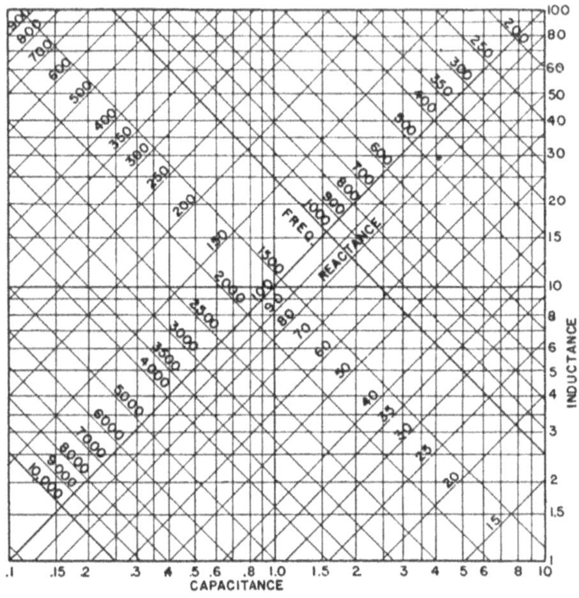

FIG. 31—FINE SCALE CHART FOR FIG. 30

line to the left until it intersects the 1000 cycle frequency line. The answer is sixty-three ohms, which is close enough for most purposes.

Be sure when the fine scale is used that the values are used correctly for the part of the main figure used. In short, the fine scale will give the numerical answer without regard to the correct number of zeros and the magnitude is determined from the large scale chart.

To cite a specific case, suppose that a one henry inductor was placed across a 120 volt, 60 cycle voltage. How much current will flow?

$$X_L = 6.28 \times 60 \times 1 = 380 \text{ ohms.}$$

As the voltage divided by the reactance gives the current, about one third of an ampere will flow at a lagging power factor.

Let it be supposed that the coil had but one ohm of resistance. This would affect the flow of alternating current but slightly, but if the coil were placed across 120 volts of direct voltage where reactance is absent, the current would be 110 amperes and the coil would promptly burn up.

If a condenser whose reactance is 360 ohms were placed across the 120 volt, 60 cycle circuit, it would draw one third of an ampere, but at leading power factor. When placed across a direct voltage it would draw no current.

Suppose that both the condenser and the inductor were placed across the circuit, in parallel, at the same time. The effect would be that no appreciable current would flow from the generator of alternating current, as the lead and lag would be the same except for sign and the equivalent would be infinite resistance and no reactance.

The sign of the reactance (lead + and lag −) is of great importance. The capacitive and inductive reactance serve to cancel one another. It also makes a difference as to how circuit elements are connected together.

With the information given up to this point it is possible to find the current through any one element but not in combination with others. Fig. 32 shows some of the simple relations of powers, voltages and impedances.

Complex Notation—There are several rather involved methods of computing the behavior of the circuit. The simplest method is known by the misleading term "complex notation."

The first principle of complex notation is that all inductive reactances shall be considered positive and all capacitive reactances, negative. Further, all such terms shall be preceded by the letter "j." The term "j" is the square root of minus one, or j times j equals minus one. The previous rules for the treatment of elements in series or in parallel will be used.

Let the inductor and the capacitor in parallel be treated by this method. As they are in parallel, the product over the sum will give the total impedance (Z) thus:

$$Z = \frac{j360 \times j360}{j360 - j360} = \frac{-129{,}600}{0} = \text{infinity}$$

NO.	CIRCUIT	VOLTAGE	POWER	IMPEDANCE
1	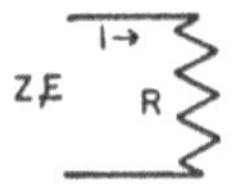	$E = IR$	$W = I^2R$ $W = \frac{E^2}{R}$	$Z = R$ $R = \frac{E}{I}$
2	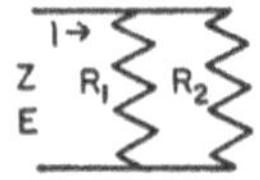	$E = IR$	$W = I R$ $W = \frac{E}{R}$	$Z = R$ $R = \frac{R_1 R_2}{R_1 + R_2}$
3	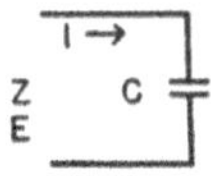	$E = IX_C$	$W = 0$	$Z = X_C$ $X_C = \frac{1}{2\pi f C}$
4	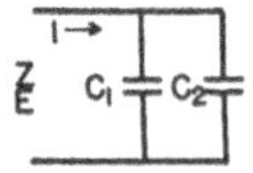	″	″	″ AND $C = C_1 + C_2$
5	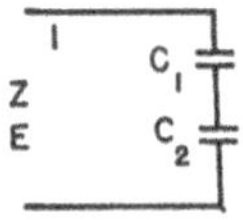	″	″	″ BUT $C = \frac{C_1 C_2}{C_1 + C_2}$
6	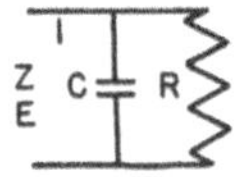	$E = IZ$	$W = \frac{E^2}{R}$	$Z = \frac{R\ X}{\sqrt{R^2 + X_C^2}}$
7	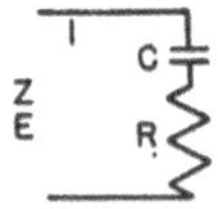	$E = IZ$	$W = I^2R$	$Z = \sqrt{R^2 + X_C^2}$
8	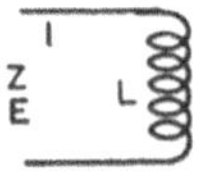	$E = IX_L$	$W = 0$	$Z = X_L$ $X_L = 2\pi f L$
9	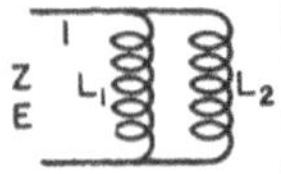	$E = IX_L$	$W = 0$	″ AND $L = \frac{L_1 L_2}{L_1 + L_2}$

FIG. 32

In other words the two in parallel have infinite impedance and no current will flow.

The coil actually had a series resistance of one ohm and it will be considered that the condenser also had some leakage which could be represented by the same power loss as a two ohm resistor in series

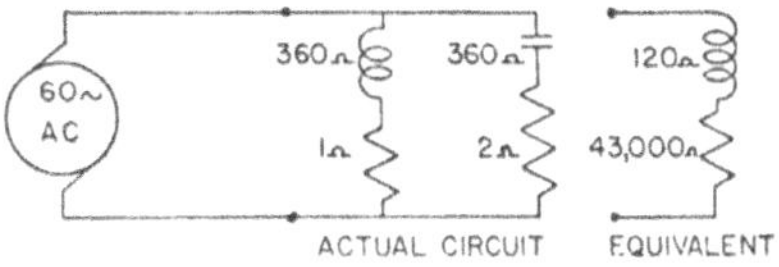

FIG. 33—EQUIVALENT CIRCUIT

with it (Fig. 33). For writing the complex expression for the inductor a term called impedance (Z) is used. It is the complex combination of the reactance and resistance of any unit or, very simply, its opposition to current flow. Thus the impedances of the two units are as follows:

For the inductor $Z_L = 1 + j360$
For the capacitor $Z_C = 2 - j360$

By multiplying these two together and dividing by the sum as before, the total impedance will be:

$$Z = \frac{(1 + j360)(2 - j360)}{(1 + j360) + (2 - j360)} = \frac{2 - j360 + j720 + 129{,}000}{3}$$

Which may be rewritten without much error as:

$$Z = \frac{129{,}000 + j360}{3} = 43{,}000 + j120$$

The answer gives at a glance the effective value of the load. The load amounts to 43,000 ohms in series with a coil (+) of 120 ohms reactance. The reactance is not yet in a form where the total impedance is given as a single number. To obtain this, it is necessary to square both terms, add them and find the square root of the sum for the answer. The impedance above is then very slightly greater than 43,000 ohms.

The impedance may be used in the simple expression:

$$E = IZ$$

This is the same expression as used for direct current circuits with the Z, or alternating current impedance, substitued for R.

The power factor does not enter this expression and it may be found separately. The power factor is:

$$PF = \frac{X}{R}$$

Where X is the reactive or j term in the final expression for Z and R is the resistance term or the first term in the expression.

Examples of Complex Notation—Series circuits are comparatively easy to solve by complex notation. The following elements are connected in series:

1. Inductor = 400 ohm reactance and 200 ohm resistance
2. Capacitor = 100 ohm reactance and 5 ohm series equivalent resistance
3. Resistor = 1,000 ohms
4. Capacitor = 100 ohms and negligible series resistance

The impedance may be written:

$$Z = (200 + j400) + (5 - j600) + 1000 - j100$$

$$Z = 1205 - j300 \qquad Z = \sqrt{1205^2 + 300^2}$$

$$Z = 1240 \text{ ohms, approximately}$$

Where parallel circuits are concerned the handling of the equations is slightly more difficult. An example will show the methods employed.

A coil whose impedance is $30 + j100$ is placed in parallel with a condenser whose impedance is $10 - j500$. The impedance of the combination is:

$$Z = \frac{(30 + j100)\ (10 - j500)}{(30 + j100) + (10 - j500)}$$

$$Z = \frac{300 - j15{,}000 + j1000 + 50{,}000}{40 - j400}$$

A common mistake is often made in the multiplication of the numerator by forgetting that the product of two positive j terms gives a negative term. If one j term is negative and the other positive, the product is positive.

When simplifying, it may be seen that the 300 term is not important in comparison to the 50,000 term so that it may be discarded with little error. Such simplifications, if done with care, save much complication. If one of the j terms were small it could be discarded as long as another remained which was large in comparison.

The equation may be simplified to:

$$Z = \frac{50{,}000 - j14{,}000}{40 - j400}$$

The resistive and reactive part of the equations cannot be separated in the present form so the whole expression should be multiplied by unity in the form of the denominator over the denominator, with the sign changed. This will give the difference of two squares in the denominator with no j term. It is very important to note that this will appear as the sum of two squares, as the squaring of the j changes the sign normally to be expected. So:

$$Z = \frac{(50{,}000 - j14{,}000)\,(40 + j400)}{(40 - j400)\,(40 + j400)}$$

$$Z = \frac{(50{,}000 - j14{,}000)\,(40 + j400)}{1{,}600 + 160{,}000}$$

For approximate solution, the 1,600 in the denominator may easily be discarded, if desired, with one per cent error. Multiplying the numerator:

$$Z = \frac{2{,}000{,}000 + j20{,}000{,}000 - j560{,}000 + 5{,}600{,}000}{160{,}000}$$

Combining and dividing everything by 1000 there remains:

$$Z = \frac{7{,}600 + j19{,}440}{160} = 47 + j120$$

This means that the combination amounts to a 47 ohm resistor in series with a 120 ohm reactive inductor. The plus sign shows that the reactive element is inductive. Slide rule accuracy is used.

The total impedance is the square root of the sum of the squares of $Z = 130$ ohms, approximately.

While the process may look complicated it is easily mastered in a short time. Various series and parallel combinations may be boiled down by combining parts of the circuit in steps. The two example problems apply in the greatest percentage of circuits which must be analyzed, particularly those associated with antenna problems on aircraft.

QUESTIONS ON CHAPTER III

1. A coil or inductor is equivalent to the hydraulic:

( ) a. Piston in a cylinder with spring centering. ( ) b. Water motor. ( ) c. Pipe with stricture. ( ) d. Pressure.

2. A condenser is equivalent to the hydraulic:

( ) a. Piston in a cylinder with spring centering. ( ) b. Water motor. ( ) c. Pipe with stricture. ( ) d. Pressure.

3. A resistor is equivalent to the hydraulic:

( ) a. Piston in a cylinder with spring centering. ( ) b. Water motor. ( ) c. Pipe with stricture. ( ) d. Pressure.

4. A resistor and a condenser in series attached to an alternating voltage will exhibit:

( ) a. Lagging power factor. ( ) b. Leading power factor. ( ) c. Always unity power factor. ( ) d. Either lead or lag.

5. A resistor and inductor in series and attached to an alternating voltage will exhibit:

( ) a. Lagging power factor. ( ) b. Leading power factor. ( ) c. Always unity power factor. ( ) d. Either lead or lag.

6. A resistor connected to an alternating voltage source will exhibit:

( ) a. Lagging power factor. ( ) b. Leading power factor. ( ) c. Unity power factor. ( ) d. Either lead or lag.

7. What is the approximate reactance of a 2 henry inductor at 50 cycles?

( ) a. 50 ohms. ( ) b. 600 ohms. ( ) c. 6000 ohms. ( ) d. 10 ohms.

8. What is the approximate reactance of a 20 henry inductor at 50 cycles?

( ) a. 6000 ohms. ( ) b. 60 ohms. ( ) c. 600 ohms. ( ) d. 50 ohms.

9. What is the approximate reactance of a 3 farad condenser at 50 cycles?

( ) a. 1 ohm. ( ) b. 1000 ohms. ( ) c. 0.001 ohm. ( ) d. 10 ohms.

10. What is the approximate reactance of a 3 microfarad condenser at 50 cycles?

( ) a. 100 ohms. ( ) b. 1000 ohms. ( ) c. 10 ohms. ( ) d. 0.001 ohm.

11. What is the approximate reactance of a 3 micromicrofarad condenser at 50 cycles?

( ) a. 1000 megohms. ( ) b. 10,000 ohms. ( ) c. 0.0 ohm. ( ) d. 10 ohms.

12. What is the inductance of a one millihenry inductor at 5000 cycles?

( ) a. 1000 ohms. ( ) b. 10 ohms. ( ) c. 30 ohms. ( ) d. 0.002 ohm.

13. What is the inductance of a one microhenry inductor at 5 megacycles?

( ) a. 10 ohms. ( ) b. 30 ohms. ( ) c. 30,000 ohms. ( ) d. 1 ohm.

14. Power factor, reactance and resistance are related as follows:

( ) a. $PF = \frac{X}{R}$. ( ) b. $PF = X \times R$. ( ) c. No relation. ( ) d. $X = R$.

15. A combination of two elements gives an impedance which is $100 - j50$. The elements probably are:

( ) a. A resistor and inductor. ( ) b. Two resistors. ( ) c. A resistor and condenser. ( ) d. A condenser and inductor.

16. Two impedances were joined in series. Their values are $40 - j200$ and $100 + j200$. The total impedance is:

( ) a. 60 ohms. ( ) b. 140 ohms. ( ) c. $140 - j400$. ( ) d. $60 + j400$.

17. An antenna has an impedance of $40 - j50$ ohms. What will be required in series with it to resonate it?

( ) a. An inductor of 50 ohms reactance. ( ) b. It is resonated. ( ) c. A condenser with 50 ohms reactance. ( ) d. A fifty ohm resistor.

18. What elements could be represented upon simplification of the following formula: $50 - j30 - 100 - j70 + 50 = Z$?

( ) a. A condenser and resistor in series. ( ) b. An inductor in series with a resistor. ( ) c. A condenser. ( ) d. A 100 ohm reactive inductor.

19. The product of two J terms gives:

( ) a. 1. ( ) b. − 1. ( ) c. j. ( ) d. − j.

20. An inductor and condenser in series will resonate if:

( ) a. The correct frequency is used. ( ) b. Enough resistance is added to make it resonate. ( ) c. The capacitor is equal to the inductor. ( ) d. They can never resonate in series.

Chapter IV

ACCESSORY EQUIPMENT

There are numerous pieces of accessory equipment on aircraft, some of which are necessary to the operation of the radio equipment. Others are used for electrical operations and indication of fuel-air ratio, temperature and so forth. Each accessory that is commonly found is treated in the following paragraphs. The principle of operation of each is covered, including schematics of the electrical connections in most cases.

Fuses—Electrical circuits must be protected against failure due to one of the elements in the circuit drawing excessive current. A fuse is a link of low melting point alloy wire in a suitable holder. When the fuse rating is exceeded for a period, the fuse wire will melt and open the electrical circuit, thus preventing damage to the unit drawing excessive current as well as to the power supply, such as a battery or generator.

Dynamotors, lamps, vacuum tubes and other equipment on aircraft usually require a much higher current when the apparatus is turned on than immediately afterwards. For this reason, fuses will have a higher rating than that of the operating load. For example, a certain transmitter used on aircraft normally has a filament drain of fifteen amperes when the filaments are hot. However, the starting current is about seventy amperes at the instant of turning on the filaments as the filaments increase resistance when heated. A twenty-five ampere fuse is sufficient for this load as the seventy amperes flow for so short a period that the fuse does not have sufficient time to melt.

A dynamotor for the transmitter drew about seventy amperes when running but at the time of starting required about 250 amperes. In this case the starting time is long enough (about a quarter of a second) for the fuse to melt unless it has a rating of about 200 amperes.

Fuses are usually made of material which oxidizes and flakes off in time, thus lowering the ampere rating of the fuse. Heating and cooling also produce a type of crystallization in fuses, thus shortening fuse life. Vibration also accelerates flaking and crystallization. Fuses on aircraft are normally replaced about every six months as a precaution against blowing out on normal loads.

Circuits to be protected by fuses are usually arranged with one large fuse to feed a certain type of load, radio for example, and this fuse is followed with smaller fuses for each smaller part of the unit installation, as in Fig. 34. The main fuse is so arranged that it will

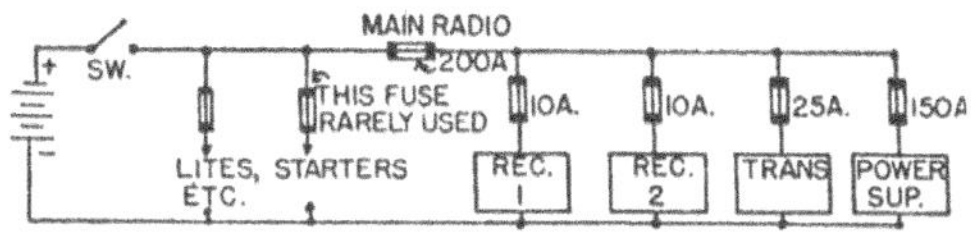

FIG. 34—FUSING

carry the sum of all the loads with a margin of safety. When one unit fails, the smaller fuse connected to that unit will normally go out first. In this way, the entire circuit is protected and if one unit fails it will not disturb the voltage supply to the other units of the system protected by the main fuse. This also simplifies trouble shooting.

An ample quantity of spare fuses is always carried on air transports.

Batteries—The batteries used on aircraft rely on a chemical action to produce power. When the battery is recharged the chemical action is reversed. The generator which charges the battery must supply a slightly higher voltage than the normal battery voltage for current to flow into the battery. If the generator is supplying a lower voltage the battery will discharge into the generator and the generator will behave as a motor, unless the usual reverse current relay is provided.

Wet storage batteries have two sets of plates per cell. One set of plates is made of lead and the other is made of lead well-pocketed with lead oxide. The solution in the battery is sulphuric acid and water. Its specific gravity, or density, is an index of the amount of charge in the battery. Specific gravities will vary somewhat, depending on the type and use for which the battery is designed. In most aircraft batteries the specific gravity at full charge is about 1.300; when discharged it is about 1.100, or slightly less.

A battery left in a discharged condition will deteriorate rapidly and it is essential to keep batteries charged. The other factor which contributes to short battery life is heating, due to high charge or discharge currents. This causes serious plate warping and will work the lead oxide out of the plates. There is only one method to charge a battery properly and that is to place it across a generator which is maintaining exactly 14.2 volts for a twelve volt battery or 28.4 volts for a twenty-four volt battery. At this voltage, the battery charging current will taper as the battery voltage rises and will thus assure a full charge and overcharging of the battery cannot occur. With small chargers, on which current alone is indicated, a safe empirical rule is to charge at a current rate not exceeding ten per cent of the ampere hour rating of the battery. As the battery approaches full charge, as determined with an hydrometer, the charge should be tapered to at least half this rate. As soon as the battery is fully charged it should be removed from the charger.

In exceptionally hot climates where batteries must be charged with room temperatures of over 90° F., they should be placed in a trough of cool running water so as to be submerged to within about an inch of the battery top.

There are two other precautions which should be observed in the care of batteries. First, the top of the battery must be kept clean and the terminals should be greased lightly. As the acid is extremely corrosive it will damage metal parts quickly and will cause burns on hands and clothing. Wool is not rapidly attacked by the acid so battery handlers often wear woolen clothing. The second precaution is to fill the battery up to the proper point with pure distilled water only. If tap water is used the battery will fail prematurely. In the aircraft non-spillable batteries the level in the cells should be up to the bottom of the filler neck and no higher. In all batteries the water level should cover the plates and will usually be about a quarter to a half inch over the plates.

Upon charge, the caps must always be removed from the cells to allow the explosive gas to escape. Sparking of wires or open lights near the battery may easily cause an explosion which will throw acid in all directions and damage the battery top.

In case acid is accidentally spilled on a person, or particularly if it gets in the eyes, it should be removed immediately with copious amounts of water, although a much better method is to apply a solution of sodium bicarbonate, also known as "baking soda."

Battery Switches—Almost every airplane has a main or "master" battery switch which is used to disconnect the electrical loads of the airplane from the battery when the plane is not in use.

Transports are usually equipped with a master switch which serves to connect the load of the airplane to the airplane's own battery or to a receptacle on the skin of the plane where a field battery (battery cart) may be attached quickly. When this switch is placed in the center position it disconnects the load from any battery.

Airplanes of the Douglas DC-3 type are equipped with a dual battery and generator system. The electrical loads on the airplane are split in two parts so that the dual system normally is separated in use. In addition to the master battery switch, there are two load transfer switches which are quite separate from the master switch. Fig. 35 shows the electrical wiring of such a system.

As shown, the two load transfer switches make connection of half the load to each battery. However, if both switches are thrown to the same side the complete load is transferred to one battery. The generators are not attached to the batteries but each generator

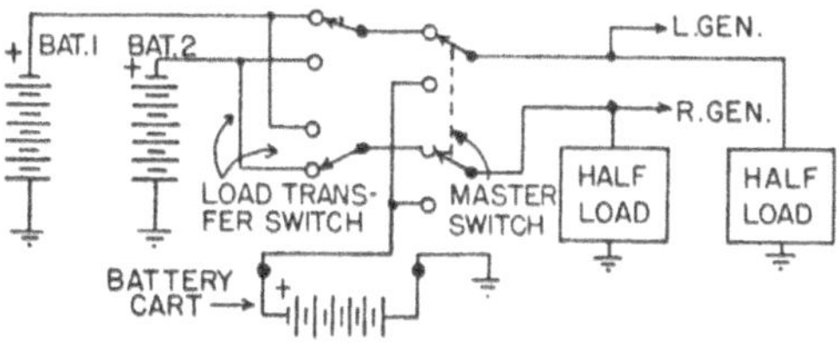

FIG. 35—DC-3 BATTERY SYSTEM

is attached to half the load. Thus if both switches are thrown to one battery both generators will charge that battery and supply the whole load. In the event of failure of one battery this assures that the system will be capable of supplying high peak loads that generators alone cannot supply.

A generator might fail, in which event the battery which is not being charged will soon run down. However, if the load transfer switches are thrown to one battery the generator will attempt to keep the one battery charged, or at least prevent it from discharging rapidly. Removal of some electric light or other electrical loads, in this circumstance, will assure continuous electrical supply.

If a battery should short-circuit, it may be removed from the circuit by throwing both loads to the good battery. In the event half of the electrical system should short-circuit to ground, it would be quite necessary that this short circuit not be transferred to the other system which is operating. In this unlikely event, half of the load system would be dead and its transfer switch would be left off.

Wiring in the cockpit is so arranged that some lights will remain in commission, as approximately half of the lights are wired to each load system. The safety factor afforded by this sort of system is of great value.

Separate generator switches are provided so that the control box and generator may be removed from the load circuit in case of generator or control box open or short circuit.

Such planes as the DC-3 are equipped with a volt-ammeter which gives load voltage readings upon pressing a button on the meter face. The voltmeter supplies valuable information in case of trouble. Low voltage indicates low batteries, insufficient charge rate or short circuits. High voltage indicates failure of part of the generator control box which will usually end in generator failure within a short time due to excessive charging current.

Failure to throw the master battery switch from the battery cart to the plane's battery is a comon occurrence. No hint of trouble will occur in daytime until the radio transmitter is used. The generator is incapable of supplying the high momentary load without a battery and the transmitter will not work. The receivers and lights will work as the generator can supply ample power for such relatively light loads. Naturally, lights will dim badly when the transmitter comes on, due to the extremely low voltage in such an instance.

Magnetos—The arrangement of a modern magneto is shown in Fig. 36. A double magnet structure containing two fixed permanent magnets are revolved through suitable gearing by the main

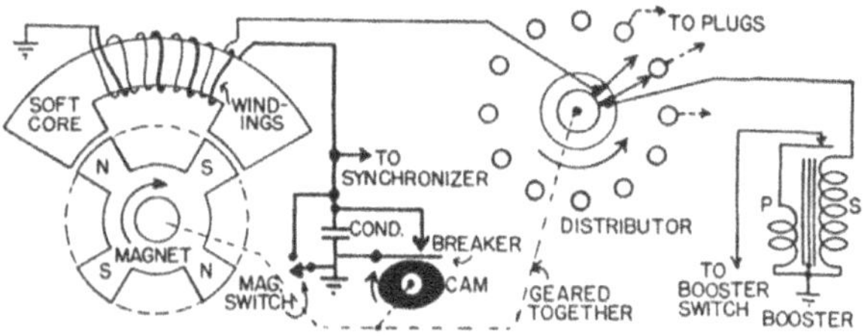

FIG. 36—MAGNETO SYSTEM

engine (one and one-eighth to one for nine cylinder engines). This sets up an alternating field of two cycles per magnet revolution in the stationary iron structure on which the primary and secondary coils are mounted.

If the magneto primary coil is open the voltage generated will be insufficient, normally, to fire the spark plugs. When the primary is closed through the breaker points a high current of about 2.2 amperes will be set up in the primary. As the secondary is open no current will flow in it. As the breaker points open the field will suddenly collapse and the rapid current decrease in the primary will induce a high voltage in the secondary of 20,000 volts, or more. The high secondary voltage will pass through the distributor to the selected plug and fire the gasoline mixture in the cylinder.

The magneto switch in the cockpit is arranged to ground the primary so that the opening and closing of the distributor points will not affect primary current. Thus, the magneto is effectively put out of service. Airplane magnetos are connected to a master ignition switch which will ground either or both of the magnetos on each engine.

It should be kept in mind that when the magneto switch is on "left" magneto (rear plugs) the electrical circuit in the switch actually is grounding the "right" magneto (front plugs). Occasionally, this becomes confusing in service and work is performed on the wrong magneto. "Left" on the switch means that the left magneto is operating and the other one is shorted out.

Capacity of the spark plug leads to their shielding will require somewhat higher magneto voltages than without shielding. Further, when the spark plug flashes, the additional power stored by the capacity will cause higher currents than without shielding and will cause more severe plug point burning.

The condenser across the points of the breaker is necessary to prevent burning of the breaker points. If the condenser short-circuits it will place the magneto out of service. If the condenser opens, the points will soon burn badly and fail to make proper contact.

There are several types of magnetos and the Bendix-Scintilla has been used for illustrative purposes. It employs a rotating magnet as described, while others use a rotating coil assembly with a fixed magnet. The principle of operation is the same however.

Boosters—Most magnetos do not have sufficient voltage developed at low engine speeds to fire the spark plugs and will not fire the plugs at 100 revolutions per minute or less. As it is usually impossible to turn the motor at 100 RPM with the starter, a booster is employed. The booster is usually a coil which has a vibrating contactor on it and produces high voltage when connected to a battery.

The booster switch is operated at the same time as the starter switch. The booster supplies high voltage to the plugs through one magneto distributor causing the engine to fire and start. The booster is often wired to the plane's battery permanently so that when the battery cart is used on the starters the booster coil will obtain full voltage from the plane's unloaded battery.

In some systems the engine timing is retarded either manually or automatically by means of a solenoid retard. In systems such as the Bendix-Scintilla, the booster is connected to another distributor arm which lags the main arm by several degrees, thus producing the retard automatically. As soon as the magneto proper comes up to operating speed the plugs fire from the magneto distributor arm in the normal manner. This booster and added distributor arm are shown in Fig. 36.

Synchronizers—It is very difficult to synchronize the engines by ear on some multi-engined aircraft for two reasons. First, the pilot is in a position where it is difficult for him to hear the engines clearly; second, two beats are heard, one between the two propellers, the other between the two exhausts. Both beats must be eliminated for perfect synchronization and it is often difficult to do this without a synchronizer.

For this reason, a synchronizer of the type shown schematically in Fig. 37 is used on modern aircraft. The magneto ground leads

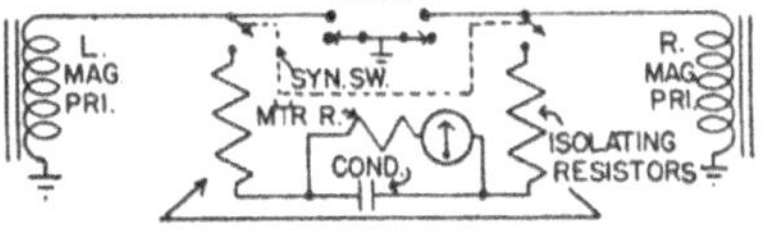

FIG. 37—DUAL SYNCHRONIZER INSTALLATION

which go to the magneto switch are lead through a pair of isolating resistors, of about 20,000 ohms value, to the synchronizer. These resistors are supplied for two reasons: First, to eliminate the possibility of the two magneto systems transferring energy from one to the other due to voltage coupling between them; second, to prevent a short circuit in the synchronizer system from grounding either of the magnetos. The resistors are located as near the magneto switch as possible to eliminate the possibility of short-circuiting of the magnetos in case of wire breakdowns.

It has been shown that as the breaker points on the magnetos open, there is a surge of voltage in the primary and secondary coil.

The primary surge delivers some voltage to the synchronizer circuit. The condenser of the circuit then charges up to a voltage dependant upon the voltage from each magneto. The charge on the condenser will be positive on one side and then positive on the other side as each magneto operates. The meter (connected through a resistor across the condenser) will then pulse from zero to near maximum scale voltage. As the two magnetos become synchronized the voltage difference across the condenser will remain constant and the needle will cease to move. Thus, the synchronized condition is shown when the meter stands still, no matter at what point on the scale.

The meter is of a rectifier type so that, in conjunction with the condenser, the pulses from the magnetos will give steady readings. If the condenser is omitted the meter will be very difficult to stop except at zero on the scale and is not as useful.

Relays—Relays are found in many places on aircraft electrical installations. One of their chief purposes is to save weight. For example, suppose that the heavy battery cables to the main engine starters were routed from the battery up to the cockpit electrical board and then back to the engines. The extra weight might amount to fifty pounds or more on large aircraft. If a relay, which is a magnetically controlled switch in the main circuit, were located between the battery and the starters, the wire length might be much shorter and the relays themselves would weigh but five pounds or so. Thus, five pounds of relays might save fifty pounds of wire with a net overall saving. The wire running to the cockpit is only large enough to operate the relay.

There are two other reasons for the use of such a relay. First, the switch to operate the starter current directly would be so large as to take up a substantial part of the switch space available in the cockpit. The small control switch for the relay occupies the space of a small lighting switch. Second, the current of several hundred amperes flowing through the wires sets up a considerable field. This field, if brought near to the magnetic compass, will cause a deviation of many degrees while current is flowing. While the starters are a poor example of this effect as they are not used in the air there are many other relay-controlled, high current circuits which are used in flight and which must, for this reason, be relay-controlled. Examples of this are the transmitting dynamotor load and heavy lighting loads.

Relays, if improperly designed or cared for, may give far more trouble than any other part of the electrical and radio circuits. Most failures are due to two factors, dirt in the contacts or incorrect spring tension.

Dirt should be cleaned from contact points with carbon tetrachloride or with rough paper slipped between the points and pulled through with the points closed. In severe cases, it may be necessary to use fine sandpaper. Emery paper is a conductor and should not be used. The use of a file will almost always cause eventual trouble with the relay and a file should never be used to clean any relay points.

Excessive bending of relay arms will usually cause trouble later. When an arm must be bent it should be bent slightly farther than is finally intended and then bent back to the correct position. The arm should be bent along its entire length and not at one point. When relay springs of the type used in small relays are well adjusted the arms should show a definite bend as the movable arm reaches

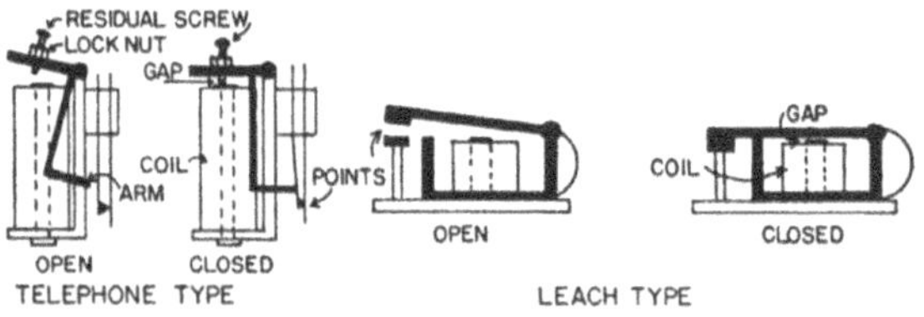

FIG. 38—RELAY TYPES

its full travel. This flexing of arms assures that the contacts will rub together slightly, which is necessary to keep the points clean.

There is one other adjustment on telephone type relays, for residual magnetism. The moving iron armature approaches the pole face when the relay is operated and it is prevented from touching the pole face by a brass screw (sometimes by a fixed air gap) which extends through the armature, as shown in Fig. 38. If the screw is so adjusted that the iron armature touches the iron pole piece the two may stay permanently stuck together because of residual magnetism remaining after the current is turned off. For this reason the brass screw should be so adjusted that the two pieces never touch. The more the air gap maintained, the better the relay and the faster the relay will operate, up to a point where the gap is so large that insufficient pull is exerted to operate the contacts with good pressure. Usually this type of relay requires a clearance between pole and armature of about four to six thousandths of an inch in operated position.

Relays are of two general classes. One class requires current to hold the relay in its operated position; the other requires current to operate the relay, at which time the current stops and if the relay is to be moved to the non-operate position, current again must flow during the time of change of the relay back to its original position. This last is called the impulse type of relay.

Fig. 39 shows an impulse relay in a diagrammatic manner. As the control switch is thrown to the "on" position, current will flow through the armature of the relay through wire 3 and the contact member (1) attached to the controlled circuit will move toward the "on" position. As it approaches the full throw the control current is interrupted in the switch circuit by the motion of the relay and inertia finishes the operation. When current flows through the "off" wire (4) the same process takes place and current is interrupted as the relay turns the controlled circuit off. The actual mechanism for accomplishing this is usually a ratchet device attached to an arma-

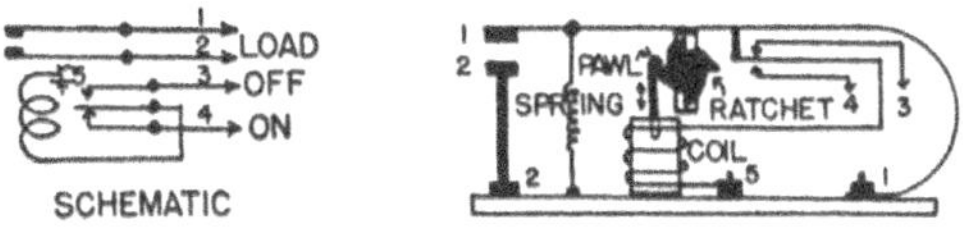

FIG. 39—IMPULSE RELAY

ture which rotates each time the relay is operated. This type of relay has the advantage of not requiring current except during operation, which conserves drain on the battery.

Fig. 38 shows a form of the more common type of relay in which the current holds the relay in its operated position. Different arrangements of contacts can be obtained. Points are closed by attraction of an armature attached through insulators to the movable contact arms. The relay is always shown on schematic diagrams in the rest, or normal, position with no current flowing through the coil. Thus, all contacts that are closed in this position will be open in the operated position.

Another special type of relay, called the solenoid relay to distinguish it from other types, is made for operation of extremely high current circuits such as for motor starters and radio transmitter dynamotors. Instead of a light armature pulling the contacts closed, a coil of wire surrounding a movable rod is supplied. The rod is attached to, but insulated from, a heavy copper plate which is sometimes plated with silver or special arc resisting alloys. The contact is large and is pulled down against two metallic plates which are insulated from one another. These two plates have contact completed between them by the plate attached to the rod through the solenoid. The plate is actually floating and is not attached to any voltage in the open position. When current flows through the coil the field attracts the rod to the relay center, causing the contactor to close the circuit. About five amperes at twelve volts will thus control currents of several hundred amperes.

The conventional types of relay described are usually designed for about fifty amperes maximum and more often for currents of but a few amperes. The impulse type of relay is sometimes found in amperage capacities up to a hundred amperes. The solenoid relays may control up to a thousand amperes or more.

There is usually considerable sparking at relay contacts, particularly when they are attached to loads which are inductive. As the current in such loads is opened the high voltage tends to create a small arc at the contacts. A small condenser and resistor are attached across the contacts to save the contacts from burning. The condenser will usually be in the neighborhood of one tenth of a microfarad and the resistor will be of one hundred ohms or so. Starting with a combination of this sort the capacity and the resistance may be varied to find the best values for least sparking. The method of determining the amount of sparking is usually to observe the sparking as values are changed little by little. Incidentally, the higher the voltage and the lower the current, the easier it will be to reduce such sparking.

Occasionally contacts will stick closed. Usually one of two things will cause this trouble. The first is residual magnetism, which will hold the core in contact with the armature. This can be remedied by adding more spring return pressure to the armature or by being sure that metallic contact is not made between the armature end and the core. The residual magnetism screw of small relays has been explained. In other types of relays using no such screw, a clearance or air gap has usually been allowed. Sometimes, due to poor manufacture or damage, the armature will not maintain this gap. The second reason for sticking is contacts arcing together by high currents when the points are closing.

Generators and Dynamotors—The operation of these machines has been described. The care of such machines is very important for satisfactory operation. Bearings in machines should be carefully examined at each inspection as these bearings may cause considerable electrical noise or slow running and heating of the machine. A lubricant, such as "Lubriplate," should be used in ball bearings as it does not melt and run, yet is not so hard as to be thrown away from the bearings.

Brushes are more important than any other one thing for satisfactory operation. If the commutator or slip ring is heavily grooved the brush is undoubtedly too hard or the brush pressure excessive, or both. Excessive brush pressure will show in very heavy brush wear. Brushes which are too soft will deposit carbon on the commutator causing sparking which, in turn, will often cause excessive radio interference from the machines. Carbon dust from soft brushes may also cause short circuits in a high voltage circuit. Many grades of brushes are made for each type of brush holder and the manufacturer will usually supply harder or softer samples than those employed until a satisfactory brush is obtained.

Commutators require more attention than any other part of the machine. The effect of brushes on the commutators has been described. Commutators which have become a red-brown color and are smooth should never be touched, as they will run for long periods when this condition exists. Usually commutators will show sparking or will become rough, due to wear. When this happens the best expedient is to take the armature out of the machine. Armatures are provided with centering holes at each end and they can be slipped into a lathe. The commutator should then be turned down until it is absolutely smooth. The finish can be obtained with fine sandpaper but never with emery paper, which contains a dangerous amount of conducting particles. After this is done, the mica separators between segments should be cut down with a thin file until they are at least five thousandths lower than the face of the commutator. The sharp edges resulting from this filing can be smoothed down with an additional sanding or the edges can be removed with a sharp knife, cutting at forty-five degrees to the edge.

After the commutator is sanded down and returned to the machine new brushes should be inserted in the holders. A piece of sandpaper, the width of the brush or a little wider, should then be placed between the brush and the commutator with its non-cutting paper back against the commutator. The paper is then pulled back and forth so that it fits the commutator. The brush will be sanded in this manner to the shape of the armature. If the sandpaper shows a broad black mark when pulled, the brush is properly fitted and the carbon dust may be blown out. Each brush should be so treated until all fit properly. The machine should then be run under no load, or a light load, until the brush runs in to final shape. This will usually require an hour or two. Large radio transmitter dynamotors, which are alternately started and stopped, need this run-in more than machines which are to run continuously. Transmitter dynamotor starting currents are high and many starts are made before the machine has been run continuously for any length of time in service.

Generators and dynamotors in aircraft service should give better than 1000 hours between service periods with visual inspection of all commutators or slip rings every few hundred hours. Frequent and miscellaneous sanding of running commutators, to get them perfectly clean, will usually result in short life between overhauls. Danger of shock is also present.

Control Boxes—Control boxes employed on aircraft limit the charge rate of the generator. The control box is placed between the generator and the generator switch which, in turn, attaches the generator to the battery and electrical system. The control box serves to limit the charging current to a sufficient charge rate to keep the battery in good condition. It also prevents excessive generator currents.

If the electrical load is forty amperes, for example, and the generator is charging at a rate of fifty amperes, the battery is actually being charged at the difference rate of ten amperes. Thus, the ammeter in the cockpit showing "charge" rate is actually showing the current supplied to the load and battery together and not to either alone. The difference between this rate and the load current gives the true charge rate.

In many automobiles, the generator is equipped with a simple relay which closes when the generator is supplying a higher voltage than the battery. No method of regulating this properly is employed except for the third brush in the generator which determines the average charge only. The third brush is set at the best guess and if the battery runs down the charge rate is too small. If the battery stays fully charged the charge rate is probably too high. Automobiles are often driven on long trips with the lights on to prevent overcharge of the battery.

Aircraft control boxes, or regulator boxes as they are sometimes called, are arranged to taper the charge current as the battery becomes fully charged and to increase the charge as the battery runs down or the load increases. A load increase merely drops the battery voltage and the drop in voltage affects the regulator to increase the charge rate.

Fig. 40 shows a control box which is used in aircraft charging systems. There are three relays employed.

The cutout relay is sometimes called the master relay. It connects the generator, attached to point G+, to the positive battery terminal attached to B+. It will do this only when the generator voltage is greater than the battery voltage. It has two coils which are interwound. If the generator is producing a voltage, current will flow through the current limiter coil and through one of the cutout relay coils to ground through a limiting resistor. The cutout is ad-

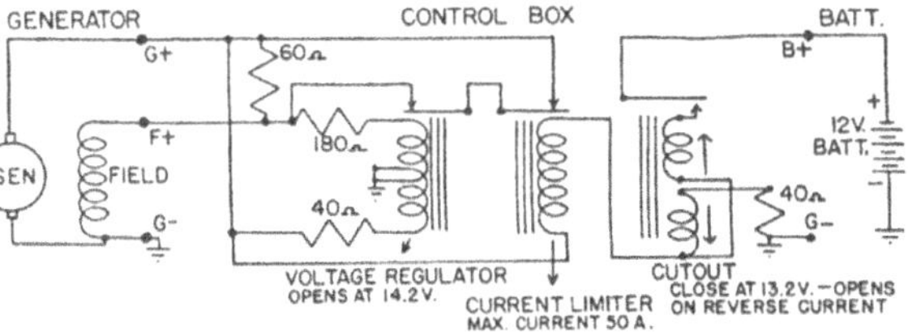

FIG. 40—GENERATOR CONTROL BOX

justed by means of a spring on its movable contact arm so that the points will close at 13.2 volts for a twelve volt battery system. The generator voltage can drop to about 12.5 volts, when not attached to a load, before the relay will open. As soon as this relay closes, the coil in series with the relay contacts will still further hold the relay closed. However, if current should flow from the battery back toward the generator, the relay will open instantly. Small arrows show the effect of such reverse current, as the field of the two coils will cancel allowing the relay to open. This relay has two functions: (1) to close the generator circuit to the battery when sufficient voltage is being generated, (2) to prevent flow of current from battery to generator.

The current limiter relay is provided to limit the maximum rate at which the generator may deliver current. Its movable arm is spring pressure adjusted to allow the contacts to open when the current reaches this maximum value. A fifty ampere generator will deliver much more current than this amount but will heat seriously and damage itself without this current limiting method. The effect of opening the relay contacts is to insert a resistor in series with the field and positive generator armature terminal. The reduction in field current causes the generator voltage, and likewise the current, to drop. This relay actually will be opening and closing if the current is tending to exceed the limits. By this pulsing, the average current will be held at the maximum permissible value.

The final relay in the system is the voltage regulator relay. The relay arm is adjusted by spring pressure to open at exactly 14.2 volts. It will again close when the voltage drops below about 13.2 volts. The essential adjustment is opening at 14.2 volts. This relay is in constant operation when the generator is in operation and limits the charge voltage to 14.2 volts, which voltage will not overcharge a twelve volt aircraft battery no matter how long it is applied. As the battery reaches full charge it will also reach 14.2 volts and no further current will flow to the battery. The two windings on the relay derive their voltage from different points. The bottom coil secures the full armature voltage through the dropping resistor. The top coil receives its voltage from the field terminal which voltage will depend on whether the field resistor is in the circuit, as determined by the voltage regulator or current limiter contacts being open. If the cutout is closed and the current limiter is not pulsing, the action of the voltage regulator is as follows: With the relay contacts closed the same voltage will be delivered to both windings but through different values of limiting resistors. Sufficient flux will be generated by both coils to open the relay and allow the field limiting resistor to come into play. This will drop the voltage to the field and upper coil and the relay will be closed by spring pressure. At this time the flux in the upper coil will again be great enough to open the contacts and cause generator voltage to drop again so as not to exceed 14.2 volts.

The proper adjustment of control boxes can only be made by careful adjustment of each of the three adjusting contact pressure springs. A test bench setup is necessary with an accurate voltmeter.

Microphones—Most aircraft microphones contain carbon grains in a sealed shallow cup. Direct current runs through the grains while the microphone is in use. A diaphragm attached to one side of the cup allows the pressure variations of the voice to compress the carbon, thus causing variation of current flow. The flow of current through the highly resistant carbon depends on the pressure.

When microphone cups of the non-sealed variety are to be refilled it is essential that the amount of carbon be exact. If there is too much carbon the microphone will pack tight and the current will be high so that whistling into the microphone will barely vary the current. If the packing is too light the carbon grains will offer very high resistance and although the current will vary widely (if any current flows at all) such packing will give poor service. Carbon must be absolutely dry to be satisfactory. Heating carbon grains to drive off water is advisable.

Microphones are equipped with switches so that the current is allowed to flow through them when the button on the microphone is pressed. Another switch, controlled by the same button, is usually employed to control relays which start the transmitter. Fig. 41 and

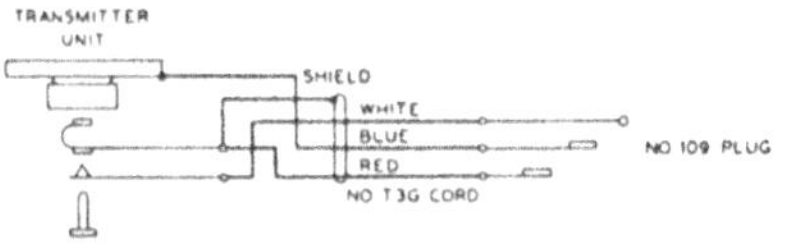

Courtesy Western Electric Co.

FIG. 41—628A MICROPHONE

Fig. 42 show two Western Electric microphone schematics. The 628A must be held vertically while the 631A may be used in any position. The cases are two and one-quarter inches in diameter by one and three-quarter inches thick, approximately.

Some microphones of the dynamic type are in use. These microphones have a small coil of wire attached to a small cone or

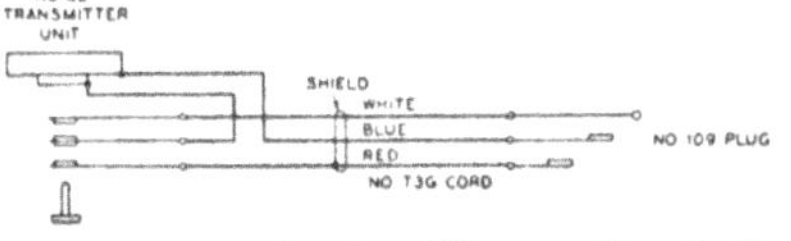

Courtesy Western Electric Co.

FIG. 42—631A MICROPHONE

diaphragm. The whole assembly is fitted into a slot in a permanent magnet. As the coil moves back and forth with the voice vibrations the coil cuts the field of the magnet and produces voltage in the coil. The output level of such microphones is much lower than that of the carbon type and preamplifiers must be used to raise the voltage to the level supplied ordinarily by a carbon microphone.

Headphones—Headphones are constructed with an electromagnet (coil of wire on a core) in a case, above which is a diaphragm separated from the core by a gap. As current passes through the windings of the electromagnet the diaphragm is alternately attracted by the alternating voice currents. The diaphragm moves the air near it and voice is reproduced.

The most satisfactory type of aircraft headphone has a limiting device which allows the diaphragm to move just to a certain point. Thus a loud static burst will not cause an excessive noise as the diaphragm stops before its otherwise greater travel, and ear fatigue from heavy bursts of noise is thereby reduced. The effect is to allow reception in otherwise difficult static conditions. Such antistatic headphones are essential in airline service.

Meters, Ammeter—The essential element for the measurement of electric currents or resistances is the ammeter with certain adaptations.

An ammeter is an instrument for measuring current. It is composed of a circular shaped magnet in which a sector has been removed. In this sector is a coil of wire on a non-magnetic spool which is the armature and it is mounted on delicate jeweled bearings. A double spring holds the armature, to which is attached an indicating needle, in the normal (or zero) position. As soon as current flows a torque exists, by virtue of the current reaction in the magnetic field. The armature attempts to turn against the spring and deflects the needle an amount dependent on the current. Thus the current may be read on the calibrated scale. A low resistance element (shunt) is placed in parallel with the winding to limit its current for heavy current measurements. By this means a given movement will indicate different current ranges with different paralleled low resistance shunts, as shown to left of switch arm in Fig. 43.

Voltmeter—A voltmeter is made in exactly the same manner but a high resistance element is connected in series with the movement (Fig. 43) so that it will definitely limit the current. Thus by knowing the resistance of the series element and the sensitivity of the ammeter a calibration of voltage can be made. Different resistance values can be used for different scales.

Ohmmeter—The third instrument made with this same ammeter is an ohmmeter, or an instrument for measurement of resistance. This is accomplished by using a voltmeter in series with voltage from a small battery, Fig. 43. When the leads to the ohmmeter are short-circuited together the meter will read full scale. This is the zero resistance indication as there is no resistance between the terminals external to the measuring instrument. The meter is calibrated for various values of resistance so that the ohmmage of unknown resistors or circuits can be measured.

Three functions are often combined in one meter for compactness. In many instances just the voltmeter and ohmmeter combination is provided. The ammeter is usually omitted, as a circuit must be broken to measure the current and this is usually difficult.

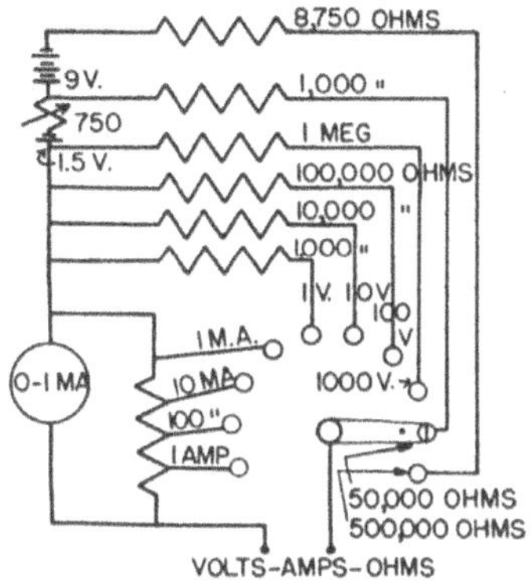

FIG. 43—VOLT—OHM—AMMETER

Voltage, and often resistance, may be measured without disturbing the electrical circuit.

It is important that the ohmmeter never be used when voltage is on the circuit as added voltage might destroy the meter. Thus the voltmeter serves its purpose on live circuits and the ohmmeter on dead circuits. It is practically impossible to service aircraft equipment without a volt-ohmmeter.

A.C. Voltmeters—Voltmeters used for alternating current circuits are usually equipped with a rectifier which is usually of the copper oxide type. As the direct current voltage produced through such a rectifier from the alternating current is not quite proportional at the lower end of the alternating current scale, the scale may be cramped as shown by the calibration. Modern meter rectifiers are almost linear however.

Aircraft measurements may be made on the 400 to 800 cycle power frequencies and audio outputs from receivers, microphones, amplifiers and the like can also be measured with the rectifier type instruments. There are several other types of alternating current instruments which will seldom be used in aircraft service, particularly as most are designed for low alternating current power frequencies.

There is one other type of meter which is used to measure alternating currents at frequencies as high as the ultra high frequencies at seventy-five million cycles (75 MC). This type of meter is the thermocouple-ammeter. As current passes through the thermocouple it generates heat at the junction made of two dissimilar metals. A direct current will flow through the two wires of the thermocouple and cause a sensitive direct current ammeter to indicate the current. Such instruments have a cramped scale near the zero end. These meters may be used to measure any alternating current or direct current. They are primarily used to indicate transmitter currents to the antenna system. Thermocouples may be mounted separately from the indicating meter.

Tachometers—Aircraft tachometers used to indicate engine speed are independent of the main electrical system. The direct current generator, with its commutator, is mounted on the engine. It contains a fixed field magnet which sets up a voltage dependent upon engine speed. Most of the magneto elements of tachometers are operated at half engine speed. The data given are in terms of engine speed and not tachometer shaft speed.

Fig. 44 shows a typical tachometer circuit. Fixed calibrated resistors are attached to the meter and when recalibration is necessary resistors of a slightly higher or lower value are substituted.

FIG. 44—TACHOMETER

Two of the standard tachometers now on the market have voltage outputs of 3.6 and 4.8 volts (direct current) respectively, at an engine speed of 1000 revolutions per minute. Thus in the two cases the magneto will develop 0.18 and 0.24 volt respectively, for every hundred revolutions per minute of the engine.

Fuel Quantity Gauges—Most modern transports are equipped with electric gauges to show the gasoline level in tanks. Usually there are a number of tank units with a selector switch to attach any tank unit to the fuel quantity gauge on the instrument panel.

Fig. 45 shows the schematic diagram of a typical gasoline gauge. Part of such a gauge is the voltage regulator which delivers a constant voltage to the measuring resistance unit over the normal range of battery voltages encountered during battery charge and discharge. The lamps in the regulator have the characteristic of diminishing resistance with decrease of current through them. This tends to lower the voltage drop through the lamps as the battery voltage goes down. This in turn causes the voltage drop to be smaller across the lamp and the voltage to remain more nearly constant on the measuring circuit.

The indicating instrument is nothing but a voltmeter that measures the voltage across part of the tank resistance unit. An arm is attached to a float in the gasoline tank and moves the slide

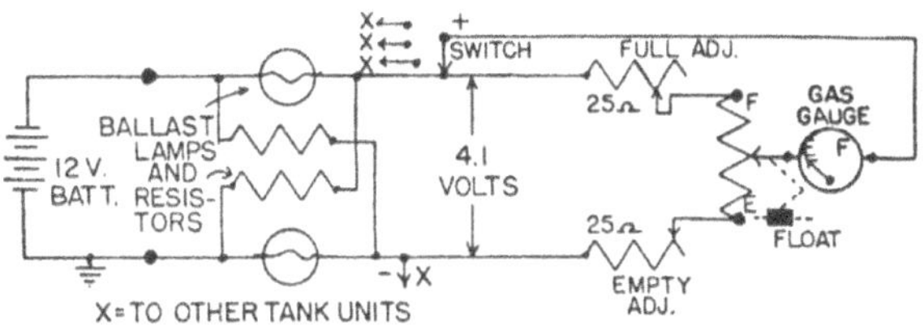

FIG. 45—GASOLINE TANK GAUGE

arm along the tank unit resistor. Mechanical linkages are so arranged that the movable arm travels over most of the resistance unit. In addition there are provided two resistors of about twenty-five ohms each which are used to electrically adjust the meter so that it reads "Full" when the float in the tank is in the "Full" position and "Empty" when the tank is empty. One resistor is used for each of these adjustments although there is some interaction between resistor adjustments.

The normal voltage delivered to the measuring circuit is 4.1 volts in one of the common units which operates on a twelve volt battery system.

Electrical Thermometers—Electrical thermometers are provided on most aircraft for reading outside air, carburetor intake, oil and other temperatures. A bulb is provided at the point at which it is desired to measure the temperature. This contains a resistor and is connected to the bridge circuit.

A wheatstone bridge operates in the following manner. Suppose the value of the four resistors W, X, Y and Z in Fig. 46 are the same. The voltage across the meter shown in the circuit will be zero as the voltage drop in W is the same as that in X. The voltage drops in Y and Z are also equal. If any one of these four resistors is disturbed in value there will be voltage across the meter.

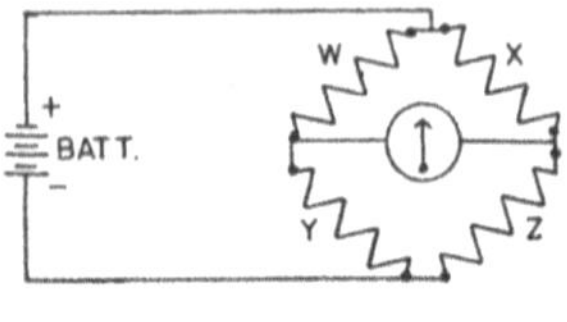

FIG. 46—WHEATSTONE BRIDGE

Thus if X were lowered in resistance or short-circuited, the meter would read the part or all the voltage across W. If resistor X were such that its resistance changed rapidly with temperature the meter could be calibrated in temperature instead of volts or amperes.

Most temperature indicators are initially set at freezing, or some point near to it, so that the voltage is just balanced and the meter will not move. As the temperature gets warmer or cooler the meter will move in the appropriate direction.

Fig. 47, a typical indicator, is nothing but a wheatstone bridge. The meter coil is actually split into two parts in some instruments, which aids in keeping the calibration more constant. Letters have

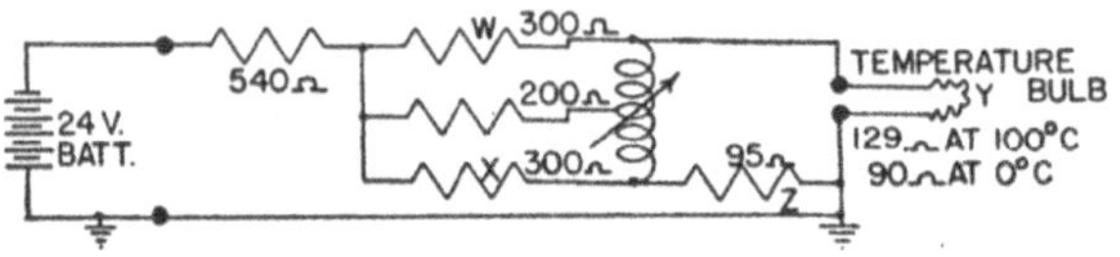

FIG. 47—ELECTRIC THERMOMETER

been placed in the figure to show its relation to the previously shown wheatstone bridge.

Engine Thermocouples—Where high temperatures are encountered another method of temperature indication is used. A thermocouple is a device which consists of two dissimilar metals in contact which will produce a voltage between them when heated. The higher the temperature the higher will be the voltage generated.

Aircraft engine head temperatures are indicated by this means and a ring of a copper alloy and iron alloy joined together is fitted under one spark plug to indicate the temperature on one particular cylinder.

The thermocouple is rather insensitive to small temperature changes and therefore is not used for temperature indications of less than several hundred degrees in range. As engine temperatures vary about 500 degrees in normal use, between cold conditions before starting and hot conditions at full power, the scales commonly used indicate up to 600 degrees.

Gas Analyzers—Gas analyzers, such as the Cambridge analyzer, have resulted in a definite improvement in operation of airplane and other engines as the mixture can be set to a predetermined and efficient value. Substantial savings in gasoline and motors can be directly traced to the use of these instruments.

The principle of operation is rather simple. A wheatstone bridge is employed with two of its elements in the exhaust gas mixture and

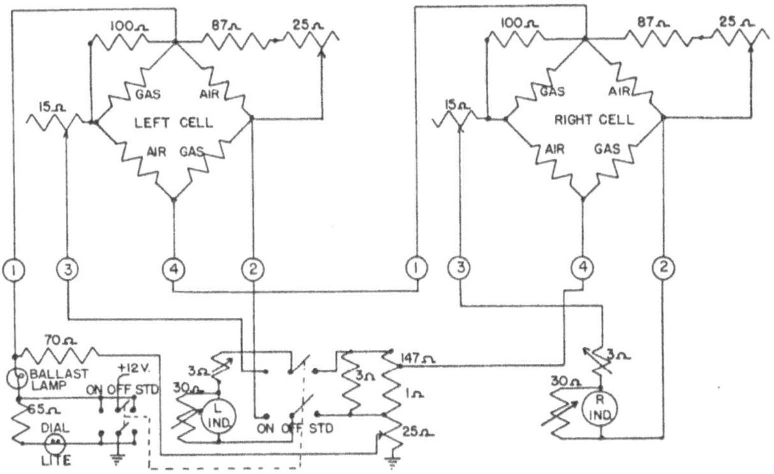

Fig. 48—Gas Analyzer for Two Engine Transport

the other two exposed to air. Current passes through the bridge and heats the bridge elements. The presence of carbon dioxide in the exhaust gas affects the heat radiation from the elements in the exhaust gas and thus allows them to radiate more or less rapidly, depending upon the amount of carbon dioxide in the gas. As the amount of carbon dioxide is directly relative to the mixture it is possible to determine the mixture of gasoline and air entering the engine.

The fuel-to-air ratio to which the instrument is calibrated is in terms of pounds of air. It so happens that the exhaust mixture has the same conductivity as air when the fuel-to-air ratio is exactly .077.

Examination of the diagram in Fig. 48 shows the operation of the instrument. A diagram has been shown for a two engine transport. The two bridges are in series with each other although actually they are independent of one another.

There are a number of adjustments which can be made on the instrument and one in particular is of great importance. A 25 ohm resistor will be seen in conjunction with a one ohm resistor. This one ohm resistor is thrown across the left engine indicating instrument by means of the switch which places the circuit in "calibrate" position. By adjusting this resistor the current through the two bridges may be set at a predetermined amount. The scale on the left indicating meter is marked with a letter "B" for such adjustment and the usual value is at a fuel-to-air ratio of .09. Once the adjustment has been made at a given voltage the ballast tube will tend to maintain this bridge voltage as the terminal voltage of the battery changes during charge and discharge.

The three and thirty ohm resistors associated with the indicating meter elements are for sensitivity adjustment of these instruments and this is usually done at the factory and need not be varied in the field. These resistors adjust the sensitivity as well as the resistance of the meter to a standard value.

It will be noted that there are two twenty-five ohm variable resistors in connection with fixed eighty-seven ohm resistors which are across each bridge. These are adjusted to bring each of the meters associated with them to the .077 value when the instrument is on but with the airplane motors not running. The meter is mechanically adjusted to the same value with no current and it is only necessary to turn on the battery and readjust the resistors so no deflection is caused electrically from this point.

The last adjustment is made with the fifteen ohm resistors at one side of each bridge circuit. This adjusts the sensitivity of the entire circuit so that the meter will read correct fuel-air ratios for any type of gasoline. In other words, the adjustment makes the meter calibrations suit the correct fuel-to-air ratios for gasolines which differ in composition.

The platinum spirals (marked air and gas) which are located in the sampling head are arranged in a block of brass so that the temperature of the four arms of the bridge will be the same at all times. As the instrument does not work on exhaust gas temperature but on heat conductivity of the gas, it is essential that all elements be arranged in this manner so that a temperature error will not occur.

On supercharged aircraft the air exposed elements are sealed in a glass container with a small amount of water vapor. The temperature of this chamber is nevertheless the same as that of the exhaust gas analyzer chamber and the two chambers are adjacent as before.

Only a small amount of cool exhaust gas passes slowly through the sampling chamber as gas at high velocity would affect cooling of the platinum spiral measuring element. Each one of the spirals has a resistance of about five ohms.

Hydromatic Propeller Control—One of the recent propellers utilizes a method of electric pump control for fully feathering the propellers when it is desired to stop the motor and propeller in the air. The normal control of propeller pitch is taken care of by a hydraulic system but when full feathering is desired an electrically driven pump is used as oil pressure is otherwise insufficient for this function.

Examination of Fig. 49 shows the propeller control schematic for two propellers. The feathering push button in the cockpit is pushed in to feather. The button will stay in due to current flow through the solenoid in the switch. Contacts on the button close the starter solenoid switch and operate the nacelle control relay of the electric propeller pump. This last relay is used to save the weight of additional heavy wires to the nacelle. The pump operates until the propeller reaches the full feathered position where the pressure rises in the hydraulic system to about 450 to 550 pounds per square inch and the pressure cutout switch causes the cockpit switch to snap out to the "off" position.

To place the propeller in operation once more, the same feathering button is pressed and held. Pressure rises to 600 pounds per square inch, which pressure causes a spring loaded valve to operate so that oil is pumped onto the opposite side of the propeller controlling piston and the electrically driven pump causes the propeller to return to normal position after which the feathering button is pulled out and the hydraulic control and governor takes hold again. If the feathering button is not released on unfeathering there is some danger of the electrical pump driving the propeller piston against the stop and doing damage. As soon as the engine starts to gain reasonable speed the automatic pitch control will finish returning the propeller to proper operating pitch.

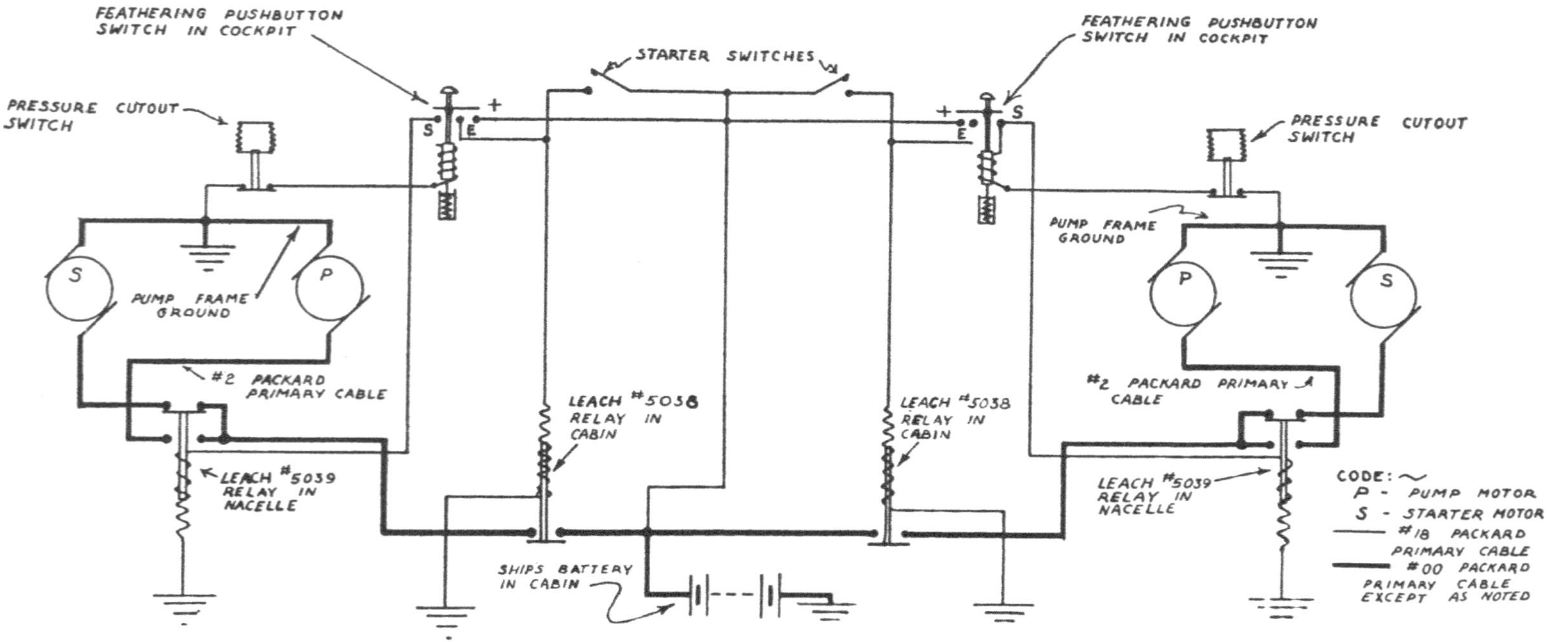

Courtesy TWA

Fig. 49—Hydromatic Propeller Control

Resistor Color Code—A standard color code has been adopted for marking resistors rather than writing the value on the unit. The body color of the resistor indicates the first figure of the resistance value. The end color indicates the second figure. The ring, or dot, on the resistor indicates by its color the number of zeros following the first two numbers.

The numbers corresponding to the colors are:

0—Black (.0 as ring or dot)	4—Yellow	8—Gray
1—Brown	5—Green	9—White
2—Red	6—Blue	
3—Orange	7—Violet	

Thus a brown resistor with an orange end and a yellow dot (or ring) is 130,000 ohms. Pure red with an orange dot indicates 22,000 ohms, the end and body having identical color in this case.

Electrolytic Condensers—Electrolytic condensers have tremendous capacity for a given space. They use two sheets of aluminum foil which are separated by a chemically treated gauze as a dielectric. A thin film of gas forms which insulates the foils from one another. If the film breaks down, due to exceeding the voltage rating, the condenser will usually heat rapidly and be of no further use. Some electrolytics are "dry" and others are "wet." The wet type will reseal a puncture and function satisfactorily again if the voltage quickly drops to a value within the voltage rating. The dry type will not reseal.

These condensers must be connected to the circuit with the proper polarity which is either designated by the lead colors or by a marking on the case. Metal cased types are usually arranged so that the metal case is the negative terminal although some have been manufactured in a metal case where the case does not form one electrode and is added for mechanical protection and better appearance.

While these condensers will be found usually in direct current circuits, there is one type that is used on alternating currents for motor starting and the like. The rating of this type is usually defined in capacity and number of seconds per minute that the condenser may be used. Sometimes the alternating current type will have a continuous rating and is very useful for correction of the power factor of autosyn indicators.

Electrolytics do not have as long a life as paper condensers, as a rule, but may give several years of satisfactory service.

Magnetic Compass—There are two types of magnetic compasses used on aircraft. The first is the aperiodic and the second the non-aperiodic. While these two compasses operate on the same principle, as described in the chapter on fields, there is a great difference in the indication of the two.

The non-aperiodic is so damped that any turn of the aircraft will cause the compass to swing past the point where it should stop and this is followed by a swing back which is gradually damped out until the compass card finally stands still. The aperiodic type swings to a new position after the plane is turned and stops at the correct point with no oscillation. By far the greatest number of compasses in use are the non-aperiodic type because the aperiodic requires, among other things, that the scale be read from above. Mounting of such an aperiodic compass is difficult, particularly as it should be mounted with the other instruments in front of the pilot on the instrument board.

Swinging the Compass—Magnetic fields are found in all aircraft due to permanent magnetism of certain parts such as meter cases, structural parts, current flow in wires, etc. For this reason the magnetic compass is usually compensated for "compass deviation" by means of two very small magnets mounted at right angles to one another and near the main magnetized compass needle. One compensator is used for North-South correction and the other for East-West correction. The plane is turned Magnetic North on a compass rose situated on the airport and the North-South compensator adjusted so that the compass reads exactly North. The plane is then headed East and the East-West compensator adjusted until the compass reads exactly East. The plane is headed South and the North-South compensator is adjusted so that the error on the South heading is halved. The plane is finally headed West and the process repeated with the East-West compensator, removing half the Westerly heading error. Following this, the plane is changed in heading by fifteen or thirty degree steps and a correction card is made showing the small error on each heading.

Northerly Turning Error—There is another serious discrepancy due to the angle that the needle makes with the horizontal. If a compass needle were located in free space in the vicinity of New York it would point Magnetic North and its North-seeking end would also be dipped some seventy-five degrees from the horizontal. In other words the needle would be within fifteen degrees of vertical.

The angle downward is called the "dip" and varies over the earth's surface. The reason for this is that the magnetic lines near the poles are steeply dipped while this dip is almost absent near the equator.

To eliminate the effect of dip, compass needles are almost universally "floated" in liquid. Thus the surface of the liquid should be at right angles to the force of gravity at all times. Unfortunately, when an airplane makes a correctly banked turn the liquid level stays parallel to the horizontal plane of the airplane. Thus the effect of turning will cause the compass to behave queerly and not indicate the correct magnetic heading. This is called the Northerly Turning Error.

If the plane is on an east or west heading and the level of the needle is not parallel to the earth's surface, the needle will tend to show the dip rather than point to North as shown on the compass scale. Thus, near New York, an error of one degree from the horizontal will cause an error of about five degrees in reading of a compass if the east-west level of the liquid is disturbed. Naturally the effect of compass liquid swirling and the liquid churning caused by change of airplane attitude will also have their effect.

Filters—A filter is used to purposely reduce alternating currents of certain frequencies. Filters may be designed for reduction of alternating currents of high frequencies while not disturbing frequencies of a low value. Other filters are arranged to allow the passage of alternating current but not of direct current.

Take the simple case of Fig. 50. A generator is supplying a simple resistive load. An inductor is placed in series with this generator and it will present an increasingly higher reactance with in-

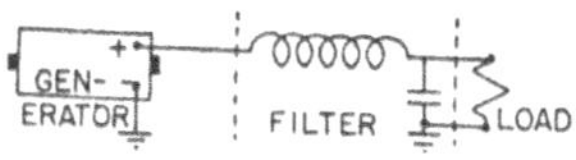

Fig. 50—Spark Filter

crease of frequency. The capacitor will represent less and less reactance to the voltage as the frequency increases. Thus the two in combination will reduce high frequencies across the load.

A practical example of the use of such circuits is found in the radio noise filters on almost all aircraft dynamotors and in some cases, in the generator circuits. The commutator shows some sparking and some ripple voltage due to the fact that segments of the commutator allow the direct current to pulse slightly. The voltage varies as the brush passes over the segment face. All of these pulsations and sparking may cause noise in radio equipment and simple filters of the type described have little detrimental effect on the production or use of direct current but successfully eliminate the higher alternating frequencies.

Another example will illustrate the use of a filter which will pass alternating current but impede any flow of direct current. Fig. 57 (A, B and C) shows the plate of a vacuum tube which is coupled to the grid of the next tube following it. The plate must be maintained at a relatively high voltage for operation and the grid must be maintained at a negative voltage. Thus a condenser placed from the plate to the grid will prevent the flow of direct current but will allow passage of alternating current. Further, the resistors will

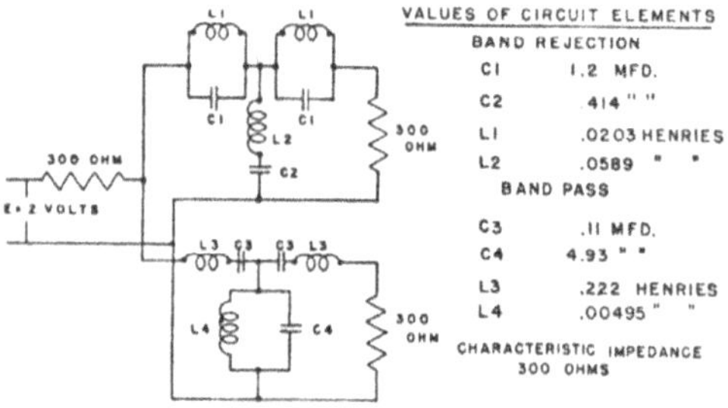

Fig. 51

allow alternating and direct current to be produced across them and the alternating voltages existing across the plate resistor will appear on the resistor in the grid circuit of the tube.

Fig. 51 shows a simultaneous radio range filter which is built in two sections. One section will pass 1020 cycles quite easily and will not pass frequencies appreciably higher or lower than this. This type of filter is known as a band pass filter. The other section of the filter will pass the audio frequency spectrum except for a portion closely centering about 1020 cycles. Thus by using one filter all but 1020 cycles is heard and by using the other filter 1020 cycles is heard.

The purpose of this filter is to separate the radio range 1020 cycle, A-N keyed frequency from the voice frequency or vice versa. The radio range may send its course signals at the same time as its voice weather broadcast and by means of the filters the pilot can select either that he desires.

Cathode Ray Oscillographs—A cathode ray tube is shown in Fig. 52. Electrons leave the filament and pass through accelerating electrodes. The tube behaves much the same way as an ordinary vacuum tube but instead of variations of the plate voltage delivering signals to a load, the electron stream passes through the plate and hits a screen of barium salts on the end of the tube and inside it. This causes a bright spot to appear. The size of the spot depends

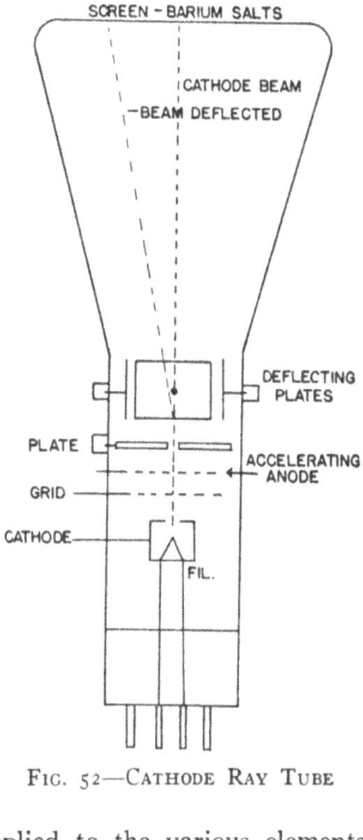

FIG. 52—CATHODE RAY TUBE

on the voltages applied to the various elements which electrically focus this spot.

After the electron stream leaves the plate it passes between two sets of plates at right angles to one another. It will be seen in the diagram that the deflected beam has been bent and this is due to the voltage on the right deflecting plate being negative and repelling the beam or the voltage on the left being positive and attracting it, or both.

If an alternating voltage were applied to the other set of

plates the beam would swing in and out. As the screen allows the light to persist for a short time a line would be seen due to the alternating voltage. Naturally this line would give little information as to the shape of the voltage wave. If the other set of plates were supplied a voltage which caused the beam to move from left to right and then suddenly return and start again, there would be seen the exact shape of the alternating current.

For this reason most oscillographs have a scanning oscillator built into them (Fig. 53). Current flows through the 34 tube until a voltage sufficient to break down the 885 mercury vapor tube exists on its plate. At that instant it short-circuits the output voltage and the cycle starts over again. Thus the wave to be viewed is spread out on the screen and will stand still if the oscillator is held at the same frequency, or some sub-multiple of the wave to be scanned.

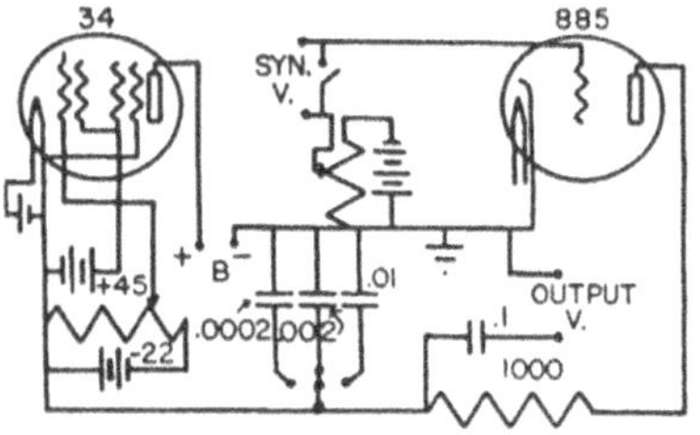

FIG. 53—TIMING CIRCUIT FOR CATHODE RAY TUBE

If it is required to hold the two frequencies in step a part of the wave to be scanned may be fed to the synchronizing voltage posts and will keep the oscillator in step by controlling the rate of the 885 breakdown through its grid circuit.

A power unit supplies 1000 volts or more for operation of cathode ray tubes. The current taken by the tube is only a milliampere or so.

QUESTIONS ON CHAPTER IV

1. Fuses must have some small resistance because:

( ) a. Current flows through them. ( ) b. Voltage drop is advantageous. ( ) c. Otherwise they would get too hot. ( ) d. Heat must be developed to melt them on overload.

2. Twelve volt aircraft batteries will overcharge on:

( ) a. 14.2 volts. ( ) b. 14.4 volts. ( ) c. 13 volts. ( ) d. Over 12 volts.

3. High charging currents will:

( ) a. Cause less gas to be generated. ( ) b. Keep the electrolyte at proper temperature. ( ) c. Shorten the life of the battery. ( ) d. Allow the use of more baking soda.

4. The use of batteries allows generators to be smaller because:

( ) a. Short peak loads can be handled by the battery. ( ) b. Storage is provided to prevent generator heating. ( ) c. Most batteries will overcharge on large generators. ( ) d. Batteries supply all the energy.

5. If lights dim badly in a plane during transmission:

( ) a. The generator is too small. ( ) b. The battery charge is too high. ( ) c. The generator is not connected. ( ) d. The generator may be attached to the circuit without the battery.

6. Magnetos require a condenser primarily to:

( ) a. Store the magnetic field. ( ) b. Prevent the breaker points from burning. ( ) c. Help start the engine at low magneto speeds. ( ) d. Prevent the voltage from becoming too high.

7. The primary magneto switch in the cockpit is closed so that:

( ) a. Primary current may flow for magneto operation. ( ) b. Synchronizers will operate. ( ) c. The engine will stop. ( ) d. The booster will operate.

8. Synchronizers are used:

( ) a. To hold the engines in step to reduce the beat noise. ( ) b. So that both magnetos will fire alternately. ( ) c. So that the number one plugs of both engines fire synchronously. ( ) d. To show condition of the magnetos.

9. Impulse relays are used because:

( ) a. Fewer relays are necessary. ( ) b. Points do not burn as badly. ( ) c. They operate better. ( ) d. Less power is wasted.

10. Contact points on relays should be filed:

( ) a. When they are overhauled. ( ) b. Only when very badly pitted. ( ) c. Only when current is low through the points. ( ) d. About every two months.

11. Residual magnetism may:

( ) a. Hold relays in non-operated position. ( ) b. Cause relays to operate faster. ( ) c. Cause relay armatures to stick down. ( ) d. Improve operation of sluggish relays.

12. Relays are shown in non-operated position on schematics:

( ) a. For standardization and ease of following circuits. ( ) b. Because this is their usual position in use. ( ) c. Because less current is used. ( ) d. Because if they were operated the contacts would be closed.

13. Brushes give least trouble of anything on motors and rarely need replacement because:

( ) a. They are made of tough metal. ( ) b. Commutators get smoother with time. ( ) c. They are made of carbon. ( ) d. The statement is untrue.

14. Commutator insulation should be undercut because:

( ) a. Mica has high resistance. ( ) b. The mica will not get carbonized as easily, thus shorting segments. ( ) c. Mica might short-circuit the segments. ( ) d. It cannot be made smooth on a lathe.

15. Control boxes limit voltage because high voltage will:

( ) a. Cause shorter life of starters. ( ) b. Cause lights to dim. ( ) c. Overcharge and ruin batteries. ( ) d. Allow batteries to run down.

16. Control boxes are arranged to be adjusted easily during flight:

( ) a. This is rarely done. ( ) b. So pilots can regulate the charge rate. ( ) c. So that time may be saved in ground servicing. ( ) d. To prevent overcharging in hot weather.

17. The voltage regulator relay is principally used to:

( ) a. Prevent the current from reversing. ( ) b. Reduce the voltage when heavy loads occur. ( ) c. Regulate discharge current. ( ) d. To prevent voltage from becoming too high or too low.

18. Microphones carbon is packed:

( ) a. Tightly to prevent looseness. ( ) b. Fairly loose for best microphone sensitivity. ( ) c. Tightly to force out water. ( ) d. With oil to assure smooth operation.

19. Headphones with the antistatic feature are used:

( ) a. To reduce all signals to a good level. ( ) b. To prevent static coming into the receivers. ( ) c. To reduce noise from loud static bursts. ( ) d. To test the compass by holding the headphone near it.

20. An ohmmeter is composed of:

( ) a. A battery, ammeter, resistor and test prods in series. ( ) b. A battery and voltmeter in series with an ammeter. ( ) c. An ammeter, resistor and test prods in series. ( ) d. A voltmeter in parallel with two test prods.

21. Tachometers of the type described rely on frequency variation for operation:

( ) a. So as to read most accurately at high speeds. ( ) b. Because less drain is imposed on the battery. ( ) c. Because direct current varies with speed. ( ) d. They generate and use direct current only.

22. Fuel quantity gauges of the type described rely on:

( ) a. Variation of pressure on the tank bottom. ( ) b. A float operated variable resistor. ( ) c. A thermocouple in the tank. ( ) d. Battery voltage being maintained exactly at 12 volts.

23. A thermocouple depends on:

( ) a. Mechanical variation of resistance. ( ) b. Generation of heat electrically. ( ) c. Generation of voltage by heat. ( ) d. The voltage of the battery being constant.

24. Electrical bridge thermometers indicate:

( ) a. Correctly with reasonable battery variations. ( ) b. Whether or not the battery is connected. ( ) c. By chemical generation of voltage. ( ) d. By use of constant resistance elements in all parts of the circuit.

25. Gas analyzers indicate:

( ) a. The amount of incoming gasoline. ( ) b. The temperature of the exhaust gas. ( ) c. The heat conductivity of the exhaust gas. ( ) d. The ratio of gasoline to air.

26. The pressure cutout switch on Hydromatic propellers:

( ) a. Keeps the pitch correct at all times. ( ) b. Releases the feathering switch in the cockpit. ( ) c. Causes the pressure to build up gradually. ( ) d. Operates the feathering control to feather.

27. Polarity of electrolytic condensers is usually observed:

( ) a. To prevent the voltage from being exceeded. ( ) b. To increase the capacity. ( ) c. To prevent spoiling the condenser. ( ) d. To eliminate the extra wire.

28. Magnetic compasses are influenced by dip when:

( ) a. The liquid level is not horizontal with the ground. ( ) b. Adjusted for best operation. ( ) c. Near the equator. ( ) d. Compensation is not correct.

CHAPTER V

TUBES AND AMPLIFIERS

Vacuum tubes behave very much like relays in electrical circuits as they are capable of controlling large currents by means of small currents. The chief difference between a tube and a relay is that a tube neither opens nor closes a contact but may cause current to change gradually or rapidly, depending on the shape of the controlling wave. Before a discussion of the type of vacuum tube which can control power it would be well to investigate the behavior of simple tubes which have but two elements.

Two-element Tubes—Two-element tubes find many applications in aircraft radio. A two-element tube is constructed with a hot filament similar to that in an electric light bulb. Sometimes this filament heats a sleeve which closely surrounds it and in this case the tube is said to have an indirect heater. Either type of filament structure is called a cathode. A plate of cylindrical or oval shape surrounds the cathode.

When a metal is heated to incandescence it will emit electrons. Electrons may be considered as particles carrying a definite negative electric charge. Some leave the filament and travel to the plate. When the plate is made more positive in voltage with respect to the cathode a heavier electron current will flow to the plate. Thus the higher the positive plate voltage the heavier will be the current through the tube. When the flow becomes excessive the plate will become a red color due to sufficient electrons hitting to heat the plate. Most vacuum tubes are not designed to have the plate show color at all but there are a few tubes which operate well up to a cherry red color.

Making the plate negative will prevent current flow. Two-element tubes are called diodes and they are used as rectifiers.

Fig. 22 shows a half wave rectifier. It will rectify but one half of the alternating current cycle. A filter is usually included between the tube and the load, as in Fig. 56, which smooths the rough rectified wave to a smooth direct current. Direct current is produced from alternating current by this means. Copper oxide rectifiers operate in the same circuits and the tube is replaced by the copper oxide element which passes current in one direction principally.

Mercury Vapor Tubes—Some diodes are filled with mercury vapor in order that the resistance of the tube will be very low. Such tubes ionize the mercury vapor when the plate is more positive than the cathode and this reduces the internal resistance of the tube very materially. The voltage drop across the tube will be about fifteen volts regardless of the current.

If the tubes are to operate below about zero Fahrenheit, it will be found that ionization may not take place and that it will be necessary to supply a heater to keep the tube envelope warm. At temperatures above 160 degrees Fahrenheit the tubes may not deionize properly and may fail in the circuit. To prevent this difficulty, the tube envelope may be reduced in temperature with the air circulation provided by a fan.

Triodes—A triode vacuum tube has a grid placed between a cathode and plate. The grid is formed of a wire mesh that will allow electrons to pass to the plate. Very few of the electrons hit the grid as it is usually negative with respect to the cathode.

The plate may be maintained at a high positive potential and the grid voltage will control the current to the plate. As the grid is made more negative a point will be reached when the plate current will cease to flow. This point is called "cutoff." Thus by varying the grid voltage the plate current may be adjusted to any desired value. It requires almost no power to operate the grid as there is normally no current flowing in the grid circuit. Thus it requires a small grid voltage to control large plate currents in vacuum tubes.

If an alternating voltage is applied to the grid, under certain conditions, the plate current will alternate in a like manner and with precisely the same wave shape. Tubes may be used to amplify voice or radio voltages at very high frequencies.

Variation of the current within a tube will accomplish no useful result unless this effect is produced in some external load. Fig. 54 (A and B) shows two such loads in the plate circuit of a triode. In the case of the resistor in the plate circuit the current flows from the battery through the plate resistor and then through the tube and back to the battery. Naturally the voltage at the tubes plate will be the battery voltage minus the drop of voltage in the plate resistor. As the plate current becomes high (grid becoming less negative) the voltage across the resistor will increase. As the plate current drops the voltage across the resistor will become smaller.

If a transformer is placed in the plate circuit the same effect will be produced in the primary of the transformer. However the transformer will show small direct current drop and large alternating current voltage drops. Thus the use of a transformer is far more

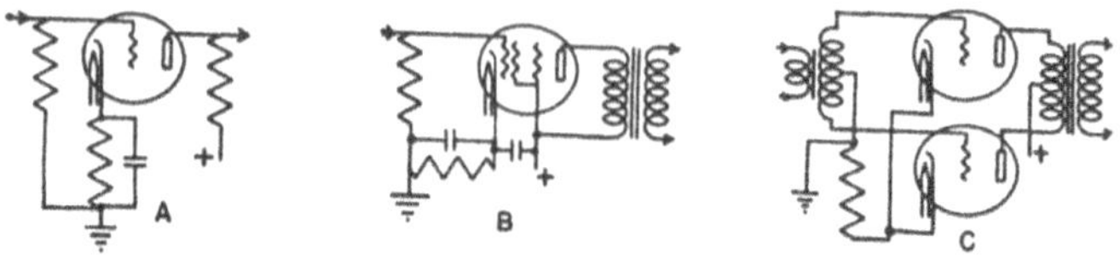

FIG. 54—AMPLIFIERS

efficient as the full plate supply voltage will appear at the tube plate. The secondary of the transformer is such as to match the resistor load. In other words the turn ratio can be such in the transformer that the maximum power can be secured from the tube. If the effective primary transformer resistance is too high, alternating maximum voltage will appear across the primary but the alternating current will be very small in value. On the other hand, if the primary resistance is small the alternating current will be high and the voltage low. In this case, most of the alternating voltage will exist between plate and cathode in the tube's plate resistance and very little power will be delivered.

The condition of maximum power production occurs when the plate resistance is equal to the load resistance of the transformer primary. Thus half the voltage produced will exist across the tube plate resistance and the other half across the load. This is always true for any generator, the internal and external resistance must be equal for maximum power transfer.

The efficiency of a tube operating as a linear Class A amplifier can never be more than fifty per cent and this is still further reduced by the fact that the grid cannot swing positive or, in other words, reach a maximum condition of plate current in the cycle where the plate resistance becomes zero instantaneously. Thus the usual biased amplifier whose plate current varies above and below some steady value will usually have an efficiency of about thirty per cent maximum. Such an amplifier is called a Class A amplifier but tubes may be operated in other ways for greater efficiencies, as will be shown later.

If the plate wave shape is not exactly the same as the grid wave shape the tube is said to distort. Operating tubes correctly in the manner described above will give the least distortion and the maximum undistorted output when the load is twice the tubes plate resistance (not equal to the plate resistance because of distortion). Distortion gives rise to harmonics of the original frequency. A tube amplifying 1000 cycles which has ten per cent second harmonic distortion will be producing not only 1000 cycles but will also produce a wave whose amplitude is one-tenth that of the 1000 cycle wave and whose frequency will be 2000 cycles. The first harmonic is the frequency itself, the second harmonic is twice that frequency, the third harmonic is three times that frequency, and so on.

The sum total harmonic distortion of all harmonics may reach a value of ten per cent without serious distortion of voice signals. The distortion will be noticeable, particularly so on music. Harmonic distortions below five per cent are not easily detected.

Triode grids are made with different meshes for the purpose of controlling the amplification constant of the tube. Tubes with fine mesh grids have higher amplification factors than those with coarse mesh. Tubes of the triode variety will have amplification factors up to about fifty. This means that the output wave will be fifty times the amplitude of the grid controlling wave with high load values. In use, the tube will never produce its full amplification as the plate load will be of relatively low value. When the plate load is twice the plate resistance, as in a Class A amplifier, the amplification of the tube will be two thirds of the tube's published amplification constant.

Mercury Vapor Triodes—Triodes used for controlling heavy loads are filled with mercury vapor. These tubes are known as thyratrons. Once the negative grid bias has been reduced to a point near zero bias, the grid will no longer have control until the plate voltage is interrupted. The voltage drop through the tube will also be nearly constant (fifteen volts) at any current within the tube's rating.

The usual way of interrupting the plate voltage is to supply alternating, rather than direct, current to the plate. The voltage of the plate is then negative for half of each cycle. If the grid is made negative, to stop current flow, the flow will be interrupted on the next negative half cycle of plate voltage.

Multi-element Tubes—There are other types of tubes used for amplification. The four-element tube is known as the tetrode. It consists of a triode, between the grid and plate of which has been placed another grid referred to as the screen grid (Fig. 55). The screen grid is usually operated at about one third of the plate's positive voltage and is grounded for alternating currents by means of a condenser to the cathode. Its function is to place a shield between the plate and the grid. A shield is necessary with high amplifications employed as the capacity between the plate and the grid may cause coupling of energy back to the grid circuit. This would destroy the control of the grid on the plate or cause the tube to oscillate violently. Tetrodes may have amplification constants of 300 or more.

Five-element tubes, or pentodes, are more common than tetrodes. In these tubes there is inserted still another grid. This "suppressor" grid is placed between the plate and the screen grid (Fig. 55). It is usually operated at zero potential or at a small negative potential. It is also grounded to the cathode through a condenser or may be connected directly to the cathode. Its purpose is to prevent the screen from attracting electrons from the plate when the plate voltage drops below the screen voltage, during that part of the alternating plate current cycle.

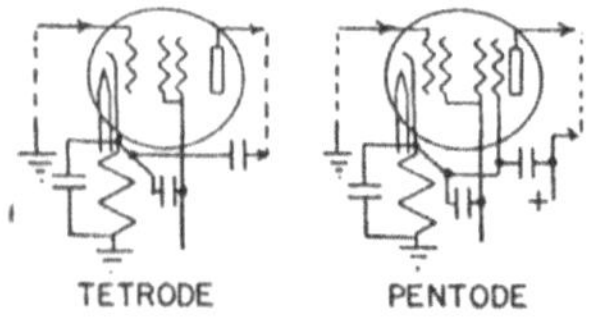

FIG. 55—VACUUM TUBE TYPES

Because of this suppressor grid the tube can produce high power outputs more efficiently than triodes or tetrodes and still maintain a high voltage gain. Such tubes may have amplification constants of as high as 3000. In power producing circuits the plate impedance, or load, is about one-fifth that of the plate resistance, as this gives the least distortion for such a tube. Whereas in triodes the amplification that is usable for power output tubes is rarely over five, pentodes will operate efficiently at effective amplifications of twenty or more.

There are other tubes which have more control grids inserted in the envelope but these tubes are essentially pentodes, tetrodes or triodes, as far as operation is concerned. Other tubes have more than one set of elements in a single envelope. Thus double triodes, two diodes and a triode, two diodes and a pentode, two diodes and a tetrode, a triode and a pentode are found in a single glass envelope and often with a common cathode.

Amplification—The principles of amplification have been described. Fig. 54 (B and C) shows audio amplifiers. Self-bias resistors are seen in the cathode circuits which raises the cathode to a slightly positive value. Thus if the grid is grounded it is effectively negative with respect to the cathode. This is a simple means of obtaining bias for the tube. The cathode resistor is by-passed by a condenser so that alternating voltages will not exist across it which would otherwise directly affect the grid. Push-pull circuits do not require a by-pass, as seen in Fig. 54C. The remainder of the circuit should be

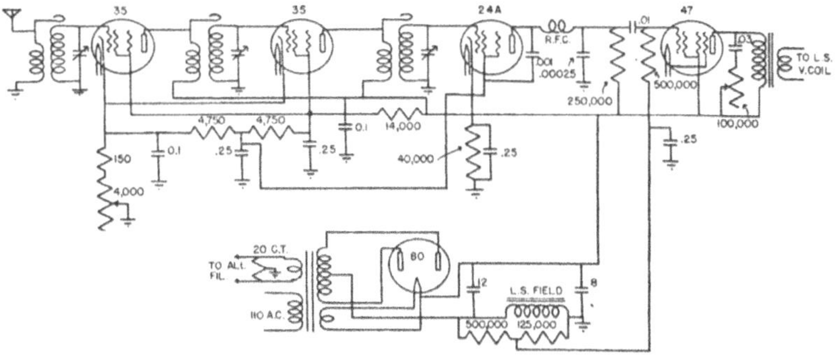

FIG. 56—TUNED RADIO FREQUENCY RECEIVER

studied for familiarization with the functions of each resistor and condenser.

The output transformer turn ratio is determined to match the output tube to the load. For example, if the output tube requires a 15,000 ohm load, a transformer is placed in the circuit whose primary impedance is 15,000 ohms at the audio frequencies to be used. The secondary circuit of the transformer contains a loud speaker whose voice coil impedance is fifteen ohms. Thus the ratio of the two impedances is 1000 to one. It has been explained that the turn ratio of the transformer determines the voltage ratio. The square of the turn ratio determines the primary to secondary impedance ratio. Thus the transformer should have a turn ratio of about thirty-three (this squared is 1000, or the impedance ratio). The transformer must also have good magnetic coupling between the primary and secondary as provided by the usual iron core.

Tuned Circuits—It was shown in the treatment of reactances that if a condenser and a coil have the same reactance at a given frequency, the impedance of the two paralleled together will be extremely high. Further, the impedance will drop rapidly when the reactances become unequal, as they very quickly will do at nearby frequencies.

Thus a coil and a condenser in parallel will act as a band pass

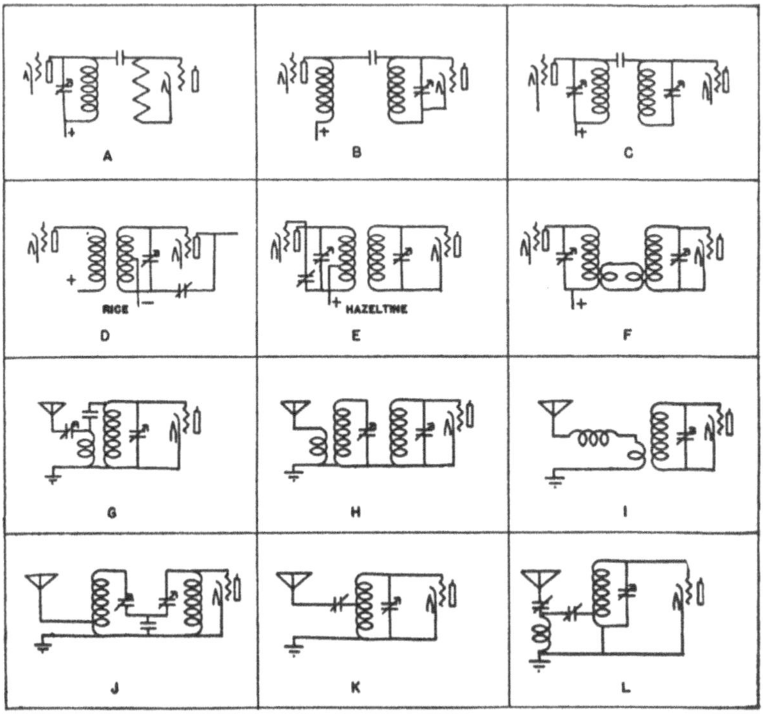

FIG. 57—COUPLING METHODS

filter for a narrow frequency range. This type of circuit is found in all radio tuned circuits. If either element is varied the frequency at which the reactances are equal will be varied. In receivers there will be found a series of coils, one set for each stage of amplification. A set of ganged variable condensers tunes all circuits simultaneously by varying the capacities of the condensers and therefore their reactance. As the reactance of the condenser changes, so will the frequency to which the circuits are tuned.

A radio amplifier can be seen in Fig. 56, in which several tuned circuits are shown. The voltage amplification with such an amplifier will vary from gains of one hundred per stage for amplifiers up to 500 KC, to gains of about thirty at 5000 KC, to gains of but one or two at 100 MC. Several forms of coupling are shown in Figs. 57A to 57F.

Class A Amplifiers—In Class A amplifiers the plate current flows at all times, even though no signal is being amplified. This is shown by the relation of grid and plate voltages in Fig. 58.

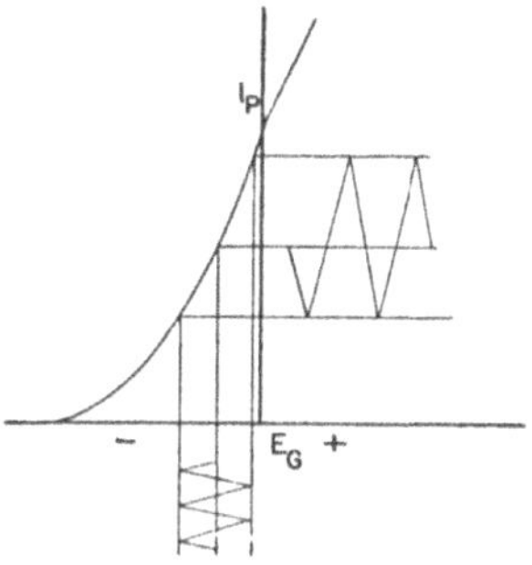

FIG. 58—CLASS A AMPLIFICATION

If a point on the grid voltage is followed vertically up to the plate current line and extended horizontally, the method of construction will be seen. The slope of the plate current curve also shows the amount of amplification. The steeper the line the greater will be the plate voltage with respect to the grid voltage change producing it.

Class B Amplifiers—Class B amplifiers operate somewhat differently, as they have their grids biased to cutoff, shown in Fig. 59A. The tube amplifies half waves instead of full waves. With no grid

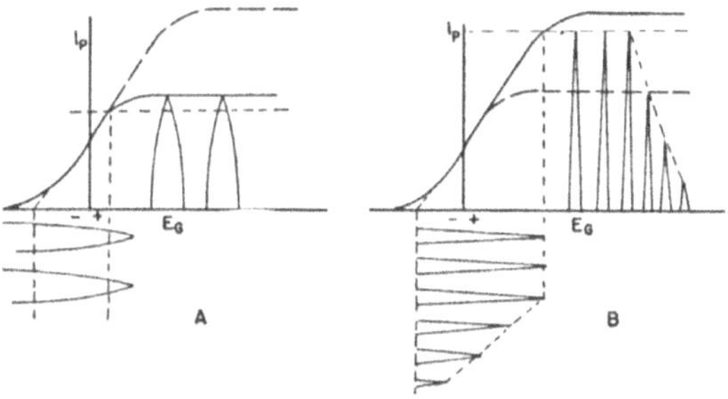

FIG. 59—CLASS B AMPLIFICATION

swing, there is no plate current flowing in the tube. The tube is operating more efficiently than a Class "A" amplifier. Efficiencies of output with respect to power supplied to the tube by the battery or rectifier, may reach efficiencies of about sixty-five per cent.

If such an amplifier is used for audio frequency amplification the distortion will amount to about thirty per cent or more. However if two tubes are used, one to supply one-half cycle and the other to supply the other half cycle, the distortion will be small and the Class B amplifier may be used. At radio frequencies, the tuned circuit in the plate circuit of a Class B tube will eliminate the harmonic distortion and the power represented by harmonics will be filtered out. The Class B grid may be positive during part of the cycle.

It must be remembered that either Class A or Class B amplifiers actually reproduce the wave shape (A) or half the wave shape (B) of the grid voltage respectively.

Class C Amplifiers—A Class C amplifier operation is employed in radio frequency amplifiers in transmitters.

A Class C amplifier is generally biased to at least twice the cutoff voltage. Efficiencies are somewhat higher than if biased to cutoff only. The efficiency of this type of amplifier may run up to eighty per cent under the best conditions.

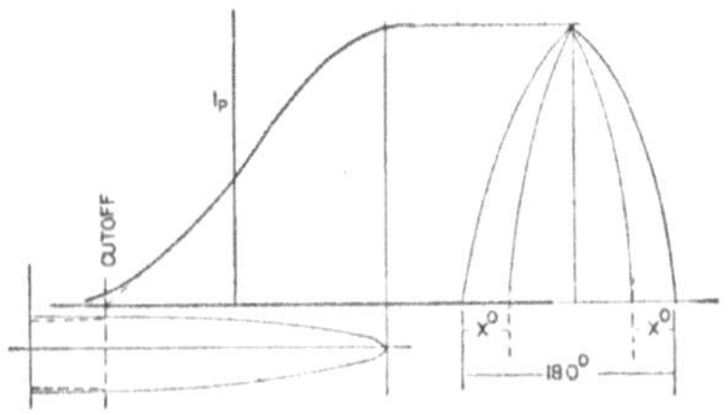

FIG. 60—CLASS C AMPLIFICATION

The grid voltage applied to such an amplifier is always sufficient to saturate the tube. In other words, no matter how high the plate voltage, the grid excitation is always sufficient to drive the plate current up to the maximum on the positive peak of every grid voltage cycle. Fig. 60 shows this.

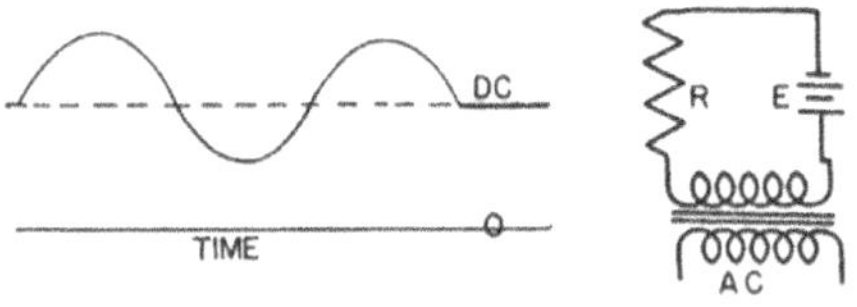

FIG. 61—ELEMENTARY MODULATION

The tube is operated so that as the plate voltage is changed the plate current will exactly vary in proportion. In other words the direct plate voltage to the tube and the output radio voltage vary in proportion. This is not true of either Class "A" or "B."

Modulation—One modulation process has already been described. This is the process employed in generating energy in a carbon microphone. A direct current flows through the microphone which is varied in amount by the pressure on the diaphragm. The steady current is thus modulated with an alternating current.

If the plate voltage of a Class C tube, considered as a resistor, is varied by placing a transformer in series with the supply, as shown in Fig. 61, the plate voltage will be varied at an alternating rate and at the frequency supplied by the transformer.

If a carrier frequency of 5000 KC were being amplified through a Class C tube and the plate supply of the tube were modulated with the voice, the carrier would vary in amplitude with the voice frequency. Thus the amplitude of the carrier is effectively being varied at a voice rate.

Unfortunately it is impossible to vary a carrier at a voice rate and produce a single carrier frequency with varying amplitude. What happens instead is that the carrier actually continues as before and two new frequencies are set up, one on each side of the carrier. The difference between these new frequencies, which are called sidebands, and the carrier frequency is exactly equal to the voice frequency.

If a 5000 KC carrier is modulated with a single voice frequency of 1000 cycles (1 KC) the radio transmitter will produce three frequencies: 5000 KC (the carrier), 5001 KC (upper sideband) and 4999 KC (the lower sideband).

Any irregular wave which is not a pure sign wave is actually composed of a number of sine wave frequencies. Thus the modulated carrier may look like a wave of varying magnitude but is actually a group of three waves being shown on a single viewing system. It can be proven that a modulated carrier actually has three frequencies present by tuning a highly selective radio receiver to each of the three frequencies. Each frequency will be found at a slightly different dial setting. The average receiver passes the carrier and its sidebands about equally well, as the percentage difference between them is extremely small.

It should be obvious from the previous explanation that once a modulated signal has been produced it cannot be amplified by anything but a Class A or B amplifier. If the signal were passed through a Class C amplifier, where the grid voltage and plate voltage have different shapes, the modulation would largely be lost as variations in grid voltage do not vary plate current proportionally.

Detection—If a modulated signal is passed through a rectifier the audio frequency will be reproduced once more. The carrier may easily be filtered out as it is very high in comparison to audio frequencies. Detector circuits are shown in Fig. 62 with diagrams of the detection process.

Diagram A in the figure shows a diode detector with a coupling

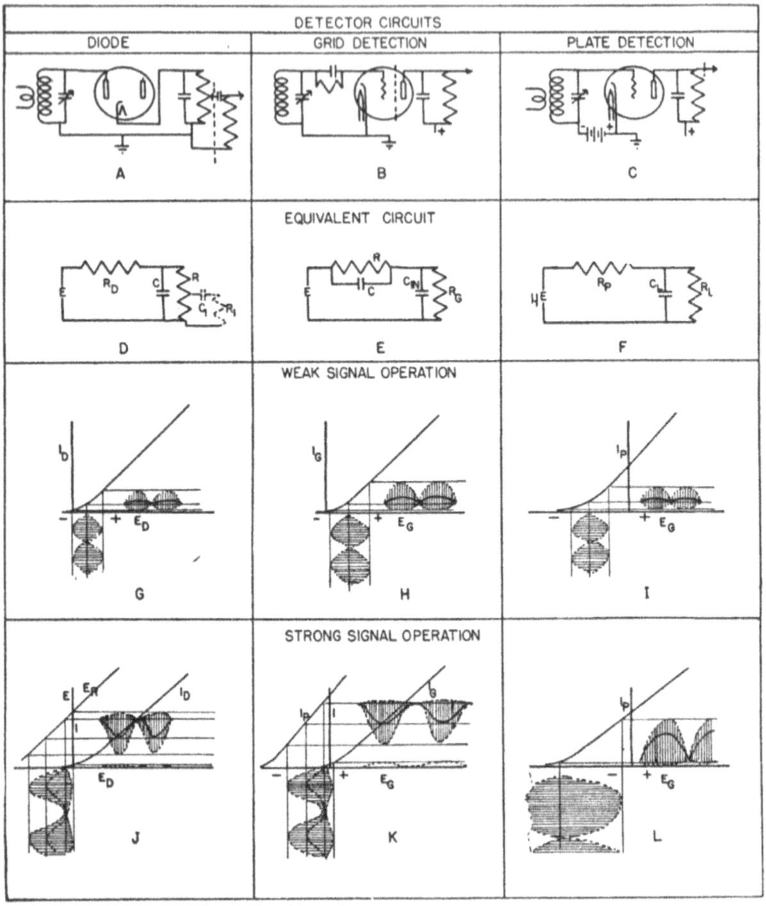

FIG. 62—DETECTION

arrangement to the following audio amplifier circuit. In diagram D the equivalent circuit is shown with R_D representing the resistance of the tube. As the tube is a rectifier, it will not exhibit constant resistance as the signal varies. The capacitor C by-passes any radio frequency so that it will not appear in the audio circuits. The ratio of the reactance of the condenser to the tube resistance will give the ratio of radio frequency voltage reduction. The condenser is usually small in capacity, representing a very high reactance to audio frequencies. Audio frequencies will not be reduced substantially by this capacitor.

The value of the coupling condenser C_1 and the audio grid resistor R_1 determine the audio response of the amplifier. If the capacitor is very large and the resistor large in value as well, the audio frequency response will be linear over a wide range. If either the resistor or the capacitor are lowered in value, the low frequency response will suffer. The effect of the radio by-pass C is to effect only the high frequency response but the effect is usually negligible below approximately 2000 cycles.

Diagram G shows the operation of the detector on a weak signal. A weak signal is one having an amplitude of less than one volt. Due to non-linearity of the diode current and voltage curve the modulated signal shown on the vertical line will be modified to the shape shown on the horizontal line. The heavy line shows the audio frequency component to be amplified.

The operation of the detector on strong signals of over a volt or two is shown in Diagram J. The signal causes a direct voltage to be built up across the load resistor R and the tube will have a substantial bias at all times.

Grid detection is shown in Fig. 62B. It is exactly the same as diode detection if the tube is considered only up to the grid. The signal is detected and audio frequency appears across the resistor, R. The grid affects the plate current of the tube and amplifies the audio grid voltage. In the diagram of strong signal detection the vertical axis is given as plate current.

Plate detection is shown in diagram C with its equivalent circuit shown below it. Bias is supplied to the grid circuit at all times by means of a voltage usually supplied by a self-biasing resistor in the cathode return. The curvature of the plate current-grid voltage curve is used rather than the grid current-grid voltage curve. The audio voltage appears in the plate circuit for the first time rather than in the grid circuit. The tube exhibits amplification at radio rather than audio frequency as the plate curvature is used. Weak signal detection results in the audio voltage increase being equal to the square of the radio grid voltage increase. In other words, when the radio signal is doubled the audio output will increase by four times. This is called square law detection. Actually any detector can operate as either a square law or a linear detector and this depends on the signal input.

Strong signal detectors are nearly linear in every case. If it is desired to have a linear detector, the audio gain following the detector must be small as this assures that the radio frequency signal must be large to be heard. If square law detection is desired the audio gain should be very large so that weak signals will give sufficient output.

Oscillators—One of the most useful characteristics of a vacuum tube is its ability to generate alternating currents at efficiencies in the order of fifty per cent. One of the simplest types of common oscillator circuits is shown in Fig. 63A. A tuned circuit is placed in

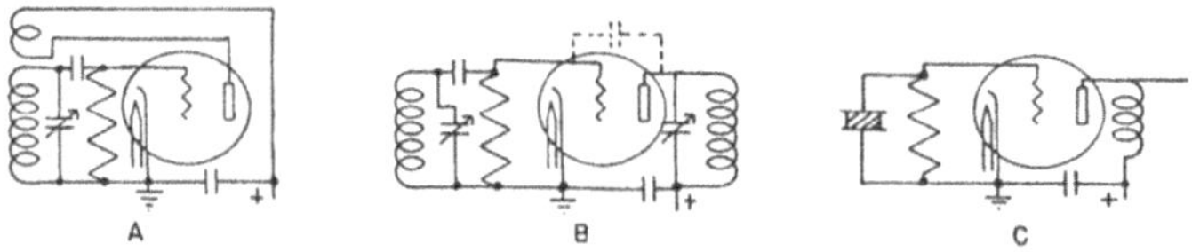

Fig. 63—Oscillator Circuits

the grid circuit of the tube and some of the energy from the plate circuit is fed back to the grid circuit.

If the grid potential is rising positively as the tube is turned on, the plate current in the coil will induce energy into the tuned circuit through its coil and cause the grid to increase in positive potential. This proceeds until the plate has reached a maximum value, due to current limitations of the particular tube. At this time the field in the coils will collapse and induce a negative voltage in the grid circuit causing the grid to reduce plate current until the tube reaches plate current cutoff. The new field due to current flow (or field change) then collapses and the tube grid starts positively once more. Thus the tube will oscillate at a frequency largely determined by the frequency of resonance of the tuned circuit.

Another type of circuit is shown in Fig. 63B. In this circuit both the plate and the grid are tuned circuits and are coupled together only by means of the capacity between the plate and the grid of the tube. Energy will be generated as before with this capacitive coupling instead of inductive coupling. The frequency of resonance will be determined mainly by the grid-tuned circuit although the plate-tuned circuit will have a noticeable effect on the frequency as well. If a coil is substituted for the plate-tuned circuit whose reactance is fairly high the circuit will oscillate as well.

In modern aircraft transmitters the frequency of the transmission must be held within .03 per cent by law. It is very difficult to build an oscillator circuit in which the coils and condensers will not change in value with temperatures and especially with humidity. Therefore a quartz crystal is used to control the frequency.

A quartz crystal will oscillate at frequencies at least as high as ten megacycles. The thickness of the crystal determines the frequency primarily. At 3000 kilocycles the thickness of the crystal will be about one sixteenth of an inch, while at ten megacycles the thickness will be a few thousandths of an inch. The width and length of the usual crystals are both an inch and have little effect upon the frequency of the crystal. Physically the crystal actually gets thicker and thinner at the rate corresponding to the frequency. Naturally such a crystal, if oscillating violently, produces appreciable heating within itself and for this reason the output of a crystal oscillator is usually somewhat limited. The usual crystal circuit will deliver a few watts of radio frequency energy.

Fig. 63C shows a typical aircraft crystal oscillator circuit and the only tuned circuit involved is that of the crystal circuit. The plate impedance of the crystal tube is usually a choke which is wound so that its capacity between windings is small. A tuned circuit may be employed in the plate circuit although it will alter the crystal frequency slightly but not enough to exceed the allowable limits.

Neutralization—It has been shown that a tube will oscillate readily if plate and grid impedances are high enough and there is coupling through the tube. The screen, in modern screen grid tubes, effectively reduces the capacity between the grid and plate so that oscillation may be prevented. Radio frequency amplifiers must not oscillate for they might produce new frequencies added to the original during amplification and would not operate properly as amplifiers when oscillating. Therefore, when using screen grid tubes it is essential to thoroughly shield the plate and grid external circuits from each other and the tube in most cases.

Triodes are often desirable on account of low cost and great reliability in comparison to other types of transmitting tubes. To amplify through a triode at high frequencies, the effect of the grid to plate capacity must be nullified or neutralized. Figs. 57D and E illustrate two methods of neutralization. The voltage at the two ends of the center tapped coil is opposite. When one voltage is instantaneously positive the other is negative and they are said to be 180 degrees out of phase. If some of the voltage is coupled through a condenser to the grid circuit from the bottom of the coil and the coupling is exactly the same as the capacitive coupling through the tube direct to the top of the coil, there can be no oscillation because these two voltages are equal and opposite and no effective coupling through tube will exist. Neutralization is often met with in aircraft transmitters and the method of adjustment will be described in the chapter on aircraft transmitters.

Doublers and Triplers—It is often desirable to produce signals at twice the original frequency. This saves neutralization as tubes will not oscillate with the plate circuit tuned to twice the grid frequency unless express coupling provisions have been made to allow oscillation. Thus a triode can be used with no great danger of oscillation. Efficiencies of doubling may amount to fifty per cent in a Class C doubler in comparison to efficiencies of up to eighty per cent with straight through amplification.

Another advantage of doubling is that the crystal used in the oscillator stage may be twice as thick at half the frequency of transmission. There is also little danger of the high transmitter output inducing high currents in the crystal circuit due to stray couplings, as the frequency of the crystal is half the frequency of transmission.

In ultra high frequency transmitters there arises a need for crystals which are too thin to be properly ground. For this reason a number of doublers may be used before the final stage in the transmitter. In some cases triplers are also used which will have efficiencies of about thirty per cent maximum.

In tuning doublers and triplers it is quite easy to select the wrong harmonic, i.e: the tube may be tripling instead of doubling. The best means to determine that the circuit is properly adjusted is to place a wavemeter near the coil. When the wavemeter is adjusted to resonance and is coupled to the tuned circuit it will apparently misadjust the circuit. Thus the wavemeter may be tuned to the fundamental second, third or other harmonic to ascertain the frequency to which the circuit is tuned. Hours of wasted effort can be saved by this simple expedient.

Power Amplifiers—It was stated that crystal oscillator circuits could deliver power outputs of a few watts. It is necessary to amplify this power to the final output power. Aircraft transmitters today have outputs of about fifty watts. Class A amplifiers are relatively inefficient and too great a number of stages are required to raise the power to the output level. Further, aircraft transmitters are required to be light in weight. For these reasons either Class B or Class C amplifiers are used. Class B amplifiers are used after modulation has taken place so as to linearly amplify the signal. If the last stage is Class C, modulation will be added at that tube's plate.

Class C amplifiers will have a power gain of about five or ten times for triodes and up to 100 times for certain pentodes. Class B amplifiers are less efficient but have power gains of from fifteen to forty times in triodes and up to 250 times gain in pentodes.

QUESTIONS ON CHAPTER V

1. Two-element tubes are primarily used as:
( ) a. Amplifiers. ( ) b. Rectifiers. ( ) c. Generators. ( ) d. Filters.

2. Three-element tubes are usually not used for one of the following. Mark which one.
( ) a. Amplification. ( ) b. Filtering. ( ) c. Rectification. ( ) d. Oscillation.

3. Mercury vapor rectifier tubes have, in comparison to vacuum tubes:
( ) a. Better amplification characteristics. ( ) b. Lower efficiency. ( ) c. Better efficiency. ( ) d. Poorer ionization.

4. With which one of the following is the self-bias usually associated:
( ) a. Cathode. ( ) b. Screen grid. ( ) c. Control grid. ( ) d. Plate.

5. If an output transformer has a turn ratio of 10 it will match two loads having an impedance ratio of
( ) a. 10. ( ) b. 100. ( ) c. 3.3. ( ) d. 20.

6. A class A amplifier usually shows:
( ) a. Little plate current with no grid signal. ( ) b. No grid current. ( ) c. High plate current until a signal is placed on the grid. ( ) d. Large variations of plate current in operation.

7. A class B amplifier generally:
( ) a. Maintains a high plate current at all times. ( ) b. Shows less efficiency than a class A amplifier. ( ) c. Reproduces half the grid wave only. ( ) d. Varies output directly with plate voltage change with constant grid signal.

8. Class C amplifiers, when in proper use:

( ) a. Show constant plate current. ( ) b. Draw no grid current. ( ) c. Operate less efficiently than Class B amplifiers. ( ) d. Do not change output with plate voltage change.

9. If a broadcasting station is radiating a carrier at 1000 kilocycles, the sidebands with 100 cycle modulation will be:

( ) a. 900 and 1100 kilocycles. ( ) b. 1,000,100 and 999,900 cycles. ( ) c. 1,000.1 and 1,000.2 kilocycles. ( ) d. Practically negligible.

10. A detector usually turns a voice modulated radio signal into:

( ) a. Voice directly. ( ) b. An audio signal and direct current. ( ) c. Another radio signal. ( ) d. A sideband.

11. The principal difference between plate and grid detection is:

( ) a. The amount of distortion. ( ) b. Rectification at high and low levels. ( ) c. Plate or grid characteristic rectification. ( ) d. The amount of rectification.

12. Oscillators that have very small coupling between plate and grid circuits:

( ) a. May not oscillate. ( ) b. Produce many harmonics. ( ) c. Are called class A amplifiers. ( ) d. Oscillate much more easily than with greater coupling.

13. Neutralization will prevent:

( ) a. Rectification. ( ) b. Oscillation. ( ) c. Modulation. ( ) d. Circuits from amplifying linearly.

14. In a properly neutralized amplifier, when the plate circuit is swung through resonance there will be:

( ) a. A substantial grid current dip. ( ) b. Little variation in grid current. ( ) c. A dip of plate and antenna current together. ( ) d. A large rise of grid voltage.

15. A crystal oscillated at 5000 kilocycles was followed by three doublers and a tripler. What was the final frequency?

( ) a. 80 megacycles. ( ) b. 120 kilocycles. ( ) c. 120 megacycles. ( ) d. 12 megacycles.

16. Power amplifiers that are to be plate modulated are operated:

( ) a. Class C. ( ) b. Class B. ( ) c. Class A. ( ) d. Class AB.

CHAPTER VI

TRANSMITTER FUNDAMENTALS

The elements of transmitter tube operation were discussed in the preceding chapter. It is particularly important that both the electrical fundamentals and the fundamentals of tube operation be understood very thoroughly, otherwise the operation and adjustment of the transmitter will not be clear.

Transmitters for aircraft service use a crystal oscillator followed by one or more buffer amplifier tubes. The buffer tube is often used as a doubler. Its original purpose was to separate the older unstable oscillators from the final amplifier so that adjustments on the latter would have small effect upon frequency. With the advent of crystal control on modern transmitters the buffer serves solely to amplify the energy for the final tube and to double the frequency.

Crystal Oscillator Adjustment—Many aircraft transmitters have no adjustment for actual tuning of the plate circuit of the crystal oscillator circuit but an untuned choke will be found with taps for rather wide frequency ranges.

Adjustment is secured by selecting a tap which will accomplish either one or both of two things. The first is maximum grid current as read on a meter in the grid return circuit of the tube following the oscillator. This assures the maximum radio frequency excitation to that tube. The second is to secure the lowest plate current in the oscillator tube which will occur near the same tap, or tuning adjustment point.

When these two conditions are obtained it will be found that a milliammeter in the grid return circuit of the crystal oscillator will show grid current in the oscillator tube. It is not particularly important that this current be a maximum and in most cases it is better to have it below maximum. The higher the current in the grid circuit, the higher is the excitation voltage on the tube and likewise the radio frequency voltage across the crystal. In some transmitters it is possible to so adjust this current as to cause damage to the crystal plate due to excessive amplitude of vibration. The usual values of safe grid current in most aircraft oscillator grid circuits do not exceed two milliamperes.

Tunable Crystal Oscillator Circuits—The adjustment of a crystal oscillator with a tuned plate circuit must be made carefully. The grid current of the tube with one direction of rotation (clockwise for example) of the plate tuning condenser will suddenly drop as the crystal starts to oscillate. When approaching this point in the other direction of condenser rotation (counterclockwise) the current will build up slowly and suddenly stop at about the same point that it started previously. It will be found possible when rotating counterclockwise, to leave the crystal oscillator in such a condition that the oscillator will deliver good excitation to the next tube but when the high voltage is removed and reapplied the oscillator will not start again. The remedy for this is in correct adjustment. The condenser should always be rotated in the direction that causes the oscillator grid current to start suddenly. Rotation should be positively continued for a small additional amount but not so much as to cause an appreciable rise in oscillator plate current nor decrease in grid current in the stage following the oscillator. Under this condition it will be found that the oscillator will start with application of voltage.

In summary, always rotate toward the point where the oscillator grid current suddenly increases and continue very slightly beyond this point.

In transmitters not equipped with oscillator grid current meters the same effect can be noted on the oscillator plate current meter. In most cases the oscillator plate current will suddenly drop as oscillation starts when approached from the correct direction of rotation. In some oscillator circuits, but not as commonly, the current will suddenly rise instead when this occurs. In any event there will be a quick change of plate current when the oscillator plate condenser is tuned in the correct direction of rotation.

Buffer and Doubler Adjustment—The buffer or doubler tube stages in transmitters are usually tuned by means of either plate- or grid-tuned circuits between tubes. In all cases the correct adjustment is secured when minimum plate current of the tube preceding and maximum grid current of the tube following the tuned circuit are obtained.

It is important that the circuit be either tuned to the fundamental frequency or the doubled frequency, as the transmitter requires. There are two ways of being sure. The first is by means of the wavemeter discussed in the previous chapter under doublers. The second is by setting the scale readings of condensers to conform to published instructions and to observe whether all plate, and especially grid, currents are within the limits given for that transmitter. Low or high grid current in a tube will indicate the incorrect tuning of the circuit associated with it, providing tubes are all in good condition and voltages correct.

Certain transmitters are equipped with broadly tuned circuits between the oscillator stage buffers and the final amplifier. A box containing these circuits for the correct frequency range is plugged into a mounting and no adjustments are necessary.

Final Amplifiers—The final amplifier is the last stage of amplification and delivers power to the antenna. The final stage is usually more difficult to adjust than any other. It involves correct adjustment of the network coupling the transmitter to the antenna, correct neutralization and correct excitation.

When the plate circuit of the final stage is not correctly tuned to the proper frequency an excessive plate current may flow which will permanently damage the output tubes or tube. It will be remembered that if the plate circuit is not adjusted to resonance its impedance will be very low. The plate circuit of the tube cannot differentiate between a low impedance due to mistuning, and an extremely heavy load. Therefore the current will climb to very high values as long as excitation is applied to the tube.

It is important when adjusting any transmitter to watch the final amplifier plate current so that excessive current will not flow for many seconds. In adjustment of a new transmitter it is possible to tune the plate circuit of the tube to resonance by applying voltage for very short periods and quickly adjust the tuning so that the plate current decreases to a reasonable value.

Thus a transmitter might have a final amplifier plate current rating in normal operation of 100 milliamperes. It might be found, upon first adjusting the plate circuit, that the current was several hundred milliamperes. By turning the transmitter on for a second or two and allowing it to cool between adjustments for ten seconds, a point will be found where the plate current is about 100 milliamperes. The adjustment may temporarily be left at this point until other adjustments before the final stage are made.

After the transmitter is adjusted for proper operation up to the final stage, the final stage plate-tuning condenser should be rotated for a minimum of plate current. This may be either more or less than the published rating. Adjustment is then made in the antenna coupling circuit to either raise or lower the plate current by adjusting the antenna load. Between any two adjustments of the antenna circuits the plate current must be dipped again for minimum plate current. The transmitter plate-tuning condenser, when finally adjusted, must be at the point of correct plate current. By reducing the coupling to the antenna the minimum will decrease but the full power output (maximum antenna current) will not be secured. Thus, at the last adjustment, the antenna current should reach its maximum value regardless of the amount of that maximum value and at the same time the minimum dip value of plate current should be at the rated value.

Antenna Adjustments—Antennas on aircraft vary in impedance from those which are fixed above the plane and have a resistance of as little as one-half ohm at certain frequencies, to those trailing wires which may have an impedance of almost one thousand ohms at some frequencies. The whole purpose of tuning the antenna circuit in a transmitter is to add an amount of reactance which is exactly equal and opposite to the antenna reactance. Thus the antenna and the antenna tuning circuits in the transmitter together will show pure resistance to the transmitter and a minimum loss will be present. The reason for this is that reactive current produces heat loss.

A common mistake is to attempt adjustment of the transmitter so that the antenna current at the various frequencies is the same. In the case of the thirty-four foot trailing wires used on some transports the resistance of the antenna is about fifty ohms on the night frequency and approximately five hundred ohms on the day frequency. Thus a fifty watt transmitter might show an antenna current of one ampere on night frequency and 0.3 ampere on day frequency and still be in correct adjustment. In both cases the final amplifier plate current might be 200 milliamperes at the minimum dip. The transmitter would be adjusted correctly and any attempt to reduce or raise the currents would result in inefficiency or excessive plate current.

Antenna circuits are easy to adjust providing (1) the plate circuit is always dipped after each antenna circuit adjustment, (2) the plate current is adjusted by means of the antenna coupling to its maximum allowable value when dipped, (3) the antenna current maximum and plate current minimum occur simultaneously. Plate current means nothing except when the plate circuit is tuned to resonance, as shown by its minimum value for that antenna load. Tuning a circuit off resonance to secure rated plate current results only in loss.

Neutralizing—A triode is employed in many transmitters for the final stage and must be neutralized. There are many ways to neutralize a transmitter but a combination of two methods will give the best results.

Upon referring to the previous discussion of neutralization it will be found that neutralization merely consists of applying a voltage from the plate circuit back to the grid circuit in such amount and phase relation that the capacitive coupling through the tube is nullified. This requires a tuned circuit with its center point grounded so opposite voltages may be secured from the two ends of the coil. It makes no difference whether this tuned circuit is in the plate or grid circuit of the final amplifier. It is usually located in the grid circuit as this simplifies coupling to the antenna by the use of a simpler plate circuit. Figs. 57D and 57E show such circuits.

When a transmitter is properly neutralized any tuning of the final circuit should not appreciably affect the grid circuit. This is the criterion of correct adjustment.

A new transmitter which is being adjusted for the first time may not be in neutralization. A power supply is attached to the transmitter after removing the high voltage lead to the plate circuit of the final tube. There will be a removable link in most transmitters to allow this circuit to be easily opened.

It should be borne in mind, when working on transmitters, that very dangerous voltages will exist. It is very easy to receive a fatal shock when working on the transmitter final stage, particularly when neutralizing it. Two precautions should be observed. The first is to use an insulated handle screwdriver for all adjustments. This screwdriver should never have the metal blade protruding through the back of the handle. The best type of screwdriver is one with its handle made of a semitransparent insulating material so that it is certain that the metal part does not extend to any point on or near the surface. The second precaution is to work on the transmitter with only one hand, especially when the back, top or bottom are open. The other hand should be kept in the pocket and not touching any grounded metal parts such as the transmitter frame. Also a rubber or wood covering should be provided on the floor. The expert service man will follow this practice while the inexpert will show his lack of experience by working on the transmitter with both hands because he knows, or thinks he knows, that the voltage at that point is not dangerous. Severe burns or death have resulted from carelessness with voltages which are not much higher than 100 volts, to say nothing of the 1000 or more volts associated with transmitters.

After removing the plate voltage link the transmitter should be turned on and all stages adjusted for proper operation. There will be no final amplifier plate current so no tuning precaution is necessary. A low range milliammeter, of about ten milliamperes full scale, should be connected from the plate return circuit of the final tube to ground. A high current meter should be tested in the circuit upon first application of voltage before inserting this sensitive meter. Up to this point there is no voltage, either on the meter or on the plate circuit. When the transmitter is turned on, the milliammeter in the plate circuit will read some current due to lack of neutralization of the stage and due to the rectification process in the final tube. The neutralizing condenser should be rotated for no plate current on this test meter. When this point is found the plate circuit should be adjusted to resonance by watching the small variation in grid current as the plate is tuned toward resonance. It may be necessary to misadjust the neutralizing condenser slightly to see this. The neutralizing condenser should be left finally half-way between the two points at which grid current change just starts to show. The test meter may now be removed after turning off the transmitter and the high voltage link reconnected.

When the transmitter is again turned on, the plate current should be dipped for a minimum with the plate condenser. The grid current of the final stage should be watched while tuning this condenser and the grid current should not vary appreciably when the plate condenser is swung through resonance. If this happens the neutralizing condenser should be adjusted in small steps until a minimum of grid current reaction occurs. At this point the minimum dip in plate current should occur simultaneously with the maximum antenna current. If there is any great disagreement between the minimum plate current and the maximum antenna current the neutralization adjustment is not correct and should be repeated.

In adjusting transmitters which have been in service for some time it will not be necessary to remove the link and go through the first part of the adjustment process. It will be found that a small readjustment of the neutralizing condenser will secure correct agreement between antenna current maximum and plate current minimum.

In some cases the neutralization will be slightly affected by the amount of load on the transmitter and the final adjustment may be made at full load.

In some transmitters there will be but one neutralizing adjustment for all of the frequencies that are selected by a quick-change mechanism. Neutralization will usually be found best for all frequencies if adjustment is made at the highest frequency to be employed. Other frequencies will not show exact neutralization but they will show reasonably close adjustment.

Final Stage Excitation, Class C—The adjustment of the final grid current (which measures the excitation) in the transmitter depends on whether the final stage is operating Class B or Class C.

If the final stage is operating Class C the excitation of the final stage must be high enough to provide sufficient excitation for the tube to be saturated at one hundred per cent modulation, where the peak plate voltage is twice the supply voltage value. Lack of excitation will cause the antenna current to remain steady or to decrease when the transmitter is modulated. The antenna current increase should be approximately twenty-five per cent when the transmitter is fully modulated and is correctly loaded. The amount of load makes some difference in Class C stages. When the load is too high the final tube may be drawing too much power from the high voltage source for one hundred per cent modulation to occur, due to limitation of the modulator supply power at the voice frequencies. The remedy is to lower the amount of the load by reducing the antenna coupling and thus the final plate current.

If the generator or supply voltage to the transmitter decreases somewhat upon modulation it is not to be expected that the antenna current will increase as much as if the voltage remains constant. The plate current to the Class C stage should not fluctuate appreciably when modulation occurs.

Final Stage Adjustment Class B—A Class B final amplifier which is amplifying a previously modulated signal is more difficult to adjust than a Class C amplifier. The amount of excitation is very important in the Class B amplifier. If it is too high the transmitter will not be capable of one hundred per cent modulation and if it is insufficient the power output will be low. In either case a certain amount of distortion will be noticeable in the received signal.

A carrier is radiated at all times when a transmitter is either modulated or unmodulated. Thus a Class B amplifier which is used to correctly reproduce a signal cannot supply more than half the full current output when unmodulated because capability in the tube must be provided for the peak current at one hundred per cent modulation (see Fig. 64 A, B and C). Thus the tube is supplying half the possible peak voltage, or one quarter of its peak output power with carrier alone. When a modulated signal is amplified by such a tube the tank or tuned output circuit will suppress most of the dis-

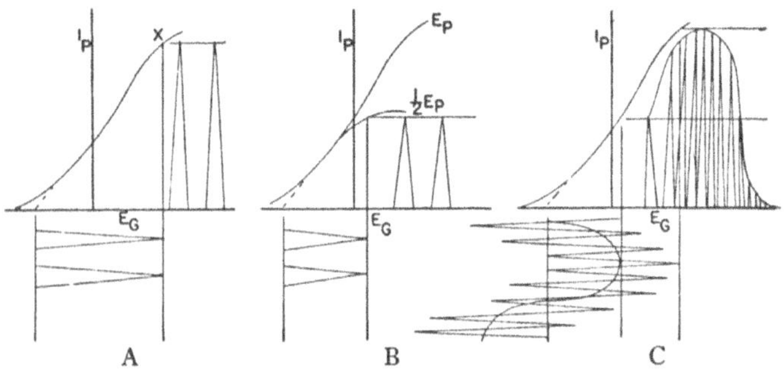

FIG. 64—CLASS B MODULATION

tortion of even a single Class B tube as the output circuit is incapable of efficiently supplying radio frequency harmonics due to its selectivity. The operation of such a linear amplifier is shown in Fig. 64C.

Grid Modulated Class B Amplifier Adjustment—A grid modulated Class B amplifier has the bias varied at an audio frequency rate and produces modulation as shown in Fig. 64C.

The plate current of a Class B linear radio amplifier, or modulated tube, is practically constant whether modulated or unmodulated as the additional supply of power necessary from the circuit when modulated (fifty per cent more output power than the carrier alone) is supplied by a change of efficiency of the tube. Class B tubes operated under these conditions will usually show efficiencies with the carrier alone of about twenty per cent. The power required to modulate the tube is very small in the case of grid modulation as the grid current (and likewise voltage) is small in comparison to the plate current and voltage. Usually the grid current is about ten per cent of the plate current and the grid power is about one hundredth of the plate power.

Class B Plate Load—The Class B linear or grid modulated radio amplifier also suffers some other disadvantages in that the plate load must be adjusted within certain limits for correct modulation to take place in the tube. Thus the operation of the circuit will be disturbed with loads that are either too high or too low. This will be shown by the increase in antenna current being greater or less than the normal twenty to twenty-five per cent rise at one hundred per cent modulation.

The bias voltage must be held within fairly close limits on Class B linear radio frequency amplifiers or grid modulated stages. If this is not done the antenna current rise will show this difficulty. Lastly, the excitation current which indicates the amount of excitation will vary even with the tube neutralized as the load is changed. Thus the excitation must be adjusted for each load condition. One of the simplest ways of changing the grid current is to change slightly the neutralizing. This will cause some change of grid current. The grid current must usually be held within about plus or minus ten per cent of the best adjustment. Higher grid currents will cause loss of power at high modulation percentages and low grid currents may cause excessive rise of antenna current and loss of efficiency.

Overall Neutralization Tests—An overall test, after adjustment of the transmitter, is to stop the crystal by shorting the crystal contact with a screwdriver so that no oscillation occurs. The antenna output should fall to zero and plate currents should all drop instantly to their static bias value, except for the oscillator plate current which may rise. If plate currents do not drop and the antenna current continues at any value, the transmitter is oscillating in some neutralized stage and should be readjusted.

The antenna current should be proper for the impedance of the antenna used and should reach a peak at the dip point of plate current of the final stage. Modulation should cause a rise in antenna current of about twenty-five per cent, or slightly less. The plate current of the final stage should not vary over ten to fifteen per cent at one hundred per cent modulation.

Individual Transmitter Adjustment Values—The adjustment of a normal aircraft transmitter has been explained. The instruction books found with most transmitters (and in company instructions) give methods utilizing these same principles although the process may be slightly modified to save time in adjustment.

The best way to gain an overall picture of transmitters is to follow through the diagrams and descriptions of the radio transmitters found in the following chapter.

The troubles commonly found in defective transmitters will be treated in a later chapter.

Pi Network Adustments—The usual configurations of a pi network is shown in Fig. 65. Such a network may be used for matching transmitters (or receivers) to antennas and the adjustment of the device is very simple.

There are two condensers in this type of pi network, one at the input (from the transmitter) and the other at the output (to the antenna).

The output condenser will determine the coupling between the transmitter and the antenna. The input condenser while also ac-

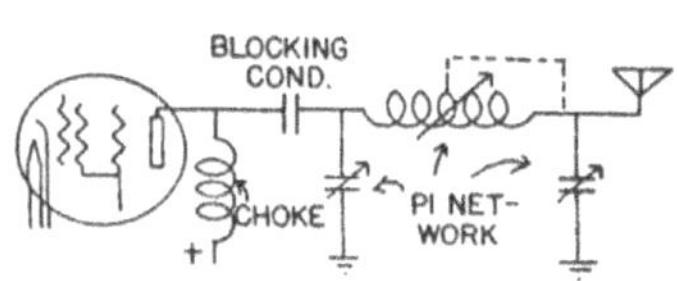

FIG. 65—PI OUTPUT NETWORK

complishing this same thing will usually be associated with the tuning of the circuit to resonance. To change the load on the transmitter the output condenser should be moved to a slightly different capacity and then the input condenser should be varied for resonance (as indicated by the dip of plate current). If the condensers come to the end of their range and the adjustment is still not correct the inductance between the two condensers should be set to a new value. If more coupling is desired the inductance should be decreased and vice versa. If this fails it may be that either the input or output condensers are too small for the purpose.

In summary, the pi network is adjusted by changing the output condenser in steps and re-resonating each time with the input condenser.

Where difficulty is found with coupling, the antenna may sometimes have to be moved to a tap on the inductor. The adjustment is always made in the manner described however.

QUESTIONS ON CHAPTER VI

1. When a crystal oscillator plate circuit is tuned properly the point of sudden current change is:

   ( ) a. To be avoided and adjustment made near to this point. ( ) b. The exact point of correct adjustment. ( ) c. A sign that correct operation has been secured. ( ) d. Indicative of efficient plate operation.

2. Final amplifier plate current will be excessive when:

   ( ) a. The antenna is too long. ( ) b. No excitation is provided. ( ) c. The plate circuit is detuned. ( ) d. The oscillator stops.

3. Antenna current on all frequencies on a fixed antenna should be the same if the transmitter is correctly adjusted because:

   ( ) a. The power output should be the same. ( ) b. This is not true. ( ) c. All antennas are matched in this condition. ( ) d. This shows best modulation.

4. Most amplifiers are initially neutralized with no final plate voltage because:

   ( ) a. There is less danger of shock. ( ) b. The plate circuit will be nearer resonance. ( ) c. High voltage causes higher grid current. ( ) d. Damage to the tube during adjustment cannot result.

5. Slight change of neutralization may raise grid current because:

   ( ) a. Energy is fed back to the grid from the plate circuit. ( ) b. Grid current is a measure of excitation. ( ) c. Oscillation occurs if there is grid current. ( ) d. Class C amplifiers are used.

6. It is possible to use a single neutralizing adjustment for all frequencies in some transmitters because:

   ( ) a. Once a tube is neutralized it is always perfectly neutralized. ( ) b. Neutralization is unimportant. ( ) c. One adjustment may give sufficiently close results. ( ) d. More than one is never provided.

7. Class B amplifiers are hard to adjust because:

   ( ) a. Small plate voltage change radically effects output. ( ) b. Excitation must be held within close limits. ( ) c. Neutralization makes them oscillate. ( ) d. Loads effect the DC plate voltage.

8. Grid modulated final amplifiers will show efficiencies of about:

   ( ) a. 8.5 per cent. ( ) b. 74 per cent. ( ) c. 50 per cent. ( ) d. 20 per cent.

9. The additional output power to supply modulation peaks in a Class B grid modulated amplifier comes from:

( ) a. Change of efficiency. ( ) b. Change of plate voltage. ( ) c. Bias voltage change. ( ) d. Excitation change.

10. Overall neutralization may be checked by:

( ) a. Removing biases. ( ) b. Removing plate voltage. ( ) c. Shorting the crystal. () d. Shorting the antenna.

11. Plate current of the final amplifier should dip at resonance:

( ) a. So that antenna current is minimum also. ( ) b. To the rated current value for best output. ( ) c. To a value giving definite downward antenna current change with modulation. ( ) d. To assure proper grid current.

12. The Pi network output condenser, or the antenna condenser, primarily is used to:

( ) a. Regulate the excitation. ( ) b. Reduce the excitation. ( ) c. Correct the modulation so it will rise. ( ) d. Regulate coupling.

13. The coil in the simple Pi network primarily:

( ) a. Regulates coupling but is not critical. ( ) b. Increases the antenna current. ( ) c. Regulates the percentage modulation. ( ) d. Offsets the tube resistance.

14. If antenna current reaches a maximum at the same tuning point that the plate current reaches the bottom of its dip:

( ) a. The transmitter should be reneutralized. ( ) b. The current should be reduced. ( ) c. Neutralization is excellent. ( ) d. The antenna should be decoupled.

15. The antenna current may rise about 50 per cent with modulation when:

( ) a. Neutralization is correct. ( ) b. The antenna load is incorrect for a class B final amplifier. ( ) c. Exact resonance is reached. ( ) d. Class B final amplifiers are properly adjusted.

CHAPTER VII

COMMERCIAL AIRCRAFT TRANSMITTERS

This chapter contains information on a great many of the present commercial aircraft transmitters with instructions on methods of adjustment of each type of transmitter. The material has been arranged in such a manner that if each description is studied in order and the reason for each adjustment is understood, a good grasp of the practical application of material which has been given in the previous chapter will be secured.

Some of the older models of transmitters have been modified by change of tube types and inclusion of special circuits. These details will become evident on comparison of the material given with the transmitters themselves. Most of the transmitters shown on the following pages are relatively new however.

A short table precedes each description giving pertinent data and approximate overall sizes. In some cases one protrusion, such as a single binding post, may exceed the dimensions shown. Space should be provided for pulling the transmitter out of its mountings as well as room to reach in the top for tube replacement in the air if this is contemplated.

Power output ratings are the same on telegraph or telephone unless otherwise specified. Battery voltages are given as twelve volts although operation from aircraft batteries at 14.2 volts is intended. This voltage will rarely be as high as twelve volts when the transmitter is operating.

25A Western Electric Transmitter—

Power output: 15 watts
Frequency range: 2800 to 6400 KC
Type frequency shift: Manual, except electrical for two closely spaced (within 2 per cent) frequencies
Number of frequencies: Retune—two per crystal holder
Power supply voltage: 12 volts
Power consumption—standby: 2 amperes
Power consumption—modulated: 17 amperes

Tube types: Oscillator 6L6, Power Amplifier 6L6, Modulators (2) 6L6

	HEIGHT	WIDTH	DEPTH	WEIGHT
Transmitter	$6^5/_8''$	$6^5/_8''$	10″ over plugs	3.5 lbs.
Power and audio unit	5″ (not incl. plug)	$7^1/_8''$	$4^3/_4''$	17.0 lbs.
Cables and microphone	....	....	....	4.5 lbs.

The Western Electric 25A radio transmitter is built in two units which may be separated from one another to suit the particular installation.

A double crystal unit can be supplied and the desired crystal is selected by a switch on the transmitter panel. The transmitter, unlike others to be described later, uses a crystal (type 5A or 5D) ground to the fundamental frequency to be transmitted except for frequencies from 5600 to 6400 kilocycles where the crystal is half the carrier frequency. No crystal heater is employed.

In itinerant aircraft, a single crystal ground to 3105 kilocycles may be used. When output at 6210 kilocycles is desired the antenna circuit is merely tuned to the higher frequency for correct operation, doubling taking place in the output stage.

The unit is simple to install. The radio frequency unit should be easily accessible to the pilot. The antenna connection inside of the plane should be as short as possible and well separated from metal parts. The receiver antenna connection is made to the transmitter so that both units may be operated from the same antenna. A changeover relay connects the transmitter to the antenna when the microphone button is pressed (relay S1).

Connections are made to the battery by means of the power cable. The black lead is connected to the positive and the white lead to the negative (ground) terminal of the battery. The cable shield braid lead is connected to a good metallic ground on the airplane structure. Other braid leads on the cables are connected to respective units. In addition, a short ground lead is connected to a phone tip which is inserted in the hole "G" (ground) in the radio frequency unit.

Crystals are placed in the transmitter and if a double crystal unit is employed the switch (D1) is thrown to the "high" (frequency). If a single crystal unit is used the switch (D1) is thrown to "low" position, or "low" frequency crystal. The schematic in Fig. 66 shows the arrangement of the transmitter. The crystal is connected to the first 6L6 tube grid and the plate circuit of this tube is broadly tuned and is non-adjustable. The oscillator draws thirty milliamperes at 220 volts. The only adjustments that can be made are in the antenna circuit of the 6L6 final amplifier where a coil and

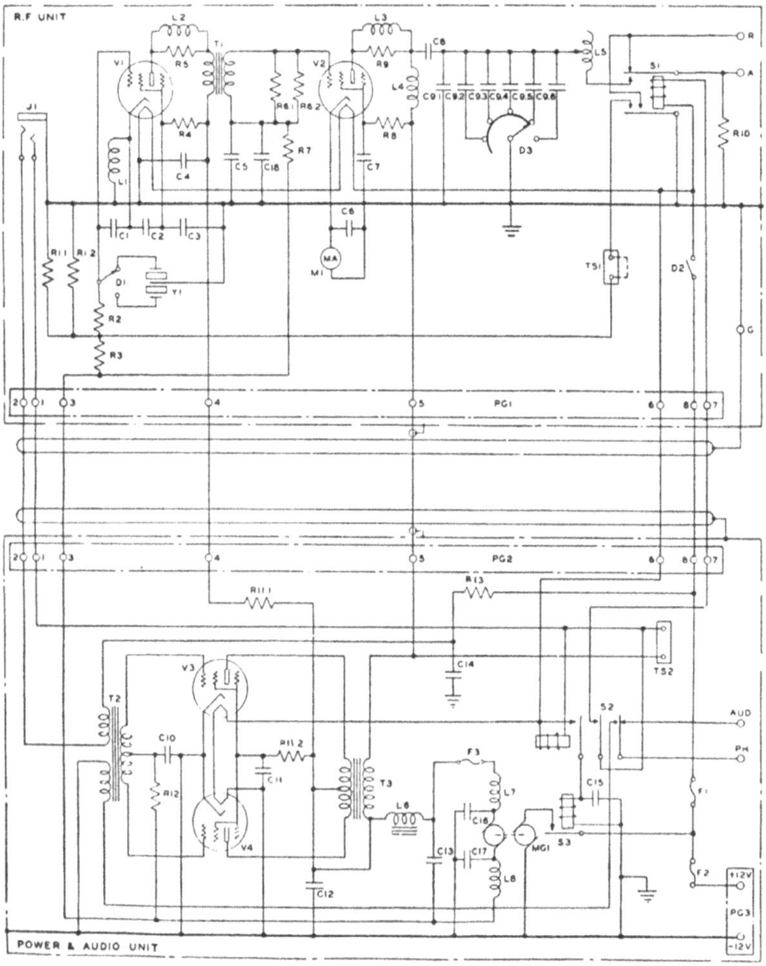

Courtesy Western Electric Co.

FIG. 66—WESTERN ELECTRIC 25A TRANSMITTER SCHEMATIC

condenser arrangement is employed for tuning. The coil is continuously adjustable and serves two important purposes. The first is to control the amount of coupling to the antenna and the second is to resonate accurately the output circuit.

When the antenna is connected to the transmitter or else to a

No. 93 Mazda lamp in place of the antenna for testing, the microphone button may be pressed and the final plate current read upon the meter on the radio frequency unit panel. For low frequencies, near 3000 kilocycles, the plate loading (condenser switch) is placed at tap 4, 5 or 6 for the highest frequencies at tap 1, 2 or 3. The tuning coil is then rotated by means of its control knob until the plate current reaches its minimum dip. This dip may not be found, in which event the plate loading tap should be placed at another point and the process repeated. The plate current should read about ninety milliamperes at 370 volts at the dip. If the minimum current is too low, the plate loading switch is rotated to the next higher point and the current redipped. If the current is too high the plate loading switch should be rotated to the next lower tap and the plate current redipped.

Occasionally it will be found that the antenna used will prevent proper tuning of the transmitter. If trouble is encountered at the high frequency it will usually be possible to tune by inserting a 100 micromicrofarad mica condenser in series with the antenna. This is especially true of long antennas or antennas with long leadins. If trouble is encountered at the low frequency end with short antennas, a small mica 100 micromicrofarad condenser may be connected from the antenna post to ground for satisfactory tuning.

It is quite possible to tune the transmitter accidentally to the harmonic frequency of the crystal. To verify the tuning it is best to listen to the transmitter at some distance (over a quarter of a mile) and be sure that loud signals are encountered on the correct frequency and that signals at half, twice or three times the desired frequency are very weak. Little difficulty should be encountered. For the low frequencies the plate loading should be about at tap 5 and the coil turns high in number, twenty or more. At the high frequency end the plate loading will be at three or less and the tuning coil should be at about fifteen or less. Some antennas may give plate loading setting and coil turn adjustments for correct adjustment at slightly different values.

Plate loading and antenna tuning readings should be recorded for the pilot's convenience although the exact antenna tuning should always be adjusted for the lowest dip in plate current, which should not be far different from the recorded reading.

The method of opening the radio frequency unit for inspection and repair requires removal of the meter and disconnecting its leads. Also the front cover, side plates and back cover are removed and, most important, the plate tuning condenser (C8) lead is unsoldered, after which the unit opens on a hinge and is readily accessible.

In cases of difficulty with transmitter operation, it is well to check the roller on the adjustable antenna coil to be sure that it is riding on and not between wires.

The power and audio unit contains the two 6L6 modulator tubes with associated transformers as well as the dynamotor, fuses and associated relays. Note that the relay S2 disconnects the audio (AUD) from the receiver to the headphones during transmission and connects the headphones so that sidetone from the microphone through audio modulators is heard.

The following voltages will be found on the connecting terminal strips for the units with the transmitter "on":

TERMINAL	VOLTAGE TO GROUND	
1	0	(microphone on)
	+12	(microphone off)
2	1.5	approx. (microphone on)
	+12	(microphone off)
3	−28	(bias)
4	+220	(oscillator plate)
5	+340	(final plate)
6	+12	(+ filaments)
7	0	(S1 control)
8	+12	(supply voltage)
Screen power amplifier	+280	
F3 (at fuse)	+380	
Audio screens	+325	
Audio plate	+370	

The following are values of the various resistors and condensers:

SYMBOL	SIZE	SYMBOL	SIZE
C1	50 mmf	R1.1, R1.2	50,000 ohms
C2	500 mmf	R2	100,000 ohms
C3, C4, C5, C6	.006 mf	R3	115 ohms
C7, C8	.002 mf	R4	150,000 ohms
C9.1, C9.2	50 mmf	R5	30 ohms
C9.3, C9.4, C9.5, C9.6	100 mmf	R6.1, R6.2	10,000 ohms
		R7	25,000 ohms
C10 to C13	special block	R8	8,000 ohms
		R9	30 ohms
C14	20 mf	R10	2 megohms
C15	.05 mf	R11.1	25,000 ohms
C16, C17	special	R11.2	10,000 ohms
C18	.1 mf	R12	5,000 ohms
		R13	600 ohms

T-30-AB Lear Transmitter—

Power output: 20 to 30 watts
Frequency range: 2900 to 6500 KC
Frequency shift: Tuning on front of radio frequency unit
Number of frequencies: Two or more (substituting other crystals)
Power supply voltage: 12V
Power consumption—standby: 1.8 amperes
Power consumption—modulated: 10 amperes (approx.)

Tube types: Oscillator 6L6, Power amplifier 6L6, Audio modulators (2) 6L6

	HEIGHT	WIDTH	DEPTH	WEIGHT
Transmitter	$4^{11}/_{16}''$	$7^{1}/_{2}''$	$5^{1}/_{8}''$	3.5 lbs.
Power Supply (GT-30-AB)	$7^{1}/_{8}''$	$6^{7}/_{8}''$	$9^{9}/_{16}''$	13.4 lbs.
Accessories, microphone, cables, etc.				2.1 lbs.

Tuning is accomplished in much the same manner as that employed in tuning the Western Electric 25A transmitter and reference should be made to the material in that description, even though the circuits differ.

A switch is provided so that the panel meter may be switched to read oscillator current in most models. An adjusting knob, or in some models a slotted shaft, will be found above the meter which should be adjusted for minimum oscillator plate current. Tuning of the output circuits is similar to that given in the reference above, after the meter switch is thrown to the P.A. position. The schematic is shown in Fig. 67.

The transmitter may be used with the G-30-AB power supply (Fig. 123) when used with the R-3-AB receiver, Fig. 124 (see R-3-AB receiver).

The transmitter draws about 225 milliamperes from the power unit at 400 volts. Other voltages to be found are:

P.A. plate	370 volts
P.A. screen	270 volts
Oscillator plate	210 volts
Oscillator screen	135 volts
Modulator plates	370 volts
Modulator screens	280 volts
Modulator cathodes	19 volts
P.A. cathode	7 volts

The values of the transmitter components are:

C1	260 mmf
C2, C9, C11	.002 mf
C4, C10	.01 mf
C3, C5	250 mmf
C6	70 mmf

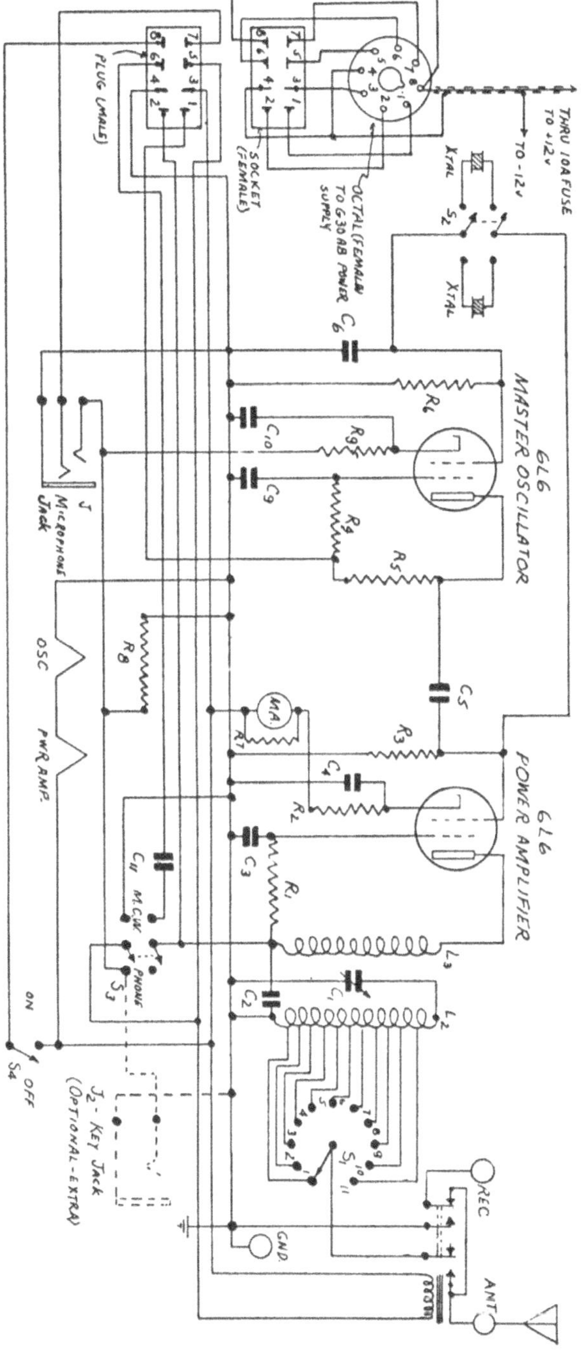

FIG. 67—LEAR T-30-AB TRANSMITTER

Courtesy Learadio

R1, R3	20,000 ohms
R2, R9	100 ohms
R4, R8	100,000 ohms
R5	7,500 ohms
R6	.5 megohm
R7	2 ohms

AVT 7B RCA Transmitter—

Power output: 20 watts
Frequency range: 2600 to 6500 KC
Type frequency shift: Electrical remote
Number of frequencies: 2 non-related. 4 related as shown
Power supply voltage: 12 volts
Power consumption—standby: 2.2 amperes
Power consumption—modulated: 14.5 amperes

Tube types: Oscillator—42, Power amplifier—807, Speech amplifier—42, Modulator—6A6

	HEIGHT	WIDTH	DEPTH	WEIGHT
Transmitter	$7^{13}/_{16}''$	$8^{7}/_{16}''$	$7^{13}/_{16}''$	11.25 lbs.
Power supply	$7^{13}/_{16}''$	$8^{7}/_{16}''$	$7^{13}/_{16}''$	19.25 lbs.

A vibrator unit is used to supply power for the plate circuits. Tube types are shown in the circuit diagram in Fig. 68.

The general arrangement of the transmitter with much pertinent data is shown in Fig. 69.

The high frequency end of the band utilizes crystals which may double frequency, if desired, to obtain the carrier frequency. Crystals ground to the fundamental may be used at any frequency. Two low frequencies within two per cent of one another and two high frequencies within two per cent of one another may be used. One popular arrangement is to equip the transmitter with a 3105 and a 3120 kilocycle crystal. Then the same crystals will also give high frequencies of 6210 and 6240 kilocycles, although this last frequency is not licensed for itinerant use.

As a small control panel is provided, the remainder of the equipment may be mounted at any convenient point. It is advisable to mount the separate power amplified plate meter in a position where it is visible to the pilot in order that correct transmitter operation may be checked.

Fig. 70 shows the two oscillator coils which must be properly connected depending upon the frequency range. The correct taps are shown in the following table and refer to the crystal frequency:

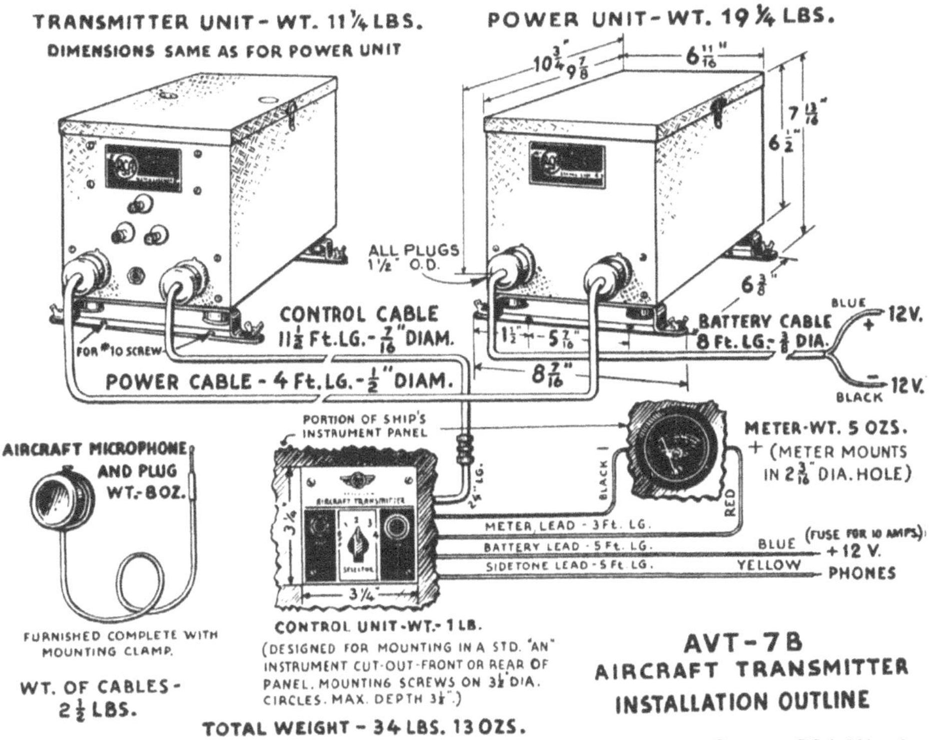

Courtesy RCA Mfg. Co.

FIG. 69—RCA TRANSMITTER

FREQUENCY IN KILOCYCLES	COIL	TAP NO.
2600 to 2800	A	1
2800 to 3500	A	2
3500 to 4000	A	3
4000 to 4500	B	1
4500 to 5000	B	2
5000 to 6100	B	3
6100 to 6500	B	4

When frequencies are near to the end of the ranges given, it may be found that an adjacent tap is somewhat superior. Crystals at 3105 kilocycles will give better results on 6210, when doubling is employed, by the use of tap 1 on coil A, although the 3105 kilocycle output will be slightly reduced.

An "off" position is provided on the frequency selector and in this position the radio transmitter microphone may be used as an interphone unit between two pilots. The radio filaments need not be turned on for this use. Selector switch positions 1 and 2 are used for the low frequencies and 3 and 4 are used for the high frequencies. There are three general arrangements that may be used in the frequency setup: (a) one low and one high frequency which are usually set up on positions 1 and 4, (b) two low frequencies within two per cent of one another and two high frequencies which are their second harmonics (double frequency), and (c) one low frequency (position 1) and two high frequencies (positions 3 and 4) one of which is the second harmonic of the low frequency and must be within two per cent of the other non-harmonic high frequency. Crystal socket position number one is for the lower frequency crystal.

Fig. 70 shows the arrangement of the high frequency (H.F.) and the low frequency (L.F.) tank coils which must be adjusted for proper operation. The number of turns to be used on each coil is shown approximately by the following table:

FREQUENCY IN KILOCYCLES	TANK COIL	APPROXIMATE NUMBER OF TURNS
2600 to 3000	Low frequency	Whole coil
3000 to 4000	Low frequency	26
4000 to 5500	High frequency	Whole coil
5500 to 6500	High frequency	11

After pressing the microphone button the power amplifier condenser is rotated for lowest dip of the plate current with no antenna connected to the transmitter. The dip that occurs at the more fully meshed condenser plate position is the correct dip if two dips can be found.

A brief description of two recommended methods of adjustment of the antenna circuits will illustrate two good methods for use with any similar transmitter. The first method supposes a fixed length of antenna of the trailing type whose length is cut to the proper operating point at the highest frequency.

The total length (feet) of the antenna and leadin to the transmitter should be about 233,000 divided by the frequency in kilocycles. The length outside the fuselage, for 6210 kilocycles, will be about thirty feet or less in most planes. For the initial ground tests a length some fifteen per cent shorter (total) should be used for adjustment.

The antenna is attached to the plane after raising the end of the antenna some ten feet above the ground on a dry non-metallic support. The transmitter is operated again and the transmitter tank condenser readjusted for the dip in plate current. The condenser

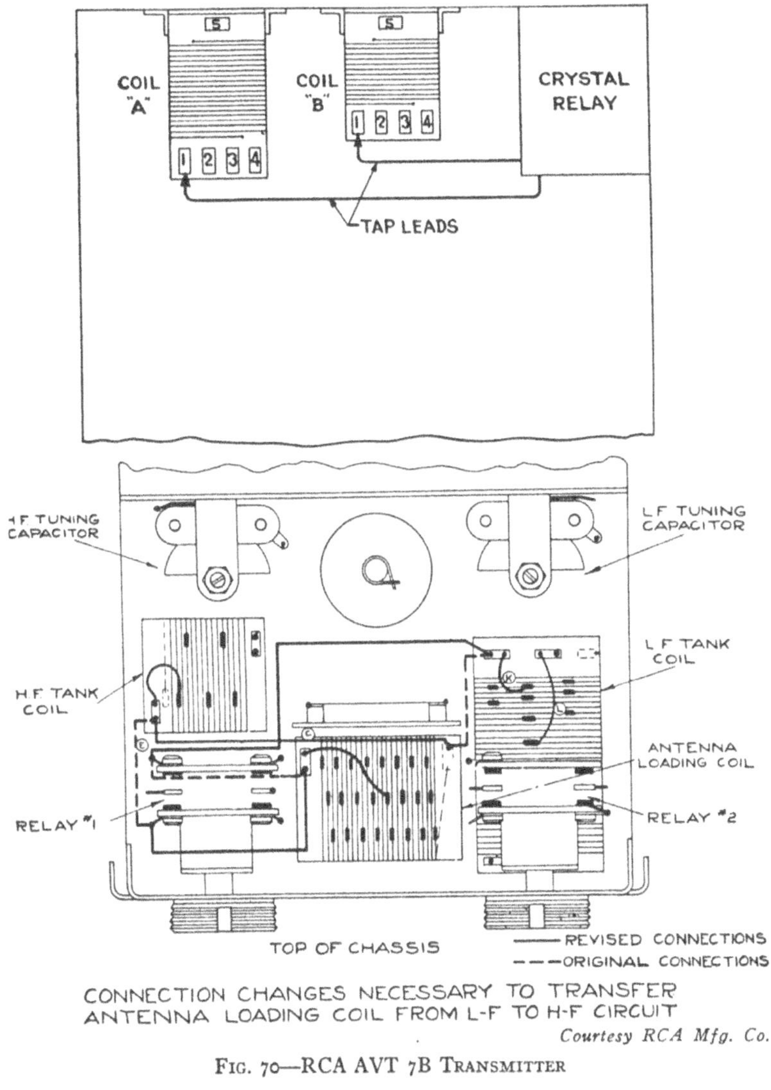

Courtesy RCA Mfg. Co.

FIG. 70—RCA AVT 7B TRANSMITTER

position will differ somewhat from that with no antenna. Now the antenna is either cut or folded back on itself a foot at a time until finally the condenser setting required is identical with or without the antenna. The antenna is electrically presenting resistance alone to the transmitter and is in correct ground adjustment.

The amount of power transferred to the antenna must now be adjusted. This adjustment is made by moving the lead which is connected from the antenna through the antenna relay to the taps on the high frequency coil, as shown in Fig. 70 (following the original dotted connections). The plate current is dipped with the output tank condenser each time a change is made until at the dip the plate current is between eighty and ninety milliamperes. A radio frequency ammeter connected in the antenna circuit will also show maximum antenna current when this adjustment is made but the meter is not absolutely necessary.

The adjustment may be made then on the lower frequency after selecting it by means of the control unit. Adjustment is made precisely as before but instead of cutting the antenna (it must not be touched) it is folded back to be fifteen per cent shorter. The taps are moved on the antenna loading coil so that the low frequency tank condenser has identical positions with or without the antenna. The load is adjusted to the same plate current limits, as before and with the antenna connected, by moving the antenna tap on the low frequency coil.

The antenna is then increased about twenty per cent in length (total including leadin) by unfolding the shortened end and the plane is taken into the air where slight adjustment of the low and high frequency tank condensers will restore minimum and practically identical plate current. The decrease in antenna length is made to compensate for the effect of the nearby ground while making ramp adjustments.

The second method of adjustment is somewhat different and the antenna is cut for quarter wave operation as before but for the low instead of the high frequency. The same formula for length, but at the lower frequency, is used. The transmitter wiring is revised, as shown in Fig. 70, so that the loading coil is in series with the high frequency tank coil. The same procedure is repeated as before but starting with the low frequency and after finding the correct length, adjusting the high frequency tuning with the loading coil for identical final tank condenser settings at dip with and without antenna.

If a reel is used with calibrated footage indicator, the same process can be used with the important exception that neither the low nor high frequency uses the loading coil. Antenna length is so adjusted that a quarter wave is used in each case. Slight length readjustments in flight will secure correct plate current so that no final tank adjustments are necessary after the first ground check.

When the transmitter is working into a quarter wave antenna, the proper adjustment is secured with the antenna reeled out for maximum plate current which should be less than ninety milliamperes at the peak. If the current exceeds this the load tap should be readjusted.

The transmitter may be connected to other types of antennas by selection of the correct leadin point from the antenna by trial, or by the addition of series coils or condensers outside the transmitter so that the transmitter may be adjusted for proper load at the plate current dip.

After adjustment the transmitter should be modulated and if the antenna current does not rise about ten to twenty per cent the loading tap should be moved so that plate current at the dip is lowered in value until upward modulation is obtained.

When adjusting two frequencies which do not differ by more than two per cent, as is permissible, some small sacrifice in output is usually encountered as adjustments are made for equally good operation at both frequencies. A small amount of detuning is necessary.

The following voltages should be found in the AVR 7B transmitter. Vibrator power supply with no-load voltages are taken by pulling out the supply cable to transmitter and holding relay No. 12 down by hand.

VOLTAGE WITH NO LOAD	VOLTAGE WITH 70 MA. FINAL LOAD	CIRCUIT
625	400	Voltage across both resistors 45 and 46
615	275	Voltage across 45 alone
600	335	Voltage from ground side of 45 to center tap of 46

Transmitter voltages with a final amplifier load of 70 ma. will be approximately as follows, using a supply battery voltage of 13.5 volts. Grid voltages of radio tubes are taken with a National type 100 choke in series with the meter lead.

CIRCUIT	OSCILLATOR (42)	POWER AMP. (807)	SPEECH AMP. (42)	MODULATOR (6A6)
Filament.....	6.3	6.3	6.3	6.3
Cathode......	...	...	−15	−5

Grid.........	−8.5	−90	...	...
Suppressor....	...	...	−15	...
Screen.......	235	250	275	...
Plate.........	290	395	270	325

8A Western Electric Transmitter—

Power output: 50 watts
Frequency range: 1500 to 6000 KC
Type frequency shift: Retune, except 2 frequency airline adaptations.
Number of frequencies: Unlimited, requiring retuning. Two in revised models.
Power supply voltage: 12 volts
Power consumption—standby: 14 amperes
Power consumption—modulated: 65 amperes

Tube types: Oscillator—205D, Doubler 205D, Power Amplifier 211D, Modulators (3) 211D

	HEIGHT	WIDTH	DEPTH	WEIGHT
Transmitter..................	17 1/4″	13 5/16″	12 3/4″	33 lbs.
Power supply (all approx.)......	10″	12″	14″	40 lbs.

A great many of the Western Electric 8A transmitters have been through modifications since their design several years ago. Many models have had quick shift added and in many cases new types of tubes have been substituted. The United Air Lines' ES271 transmitter is typical in this class and is given in the following section.

The schematic shown in Fig. 71 shows the simplicity of the transmitter. The transmitter uses a 205D oscillator followed by a 205D doubler which feeds the final Class C neutralized amplifier which is a 211D tube. The three modulators are 211D tubes as well. Western Electric 261A tubes have usually been substituted for the 211D tubes, with no change in operating characteristics.

Typical meter readings are as follows, with currents at the low frequencies (1500 kc) first and at the highest frequency (6000 kc) last (see test bench setup—Fig. 72).

CIRCUIT	JACK	NORMAL CURRENT
Oscillator grid..........	J1	.4 to 2.0 milliamperes
Oscillator plate.........	J2	40 to 42 milliamperes
Doubler grid...........	J3	3 to 6 milliamperes
Doubler plate..........	J4	40 to 47 milliamperes
Final amplifier grid.....	J5	15 to 10 milliamperes
Final amplifier plate....	J6	90 to 140 milliamperes
Audio amplifier grid.....	J7	0
Audio amplifier plate....	J8	160 to 125 milliamperes

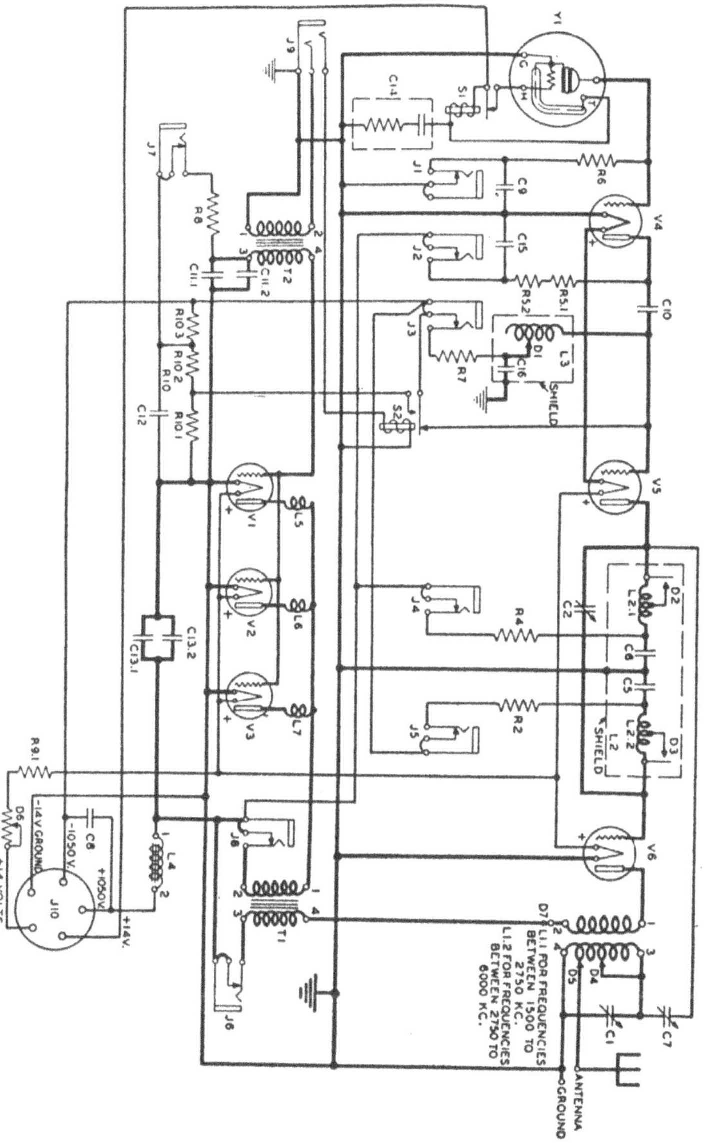

FIG. 71—WESTERN ELECTRIC 8A TRANSMITTER SCHEMATIC

Courtesy Western Electric Co.

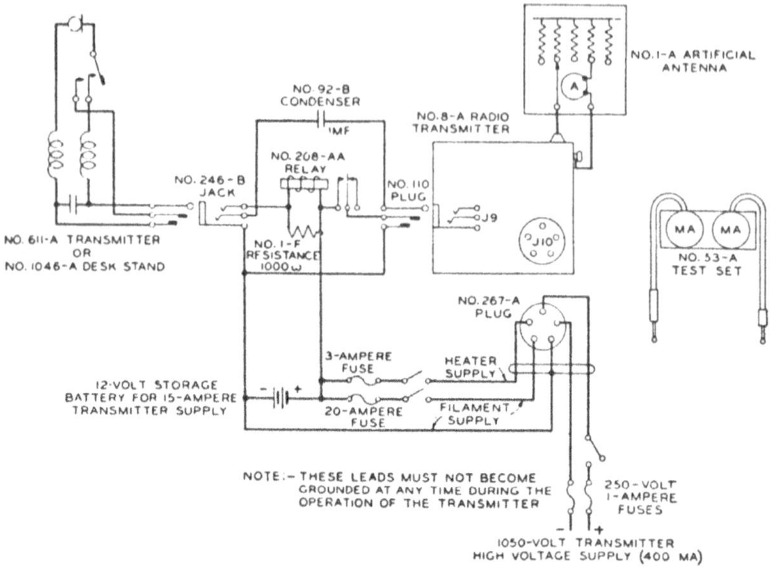

Courtesy Western Electric Co.

FIG. 72—TEST BENCH SETUP—WESTERN ELECTRIC 8A TRANSMITTER

The readings above are taken with the following conditions existing:

211D filament voltage.............. (Note R9.1 and R9.2 adjust filament voltage)	10 volts
High voltage supply...............	1050 volts
Grid bias (S2 operated)............	53 volts
Final tube bias (S2 not operated)....	90 volts
Total plate current (S2 operated).....	340 milliamperes
Total plate current (S2 not operated).	170 milliamperes
Minimum allowable final grid current for proper operation............	8 milliamperes

The components in the 8A transmitter are:

PART NUMBER	VALUE	PART NUMBER	VALUE
C5, C6, C15.........	.01 mf	R2, R7.........	8,000 ohms
C8, C13.1, C13.2.....	1 mf	R4.............	16,000 ohms
C9, C16.............	.008 mf	R5.1, R5.2......	9,000 ohms
C10................	.002 mf	R8.............	24,000 ohms
C11.1, C11.2.........	.1 mf	R9.1...........	0.16 ohm
C12.................	.006 mf	R9.2...........	0.14 ohm
C14......	.1 mf plus 130 ohms	R10.1 to R10.3..	525 tapped at 150, 175 and 200 ohms

ES 271 United Airlines Transmitter—

Power output: 50 watts
Frequency range: 1500 to 6500 KC
Type frequency shift: Remote mechanical, push cable
Number of frequencies: 2
Power supply voltage: 12 volts
Power consumption—standby: 14 amperes approximately
Power consumption—modulated: 65 amperes

Tube types: Oscillators (2) 807, Power Amplifier 261A, Modulators (3) 261A
(807 is RCA type; 261A is Western Electric)

Sizes and weights similar to Western Electric 8A

This transmitter is a revision and improvement of the Western Electric 8A transmitter. The original case size was maintained and Fig. 73 shows the schematic of the transmitter.

Two RCA 807 tubes are used, one for the day and one for the night oscillator and doubler. The upper one in the diagram is the night, or low, frequency tube with its crystal. A mechanically operated quick shift serves to place the transmitter on the proper one of two pretuned frequencies.

The grid condenser C1 of the modulated amplifier must always be adjusted first, as another condenser C2 is paralleled with it for tuning the lower frequency channel. Separate output coils are provided, whose taps are adjusted, and connect the transmitter to the antenna.

Three Western Electric 261A tubes are used in parallel to modulate the final amplifier which is a tube of the same type. The modulator tubes are operated Class A and the microphone drives them directly.

Neutralization is provided through C7 and is adjusted by any one of the methods in the previous chapter.

Plate current readings for normal operation are shown in the following table with their jack number:

CIRCUIT	JACK	AVERAGE CURRENT
Oscillator grid.	J1	1.5 milliamperes
Oscillator screen grid.	J3	5 milliamperes
Oscillator plate.	J4	80 milliamperes
Power amplifier grid	J5	25 milliamperes
Power amplifier plate. . . .	J6	140 milliamperes
Modulator plates.	J8	240 milliamperes

In addition the filament voltage may be measured at J7.

The transmit-receive relay operates a pair of contacts which short-circuit a part of the 807 filament dropping resistor while transmitting, assuring that with the temporarily low battery voltage, 6.3 volts will be maintained on the 807 tubes.

The value of the parts shown on the schematic of Fig. 73 are:

SYMBOL	VALUE	SYMBOL	VALUE
C1	50 mmf	R2	5,000 ohms
C2	150 mmf	R4.1	2,500 ohms
C8, C13.1, C13.2	1 mf	R4.2	3,000 ohms
C11	.1 mf	R6.1, R6.2, and R12.2	100,000 ohms
C12, C24	.006 mf	R8	24,000 ohms
C17, C20	.005 mf	R10	100 ohms
C18	.5 mf	R11.1, R11.2	30 ohms
C21	125 mf	R13	40,000 ohms
C23.1, C23.2, C25.1, C25.2	.01 mf	R14	40 ohms
C28.1, C28.3	5 mmf	R15	4.27 ohms, tapped

AVT 12B RCA Transmitter—

Power output: 50 watts, telephone. 90 watts, telegraph
Frequency range: 2600 to 6500 kilocycles
Type frequency change: Electrical, remote
Number of frequencies: 4 related or 2 non-related
Power supply voltage: 12 volts
Power consumption—standby: 6.4 amperes
Power consumption—modulated: 62 amperes

Tube types: Oscillator 837, Power Amplifier 803, Modulator 42

	HEIGHT	WIDTH	DEPTH	WEIGHT
Transmitter	11 7/8″	9 3/4″	10 1/4″	21 lbs.
Power Unit	12″	7 1/2″	11 3/8″	50 lbs.
Control box and misc.				9 lbs.

The AVT 12B RCA transmitter is somewhat similar in operation to the AVT 7B, although the tubes used and the output are quite different. Many adjustments are made in a similar manner.

Tuning in the initial stage of the transmitter is different from the AVT 7B. A wrench is supplied with transmitters which will unlock the lock nuts on condenser shafts. The plate lead to the top of the large 803 tube should be removed and as this lead and others have 1700 volts on them in operation, care should be exercised in touching any internal part. Filaments should be allowed to heat for at least forty seconds before applying any voltage to the plates by pressing the microphone button.

An insulated case meter is plugged into the "OSC-MA" jack, which is in the oscillator cathode circuit. Starting with the low frequency first (Position 1), the screwdriver end of the tuning wrench is inserted in the L.F. condenser slot and turned until a sharp drop in current is noted. From the point at which the current is minimum, the condenser is rotated to unmesh the plates until the current increases about five per cent, at which point the adjustment is correct for the frequency. The control switch is moved to Position 2 and, if another crystal is employed and adjustment is correct, the current is within five per cent of the reading on crystal 1. If the current is greater, readjustment is made as before and the current measured on

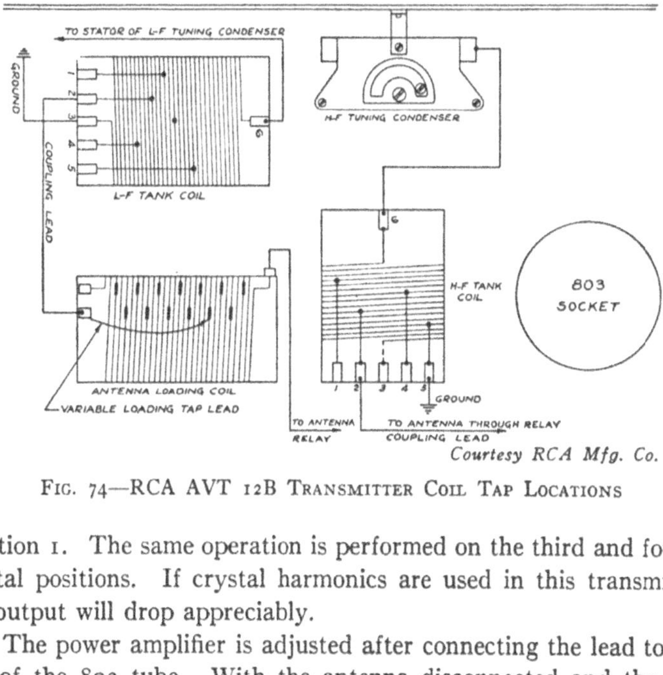

Courtesy RCA Mfg. Co.

FIG. 74—RCA AVT 12B TRANSMITTER COIL TAP LOCATIONS

Position 1. The same operation is performed on the third and fourth crystal positions. If crystal harmonics are used in this transmitter the output will drop appreciably.

The power amplifier is adjusted after connecting the lead to the top of the 803 tube. With the antenna disconnected and the test meter plugged into the "PA-MA" jack, the adjusting wrench may be placed on the large tuning condenser shaft marked L.F., in the front compartment of the transmitter. (Plate voltage must not be left on the transmitter more than about ten seconds while it is out of adjustment.) During the time power is on, the condenser should be rotated to the dip which will be at less than fifty milliamperes. The same is done with the H.F. condenser and in both cases the companion frequency is adjusted to about the same plate currents utilizing small changes in the oscillator adjustment if necessary. The higher final amplifier current of a frequency pair should not be more than fifty per cent greater than the one on which adjustment is made.

The final amplifier is adjusted very similarly to the adjustments described for the AVT 7B. A diagram of the coil locations is shown in Fig. 74. A schematic is shown in Fig. 75.

Tube voltages and currents are shown in the following table (see Fig. 76):

TUBE	SUPPRESSOR VOLTAGE	SCREEN VOLTAGE	PLATE VOLTAGE	PLATE CURRENT
803	−50	300 to 600	1700	120
837	....	275	480	60
42	....	300	280	no jack

For harmonic operation, the 803 plate current should not exceed 87 milliamperes.

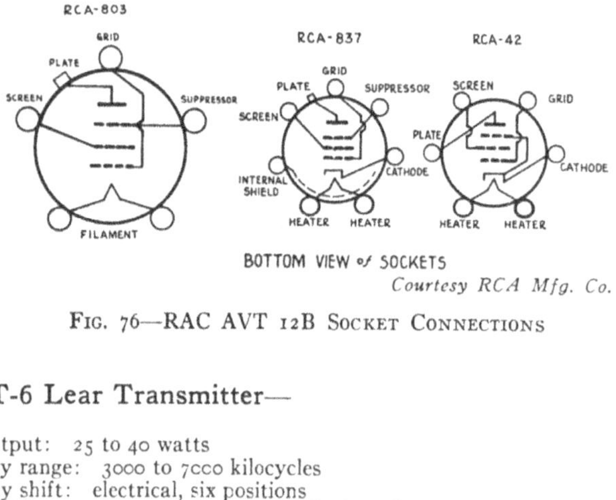

Courtesy RCA Mfg. Co.

FIG. 76—RAC AVT 12B SOCKET CONNECTIONS

UT-6 Lear Transmitter—

Power output: 25 to 40 watts
Frequency range: 3000 to 7000 kilocycles
Frequency shift: electrical, six positions
Number of frequencies: six frequencies in four bands—two crystal bands may contain two frequencies each with two per cent separation.
Power supply voltage: 12 volts
Power consumption—standby: 2 amperes
Power consumption—modulated: 12.5 amperes

Tube types: Oscillator 6L6, Power amplifier RK-39, Modulator—(2) 6L6

	HEIGHT	WIDTH	DEPTH	WEIGHT
Transmitter	$4\frac{3}{4}''$	$8\frac{15}{16}''$	$17\frac{1}{4}''$	24.1 lbs.
Control and accessories	$4\frac{1}{8}''$	$4\frac{1}{8}''$	$2\frac{3}{4}''$	5.3 lbs.

Separate condensers are supplied for each of the four bands and may be adjusted after selecting the proper coil taps for the frequencies to be used. The wiring diagram, Fig. 77, shows the layout in a very simple manner and includes the circuits of the control equipment.

It will be noticed that a relay is provided which will switch the transmitter between two antennas. Thus a fixed antenna with a tuning unit, adjusted with its tuning unit to appear to have the same impedance as the trailing wire used in flight, may be employed.

W4A Transmitter—

Power output: 50 watts
Frequency range: 2800 to 5900 KC
Type frequency shift: Remote electrical, special dial control
Number of frequencies: 8 in two non-related groups of 4 each, two per cent wide
Power supply voltage: 12 volts
Power consumption—standby: None
Power consumption—modulated: 65 amperes

Tube types: Oscillator 801, Doubler 801, Power amplifier (2) 203A, Modulator 801 (210's may be used for 801's with no changes)

	HEIGHT	WIDTH	DEPTH	WEIGHT
Transmitter	11 1/2"	20 1/2"	15"	28 lbs.
Power Unit (for receiver too)	11 (over shocks)	15 1/4"	12 7/8"	58.5 lbs.

(See W4A receiver also)

The W4A transmitter has been manufactured by both Western Electric and RCA. The output is fifty watts on any of eight frequencies which must be divided into two groups of four frequencies. The total frequency band of each group is about two per cent of the average frequency.

Control unit, power unit and a schematic layout of the complete W4A equipment are shown in Figs. 78, 79 and 80.

Location of the equipment in a typical installation is shown in Fig. 81.

Crystals are inserted in the crystal selector jacks on the front of the unit. RCA utilizes the Positions 1 to 4 for night frequencies and 5 to 8, inclusive, for day, or highest frequencies. The Western Electric transmitter utilizes Positions 1 to 4 for the high frequencies and 5 to 8, inclusive, for the low frequencies. Otherwise the units are about the same.

The operation of the crystal selector arm is accomplished electrically by means of three wires and the common ground return. The switch which operates the receiver and transmitter selector simul-

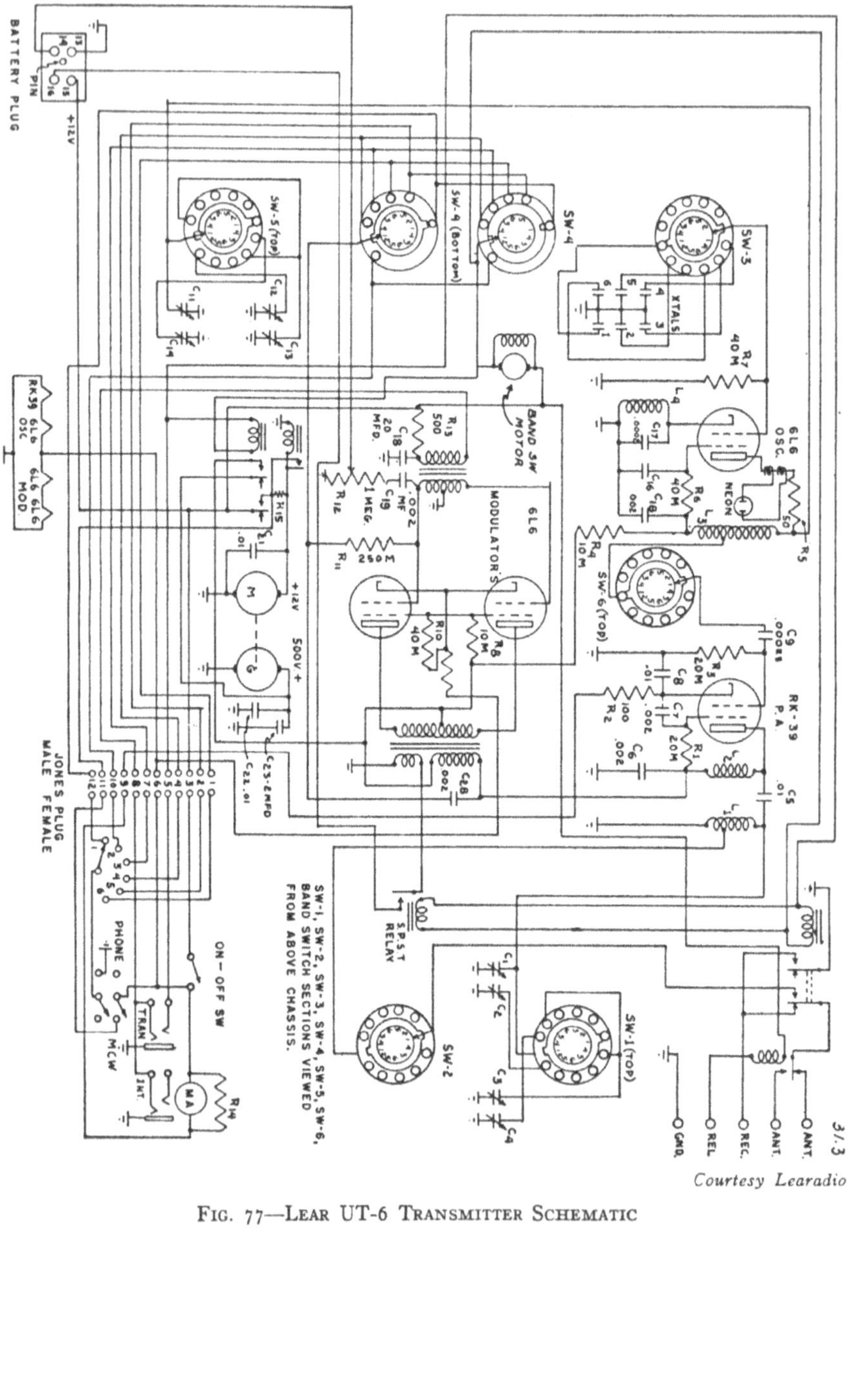

Courtesy Learadio

FIG. 77—LEAR UT-6 TRANSMITTER SCHEMATIC

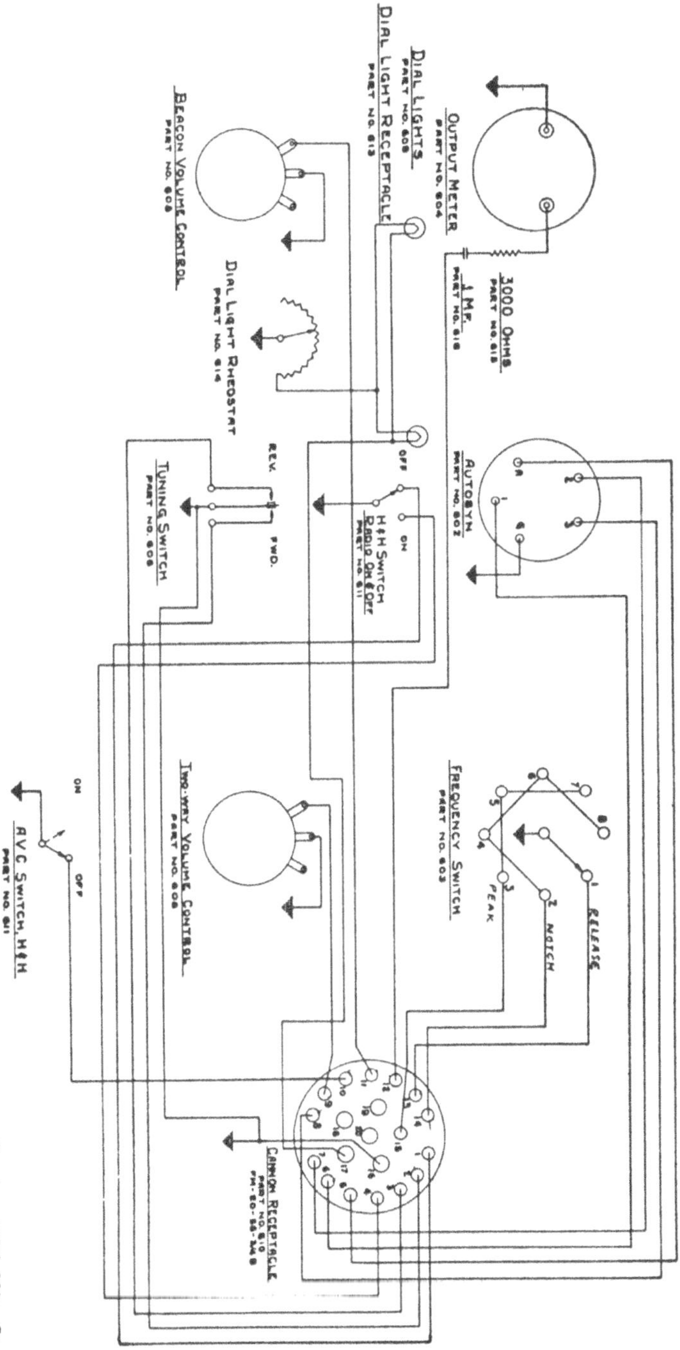

Fig. 78—W4A Control Unit Schematic

Courtesy RCA Mfg. Co.

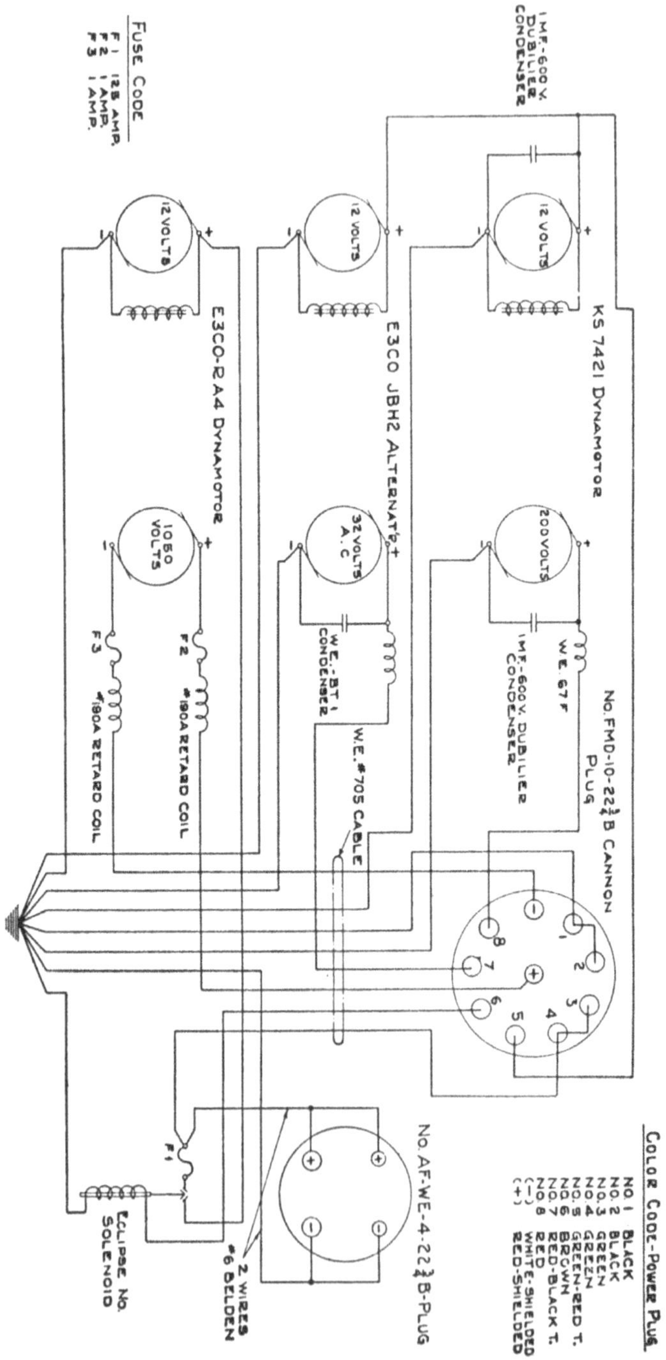

FIG. 79—W4A POWER UNIT SCHEMATIC

Courtesy RCA Mfg. Co.

taneously from the same wires controls functions in the following manner. A solenoid is arranged to notch the arm around by steps from Position 1 to 8 by successive impulses. When selecting Point 1, a second solenoid operates and pulls the retaining ratchet back, allowing a spring to return the selector to Position 1. As the arm reaches Position 1 the arm shaft rotation breaks contact through the releasing solenoid and it ceases to draw current.

This same impulse contact feature is also employed for the notching assembly which rotates the arm by steps from Position 1

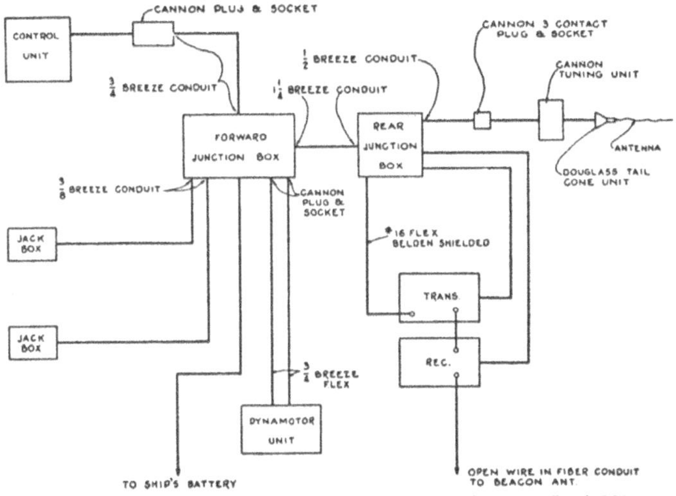

Courtesy RCA Mfg. Co.

FIG. 80—W4A EQUIPMENT LAYOUT SCHEMATIC

to Position 8. A cam, shown on the wiring diagram of Fig. 82, alternately closes the notching solenoid circuit to one control wire or the other of the control switch. These two wires are referred to as "peak" and "notch," with reference to either the peak or valley (notch) of the cam on the selector arm shaft. Thus, if the cam is at any "peak" position, contacts are closed to the "notch" wire, breaking the circuit that caused the cam to rotate to this step. It is therefore necessary, in order to cause the arm to rotate, to alternately make contact with a peak wire, then a notch wire, then a peak wire, etc.

The control unit schematic, Fig. 78, shows the switch. It has three contact discs, one each for peak, notch and release wires. The cam on the top of the crystal unit operates the final amplifier shifting relay and the trailing wire-tuning unit in the day position.

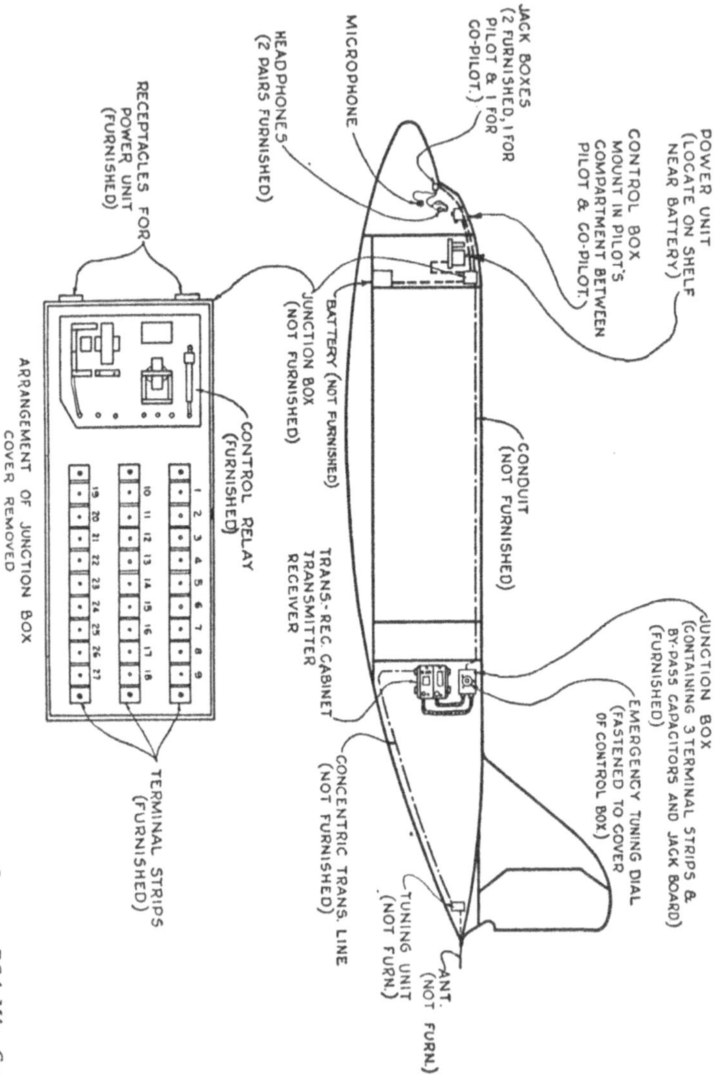

Fig. 81—Typical Installation—W4A Equipment

Courtesy RCA Mfg. Co.

Fig. 83 shows a complete wiring diagram of a W4A installation. It should be noted that no voltage is supplied to the transmitter filaments until the microphone button is pressed, which also starts the

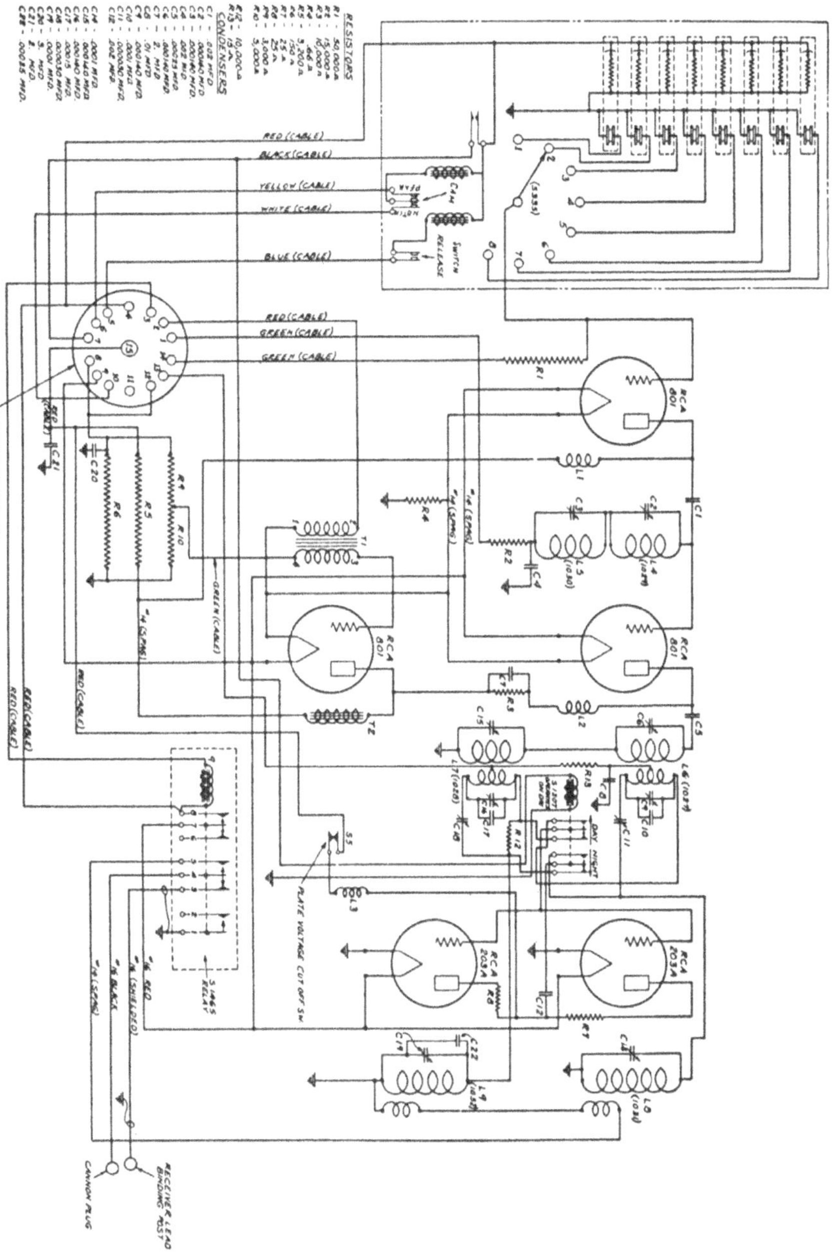

Fig. 82—W4A Transmitter Schematic

Courtesy RCA Mfg. Co.

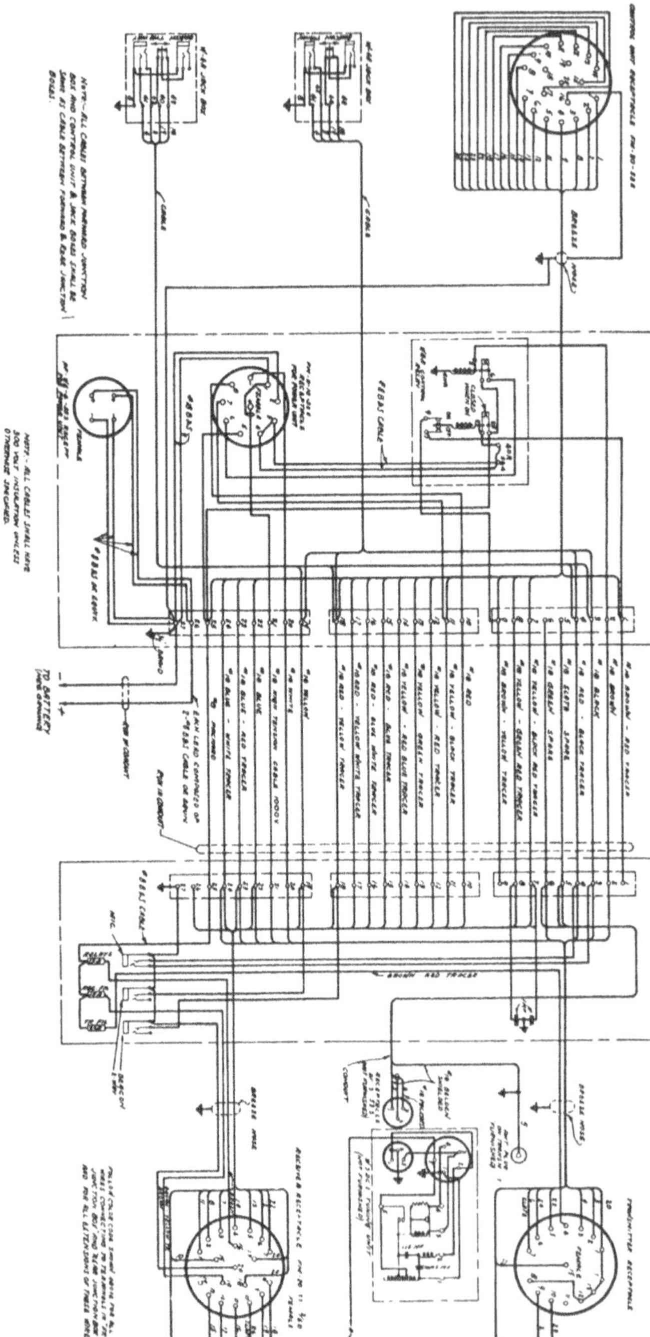

Fig. 83—W4A Complete Installation Schematic

Courtesy RCA Mfg. Co.

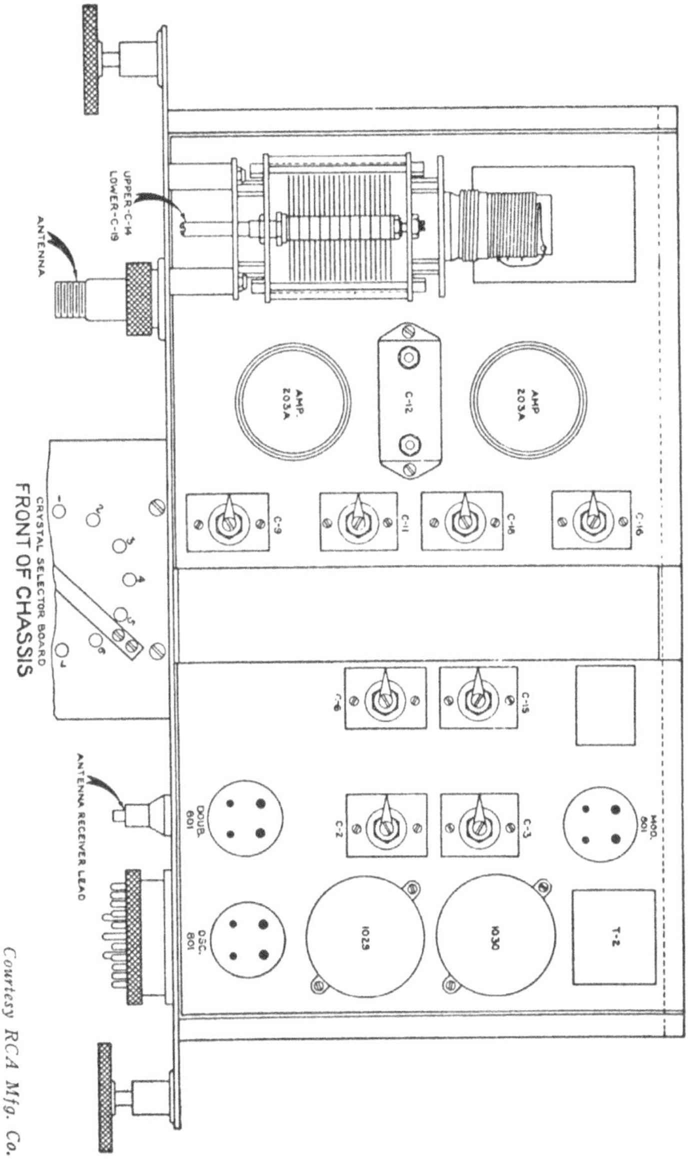

FIG. 84—PLAN VIEW OF TRANSMITTER

Courtesy RCA Mfg. Co.

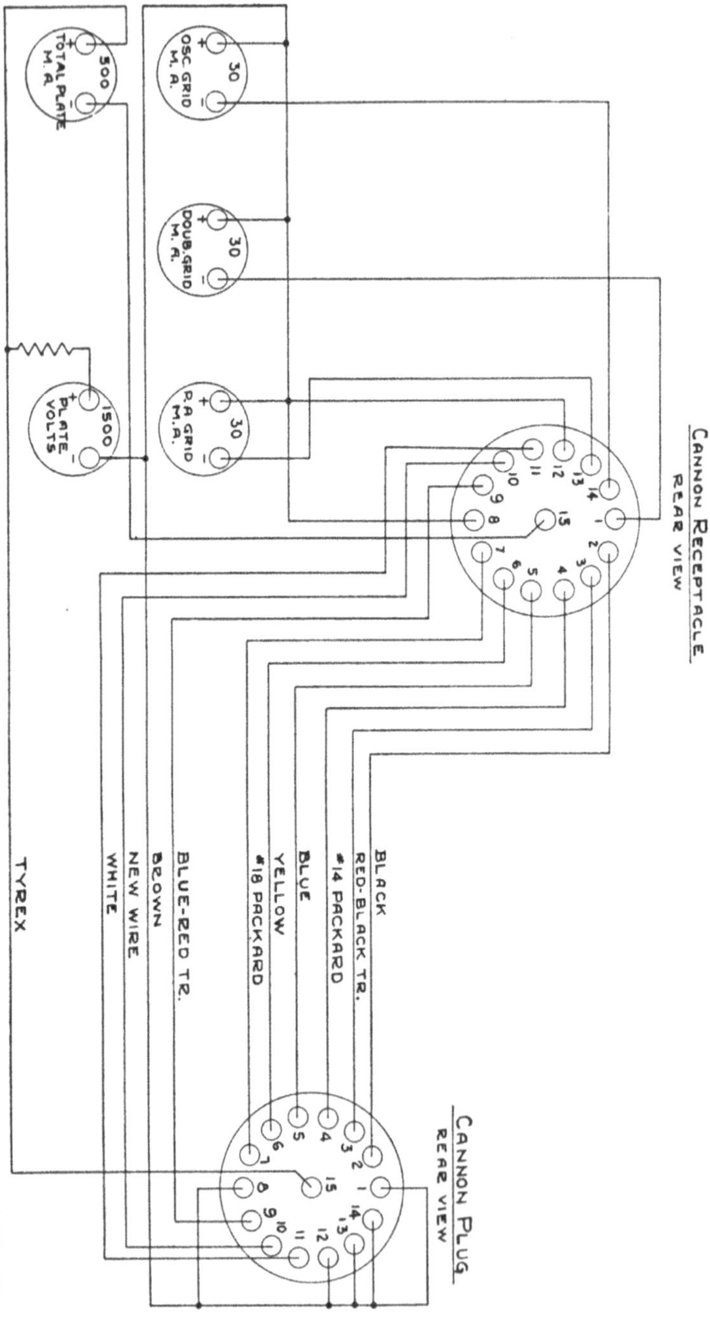

Fig. 85—W4A Test Kit Schematic

Courtesy RCA Mfg. Co.

main dynamotor. No standby switch to preheat filaments is necessary as in most transmitters.

The arrangement of the tuned radio frequency circuits in the W4A transmitter is novel in that the day and night tuned circuits are permanently connected in series. While the transmitter covers the same band as the W4A receiver, the coils in the transmitter must be adjusted to the proper inductance before initially tuning the transmitter. RCA provides taps on the coils for this purpose. It is especially important that the night and day oscillator condensers be fairly well meshed or either one of the crystals may not start oscillating properly. It is also important to check the two oscillators after adjusting either one.

The day (highest frequency) trimmers are C2, C6, C9 and C14 (Fig. 84). The night (lower frequency) trimmers are C3, C15, C16 and C19. Locks are provided on each trimmer at the shaft nut, reached through the base panel or through the front panel for C14 and C19. Day trimmers are adjusted first. After the night trimmers are adjusted the day operation should be rechecked. The test kit of Fig. 85 is used in adjustment and is inserted between the transmitter plug and connector receptacle. When the oscillator adjustment C2 is correct its grid current will be below two milliamperes. The peak current point should not be used and the condenser should be rotated on the oscillating side of the peak current for a reasonable grid current of about one half to three quarters of the maximum grid current. This will correspond to nearly maximum doubler current. Adjustment of the next trimmer, C6, is made at the same time that C16 is adjusted for maximum final amplifier grid current. Finally the output circuit is adjusted with C14 for maximum antenna current. Neutralization can be accomplished with C18, as explained in the previous chapter.

The same operation is repeated for the night trimmers in the order given above. Day frequency should be realigned.

Dial plates are affixed to the chassis for guidance in tuning. Care must be taken to tune the circuits to the desired frequencies and not accidentally to half or double the correct frequency. The crystal plate circuit is tuned to the crystal frequency which is half the operating frequency. The next tube is a doubler and its circuit, as well as that of the power amplifier grid, must be tuned to the operating frequency which is twice the crystal frequency.

It is very important that the grid current of the power amplifier does not exceed about ten milliamperes. The last stage is a Class B amplifier and high grid current (over 10 ma.) will cause much distortion. If the grid current cannot be brought down to this value with a load correctly adjusted, the excitation must be reduced by either slightly detuning the grid circuit, slightly changing the neutralizing condenser (particularly helpful on day frequency) or by adding a two watt resistor across the secondary of the power amplifier grid circuit to lower the current to the proper value (10,000 ohms or less as a rule).

Class B output stages, as in this transmitter, are far more critical to correct antenna adjustments than Class C modulated output stages.

The load to the transmitter must be about thirty to fifty ohms and non-inductive. If the load is inductive it will be found that correct modulation will not take place and a tuning network external to the transmitter will be required. If the load is too high or too low in resistance a tuning network is required in the output lead to bring the load to the correct value. Usually the transmitter feeds a concentric transmission line to the antenna tuning unit, as it was designed for such operation.

13C and 13CB Western Electric Transmitters—

Power output: 50 watts
Frequency range: 2000 to 6500 KC, 13CB covers 325 to 500 KC also.
Type frequency shift: Mechanical remote through geared tachometer shaft.
Number of frequencies: 3 non-related
Power supply voltage: 12 volts
Power consumption—standby: 6 amperes, approximately
Power consumption—modulated: 65 amperes, approximately

Tube types: Oscillator 205D, Doubler 282A, Power Amplifier (2) 282A, Modulator 205D, Ballast Lamps (1) 113A, (3) 112A
(All tube types are Western Electric)

	HEIGHT	WIDTH	DEPTH	WEIGHT
Transmitter (with mount)....	10$^1/_8$″	13$^3/_4$″	21$^3/_4$″	40 lbs.
Coils and crystals per channel..............................				2.5 lbs.
Same for low frequency channel (13CB)......................				3.25 lbs.
13CB loading coil..				5.1 lbs.

The 13C transmitter is a three frequency transmitter. The 13CB transmitter is a modification which will allow operation of one or more channels in the band from 325 to 500 kilocycles and also provides for keying for continuous wave transmission.

The transmitters utilize a crystal oscillator tube, followed by a doubler which is used in all cases and excites the output stage consisting of a pair of Western Electric 282A tubes. The crystal frequency is half the carrier frequency except in the 300 to 500 kilocycle band. The screen grids of the doubler and final tubes are all modulated together. Fig. 86 shows the schematic of the 13C transmitter, Fig. 87 shows a modification for one airline's service.

No tuning, except the output antenna coupling transformer, is required for this transmitter. The correct interstage transformer must be plugged in to the interstage receptacles. The interstage transformers that can be supplied are:

NUMBER, WESTERN ELECTRIC	FREQUENCY RANGE, KILOCYCLES
271K	300 to 500
271C	2250 to 2800
271B	2800 to 3500
271F	3500 to 4200
271E	4200 to 4900
271A	4900 to 5700
271D	5700 to 6500

The final stage tuning inductors, which have a series condenser attached to them, cover the following approximate bands:

ANTENNA TYPE	CARRIER FREQUENCY IN KILOCYCLES	SERIES CAPACITY, MMF	WESTERN ELECTRIC TUNING COIL
Nose to tail 40′ fixed...	2000 to 3000	4000	7C
	3000 to 4000	4000	7B
	4000 to 5000	250	7B
	5000 to 6000	250	7A
	6000 to 6500	125	7A
Seventy ohm transmission line with tuning unit	2000 to 3500	250	7B
	3500 to 6500	250	7A

In the case of the seventy ohm line, a 1000 mmf condenser must be attached between the antenna post contact of relay S6 and ground.

Occasionally it will be found that the series condensers supplied will not give correct operation and a 62.5 mmf condenser may be used. The condensers are all of the PL6 Dubilier type except for the 1000 mmf condenser which is of the PL9 type. The low frequency band for 300 to 500 kilocycle operation requires a special output coil as well as a loading coil, as shown in the schematic of Fig. 88.

A special two scale test milliammeter on a standard telephone plug is used in adjusting the 13 type transmitter. An antenna current thermocouple is provided in the transmitter which should be short-circuited at the meter connection when not in use. A test bench setup shown in Fig. 89 can be used for initial adjustments.

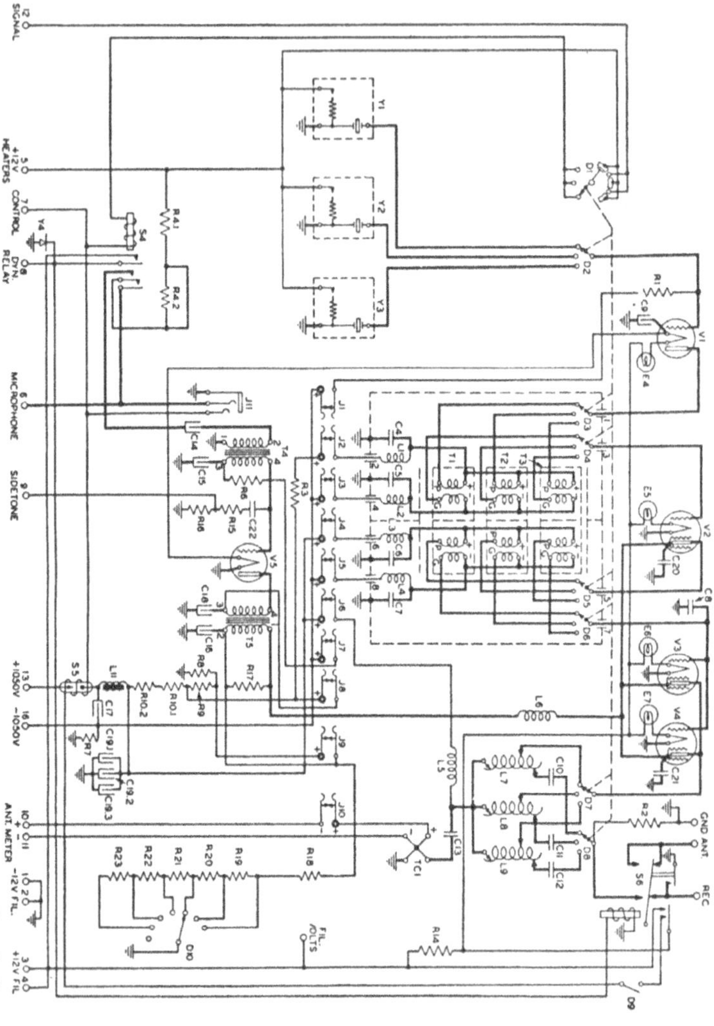

FIG. 86—WESTERN ELECTRIC 13C SCHEMATIC

Courtesy Western Electric Co.

The Number 1 channel is used for the lowest frequency while the Number 3 channel is used for the highest frequency, as there is a slight advantage in so doing.

The milliammeter is plugged into the "2nd A plate circuit" jack for initial alignment. The cover must be on the transmitter while making adjustments in order to close the safety switch so that output

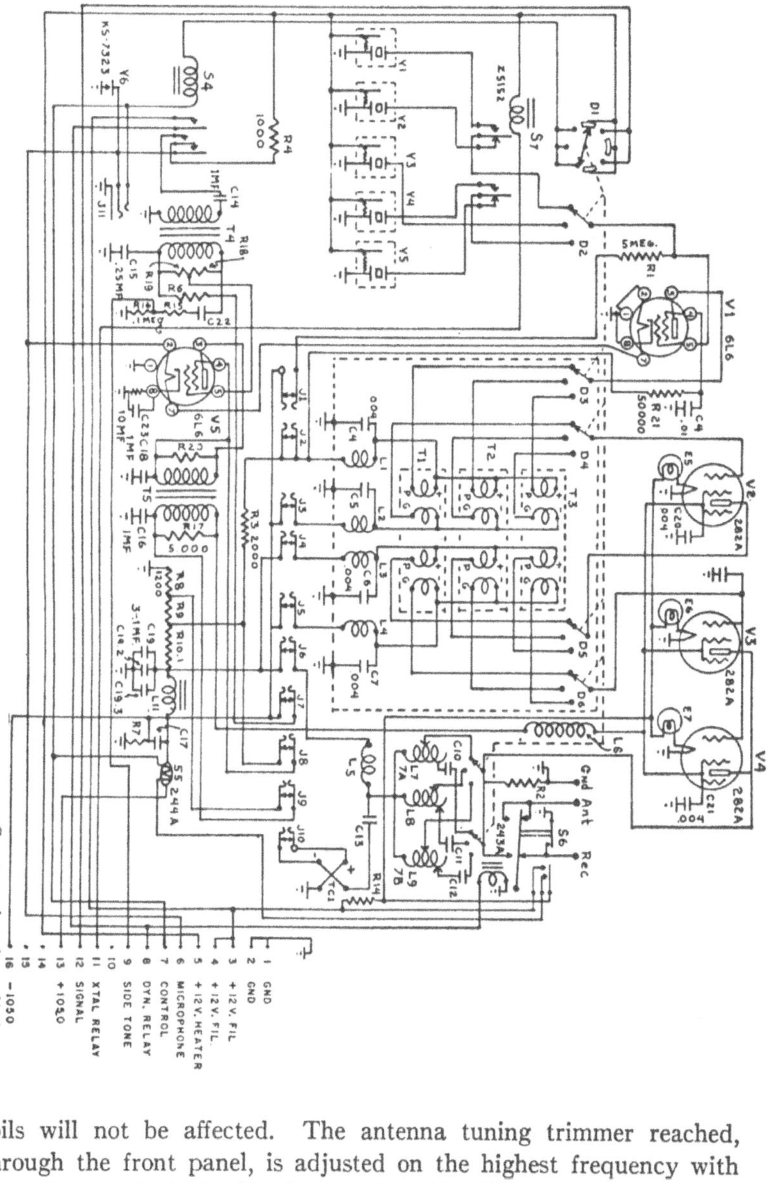

FIG. 87—MODIFIED 13C TRANSMITTER

Courtesy American Airlines

coils will not be affected. The antenna tuning trimmer reached, through the front panel, is adjusted on the highest frequency with the antenna attached after first turning the adjustment as far to the right, clockwise, as it will go with the antenna clip on the 7-type coil on the rear turn of the tuning coil and with the plate clip at about

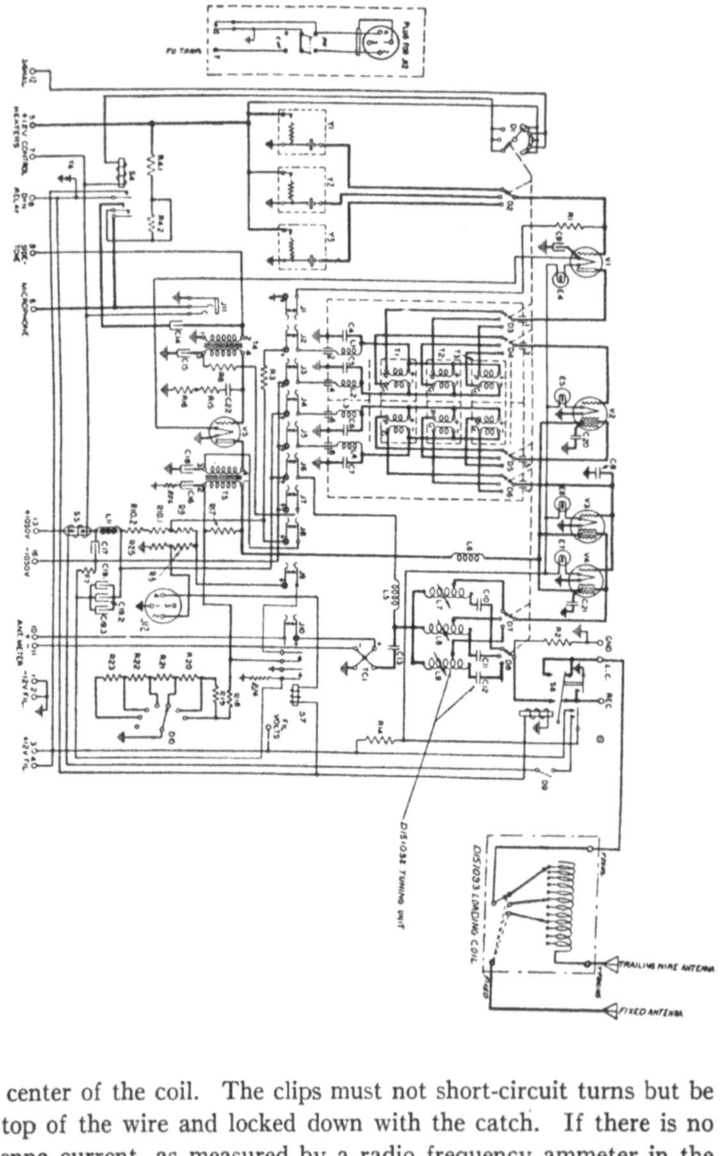

FIG. 88—WESTERN ELECTRIC 13CB TRANSMITTER

Courtesy Western Electric Co.

the center of the coil. The clips must not short-circuit turns but be on top of the wire and locked down with the catch. If there is no antenna current, as measured by a radio frequency ammeter in the

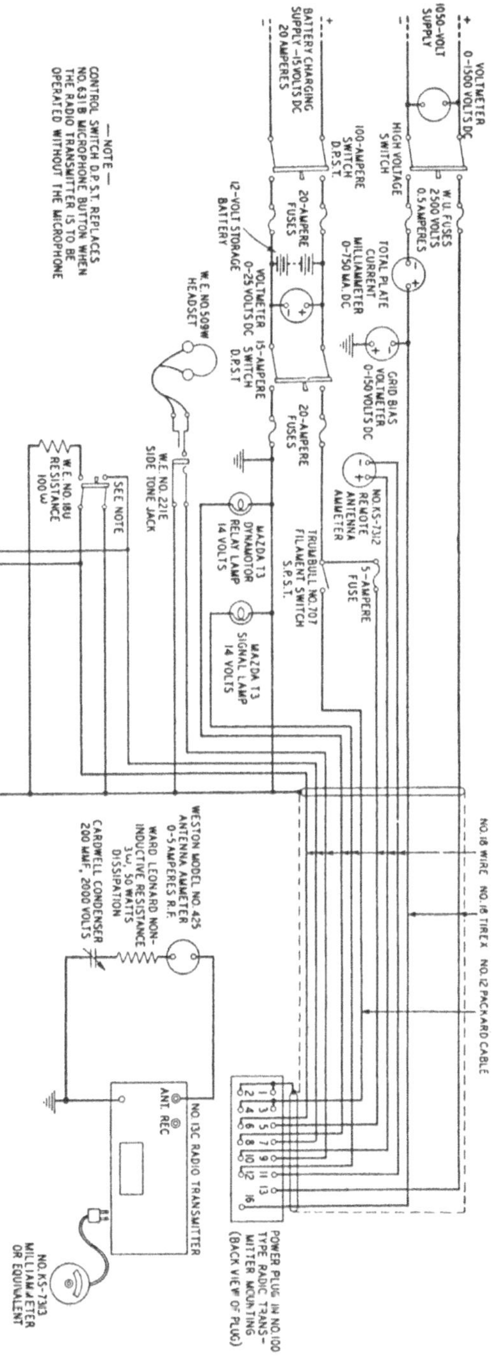

FIG. 89—TEST BENCH SETUP FOR 13B TRANSMITTER

Courtesy Western Electric Co.

antenna circuit, the transmitter is instantly turned off by releasing the microphone button, then moving the antenna tap forward a turn at a time, until maximum antenna current has been passed. The screwdriver adjustment in the end of the coil is then used to trim for exact maximum antenna current. If the antenna clip reaches the plate clip, both are moved forward together until maximum antenna current is secured.

After the best point has been found, the milliammeter is plugged into the "SCR" jack and the screen current measured and adjusted to approximately twenty milliamperes with the switch D10 at the bottom of the antenna coil compartment. If the antenna current does not rise upon modulating the transmitter, the plate clip is moved one turn nearer the front of the transmitter or until upward antenna current of at least twenty per cent is secured upon modulation. The trimmer on the antenna coil must be retrimmed after each adjustment.

If no satisfactory point can be found the next smaller, or next larger, condenser may be substituted on the antenna tuning coil and adjustment tried again.

When the transmitter is in proper adjustment the current readings should be approximately as shown in the following table:

CIRCUIT	APPROXIMATE VALUE
Oscillator grid	1 ma.
Oscillator plate	32 ma.
1st amplifier grid	12 ma.
1st amplifier plate	45 ma.
2nd amplifier grid	60 ma.
2nd amplifier plate	175 ma.
Screen	20 ma.
Audio grid	0 ma.
Audio plate	32 ma.
Plate supply voltage	1050 v.
Antenna current	1 to 6 amps.
Filament voltage (transmitter on)	11 v.
Grid bias (grid jack to ground)	−27 v.
Screen (screen jack to ground)	50 v.
Audio plate (audio jack to ground)	380 v.

The schematic in Fig. 86 has the following values of units: V1 and V5 are 205D tubes, V2, V3 and V4 are 282A Western Electric vacuum tubes, E4 is 113A ballast lamp and E5, E6 and E7 are 112A ballast lamps.

C4, C5, C6, C20, C21, C13	.004 mf	R1, R6, R16	100,000 ohms
C8	75 mmf	R2	2 megohms
C9, C22	.01 mf	R3, R22	2,000 ohms
C10, C11, C12	as in table	R4.1, R19	500 ohms
C14, C16, C17, C19.1, C19.2, C19.3, C18	1 mf	R4.2	1,595 ohms
C15	.4 mf	R7	65 ohms
		R8	1,200 ohms
		R9, R17	5,000 ohms
		R10.1, R10.2	2,350 ohms
		R14	0.3 ohm
		R15	0.5 megohm
		R18	2,500 ohms
		R20, R21	1,000 ohms
		R23	3,000 ohms

The following parts are changed or added in the 13C to convert to a 13 CB radio transmitter:

C10, C11, C12	.05 mf, part of tuning unit	R24	2,000 ohms
C13	.05 mf (changed)	R25	750 ohms
		R26	20 ohms
		R17	2,000 ohms (changed)

17D Collins Transmitter—

Power output: 100 or 175 watts
Type frequency shift: Electrical telephone dial control
Number of frequencies: 10 non-related frequencies
Power supply voltage: 12 volts
Power consumption—standby: 6 amperes
Power consumption—modulated: 81 amperes (100 watt). 107 amperes (175 watt)

Tube types: Oscillator C100D, Buffer 807, Doubler (2) 807, Power Amplifier C120, Audio Amplifier (2) 6F6, Modulators (2) C120, (C type tubes are Collins types)

	HEIGHT	WIDTH	DEPTH	WEIGHT
Transmitter	14″	20″	13½″	59.5 lbs
100 watt power supply				47 lbs.
175 watt power supply				57 lbs.

The 17D Collins transmitter is capable of either 100 or 175 watts output depending on the output tube and power supply.

Fig. 90 shows the heart of the quick-change "autotone" which tunes the transmitter to a dialed frequency in about three to four seconds. The system tunes necessary circuits and selects coil taps automatically, once the correct tuning has been initially established. There are two motors that may be seen in the figure and these accomplish two separate functions. One motor rotates the drums, seen protruding above the panel, in a direction selected by the control relays for that motor. The other motor runs a stop along the face of the drum to the proper position. These stops are hidden behind the drums in Fig. 90 but may be seen better in Fig. 91.

Ten lugs protrude from each drum and each lug is on a separate disc which, in turn, is separated from adjacent discs by a keyed washer. When tension is removed from the stacked discs and washers these lugs may be rotated to any position desired.

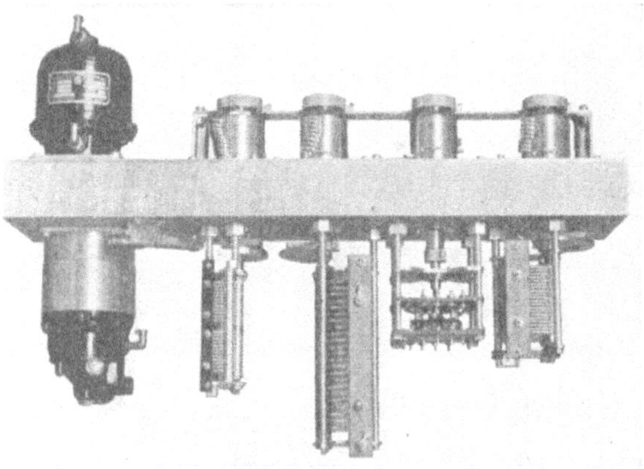

Courtesy Collins Radio Co.

FIG. 90—COLLINS 17D TRANSMITTER "AUTOTUNE"

Each condenser and switch is equipped with a coil spring which holds its drum lugs against the stops. Thus condensers and coil switches may be set to the best point of tuning and the lug placed against the stop. The lug will press against the stop by the coil spring pressure.

Assuming that the transmitter has been tuned separately to a number of frequencies and the drums tightened, the mechanism of shifting is followed in the sequence below:

1. The drum drive motor, through suitable gears, rotates each drum against the spring tensions on each shaft, so that all lugs are well removed from the stop position. This is called the "home" position.
2. The stop drive motor then turns the threaded shaft, upon which each stop is located, to a new position that was selected by the control dial.

3. The drum drive motor then rotates the drums again, in the reverse direction from Step 1 above, and as each lug touches the stop it ceases rotation and is held in position against the stop by the spring on its shaft.

The circuits controlled by the "autotune" are those that would be tuned by hand were the transmitter a non-quick shift transmitter

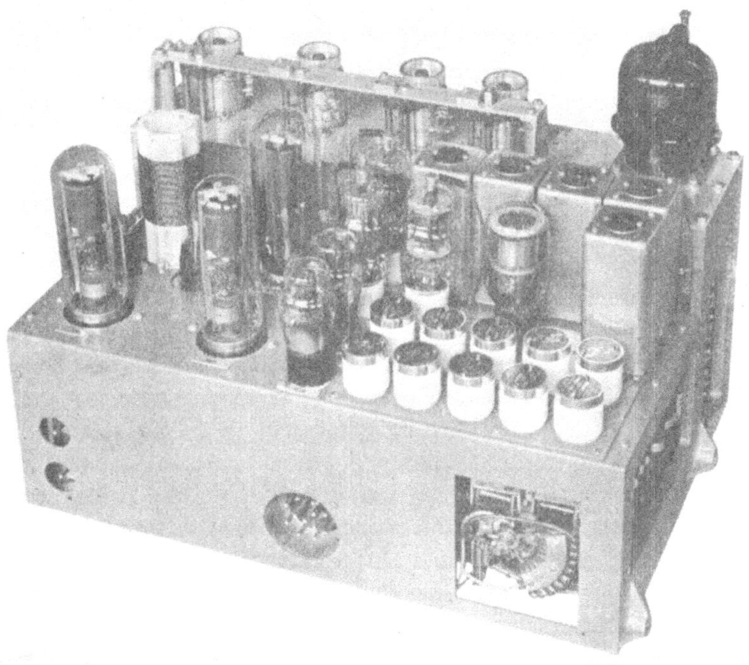

Courtesy Collins Radio Co.

FIG. 91—COLLINS 17D AIRCRAFT TRANSMITTER

of conventional design. No trimming is necessary as each condenser is accurately set to within a small part of a degree.

A description of the electrical control performing the above functions for dialing of a single number follows. The operation may be traced on the schematic diagram of Fig. 92.

1. Digit 2 is dialed by the operator either remotely or on the transmitter's dial.
2. As soon as the dial starts to be pulled to the finger stop, the dial's "off normal" springs close and operate a "slow

release" relay 12. This relay breaks the circuit to return relay 8 and to the minor switch release magnet 18.

3. The minor switch arm 20 is stepped up to contact 16B as the dial is released by the operator's finger and returns to normal by spring operation. This gives the two impulses necessary through the pulsing contacts on the dial to magnet 17.
4. As the minor switch is moved up to magnet 17, it mechanically operates the "off normal" contacts shown below it at 19 and opens one pair of contacts, thus opening the dynamotor control solenoid circuit. The other pair of contacts closes and applies power to homing relay 9 and the minor switch arm 20.
5. Relay 11 now operates and holds relay 12 operated by energizing its second coil. Number 12, being a slow release relay, has had no opportunity to open due to its opening time lag.
6. Homing relay 9 operates and runs motor 1 toward the home position.
7. As the condenser drum drive motor shaft reaches the "home" position it mechanically operates switch 7. Contacts A and B of switch 7 close and apply power to the field of motor 2.
8. Selector switch number 3 is geared to motor number 2 and the minor switch has already applied voltage to its second contact (16B). This contact has placed voltage on the second contact of switch 3, so that current will pass to the half sector (5A) touching the second contact (4B) and apply voltage to the number 12 wire leaving the unit. This will operate relay 11 and cause motor 2 to turn in the proper direction so that the contact (4B) will eventually reach the open sector (6). Note should be taken of the fact that had switch 3 been resting on contact 1 (instead of contact 5) relay 10 (rather than 11) would have closed so the motor would rotate selector 3 in the opposite direction so contact 4B would have eventually been reached by the open sector (6) and allowed relay 10 to open, stopping the motor 2. If, due to inertia, the switch 3 runs by its position the number 2 motor will reverse by the relay 11 closing and the insulated sector will come to rest under contact 2.
9. The sequence is finished as far as a tap selection is concerned. The second crystal (not shown) is in the circuit, as well as other coil taps having been selected.

10. Relay 12 will now release about one tenth of a second after both relays 10 and 11 have ceased operation. When relay 12 releases, relay 8 will operate and motor 1 will run from the "home" position toward the "operate" position and all lugs will stop against their stops previously selected by the number 2 selector motor.
11. Switch 7 will be operated by the motor shaft when the motor reaches the "operate" position and stop motor 1. The release magnet 18 is also operated and the minor switch release magnet allows the switch to drop to the normal position. The minor switch arm operates switch 19, which restores the circuit for the dynamotor solenoids so that the transmitter may be used.

Several facts are now evident. Relays 8 and 9 control the direction of rotation of motor 1. Relay 9 always operates first to return the drums to "home." Also relays 10 and 11 control the direction of rotation of motor 2. The correct relay is closed to make selection in the proper direction as chosen by the proper half sector of switch 3. The system is quite simple and after tracing through the wiring the operation will become apparent. Naturally it is not safe to close by hand any of the four relays mentioned just to "see what happens."

A test receptacle is provided on the front panel to which a cord plug attached to a number of meters may be inserted. The dummy plug shown normally short-circuits certain connections, as shown by the dotted lines on the schematic. The following meters, with the normal current and voltage values, may be used for 100 watt operation:

BETWEEN TERMINALS	CIRCUIT	FULL SCALE METER VALUE	NORMAL VALUE
3 and 12	Oscillator current	100 ma.	50 ma.
10 and 12	First amplifier plate current	50 ma.	30 ma.
4 and 12	Second amplifier plate current	300 ma.	175 ma.
1 and 8	Final amplifier plate current	300 ma.	155 ma.
2 and 12	Speech amplifier plate current	50 ma.	35 ma.
9 and 8	Modulator plate current	400 ma.	50 ma.
	Same with steady tone (100% mod.)	....	150 ma.
7 and 11	Battery voltage	15 volts	12 volts
8 and 11	Plate voltage	1500 volts	1025 volts
5 and 11	Second amplifier grid current	30 ma.	8 ma.
6 and 11	Final amplifier grid current	100 ma.	55 ma.
	Final amplifier grid current, no plate current	100 ma.	75 ma.
12 and 11	Low voltage supply	1000 volts	425 volts

It is possible to equip the test meter with switches to reduce the number of meters necessary. For example, both high voltages can be read on a 1500 volt meter with a switch sufficiently insulated for high voltage to switch between the circuits. A switch associated with terminal 12 might be used to short-circuit all but one circuit at a time of the four and to connect that one circuit to a 300 milliampere meter. Another 400 milliampere meter might be used for the two currents associated with terminal 8, using a switch to select the proper circuit and short the other one to terminal 8. Likewise grid currents can be switched for terminals 5 and 8 to a 100 milliampere meter. Thus four meters would suffice, assuming that a 12 volt reading is not required.

Tuning is accomplished after allowing filaments to warm up. Approximate current readings are given in the last column of the table above. High voltage should be disconnected from the final tube.

The small plug-in coil associated with the frequency selected is tuned after substituting an adjusting can provided with two holes in its top for the regular can. Two frequencies are adjusted in each can with an insulated screwdriver. The lock is tightened after adjustment and then the regular can is replaced. The second amplifier grid current should be about 6 to 10 milliamperes.

Adjustments of the following circuits require that the lugs be set on the drums for the proper tuning point. The large hexagon nut behind the knurled knob is slightly loosened while holding the knob in the fingers at the correct setting. The lug desired may be slipped against the stop and the nut retightened. No lug should ever be left in a position in the vertical path of the stop at "home" position or the mechanism will jam.

Second amplifier adjustments are made with the knob as the plate current is dipped for a minimum. Maximum grid current will result on the final amplifier.

It is well to insert a 2500 ohm, 200 watt resistor in series with the high voltage to the final tube in initial adjustments. Current readings will be abnormally low until the resistor is removed. The output is resonated approximately for plate current minimum by means of condenser B, with the antenna disconnected. The proper tap on the output coil (70) must be connected to the tap switch and the connections shown are normal with the lower numbers (1 and 2) on the tap switch used for the highest frequencies. The antenna, or a dummy, is next connected to the transmitter. The loading coil (71) tap is placed where the final plate current is a maximum. The correct value of final amplifier plate current is then obtained by adjustment of condenser D and proper selection of condensers 60, 61 or 62. The plate tuning must be resonated finally each time with knob B. The tuning process is repeated for each frequency.

It is very important to use a light vegetable oil on the mechanism so that proper operation will be assured at very low temperatures. Motors and other parts are serviced in the routine manner for such equipment. It will be found that the whole relay system may be removed on a plug for servicing. The schematic diagram (Fig. 92) shows the grouping of units on removable plugs.

27A Western Electric Transmitter—

Power output: 135 watts
Frequency range: 2,000 to 10,000 KC
Type frequency shift: Electrical remote
Number of frequencies: 10 non-related
Power supply voltage: 12 volts or 24 volts
Power consumption—standby: 15 amperes or 9.5 amperes
Power consumption—modulated: 77 amperes or 40 amperes

Tube Types: Oscillator 307A, Doubler 307A, Final Amplifier 322A (or 803), Audio Amplifier 1E5G, Driver 300A, Modulator (2) 830B. (All Western Electric tubes except 1E56 and 830B)

	HEIGHT	WIDTH	DEPTH	WEIGHT
Transmitter	$17^3/_8''$	$17^5/_{16}''$	$14^{11}/_{16}''$	55 lbs. (1 freq.)
Add per frequency				1.4 lbs.
Power unit				49 lbs. (approx.)

The frequency selection is entirely electrical. A large drum accommodates the crystal, intermediate tank coils and final output coils. These engage with the voltage supply leads and tube leads when the channel is rotated electrically into position. Figs. 93 and 94 show the general arrangement.

Electrically, the transmitter is very simple (Fig. 95), employing a Western Electric tube complement of one 307A oscillator, followed by a 307A doubler tube and uses a final amplifier, Class C, tube of the 322A type. It is possible to use an RCA 803 for the final tube, if desired, but the power output is about 100 watts with this tube. The modulator system consists of an RCA type 1E5G amplifier, Western Electric type 300A driver and two 830B modulators in push-pull Class B.

The transmitter will operate on either 12 or 24 volts with a dynamotor change only. The medium high voltage dynamotor is geared to the frequency shift mechanism through an electric clutch

Courtesy Western Electric Co.

FIG. 93—ROTATING COIL ASSEMBLY—27A TRANSMITTER

and operates a fan as well as supplying the plate voltage for the low plate voltage tubes. An additional external dynamotor unit supplies 1000 volts for the final modulator tubes.

TA-2D Bendix Transmitter—

Power output: 100 watt
Frequency range: 300 to 600 and 2800 to 18,000 KC
Type frequency shift: Electrical remote, tap switch control
Number of frequencies: 8 non-related
Power supply voltage: 12
Power consumption—standby: 5 amperes approximately
Power consumption—modulated: 70 amperes approximately

Tube Types: Oscillator 807, Doubler 807, Power amplifier 803, Audio Amplifier 801, Modulators (2) 830B

Courtesy Western Electric Co.

FIG. 94—REAR VIEW—27A TRANSMITTER

	HEIGHT	WIDTH	DEPTH	WEIGHT
Transmitter..................	16″	10 1/4″	18 1/2″	39.5 lbs.
Power supply.................	9 1/2″	14″	14″	50 lbs.
Antenna tuning unit...........	14 3/4″	11″	14 1/8″	25 lbs.

No tuning adjustments for the oscillator and doubler plate circuits are provided as coils are supplied with the transmitter for the particular frequencies of operation.

The output stage is tuned in the following manner:

1. Place the transmitter in operation by turning in on and selecting frequency Number 1.
2. With a patch cord, connect jacks J5 and J2 together. The microphone is plugged into J4.
3. Set the plate tap on the output coil selected to the front of the coil and the antenna tap about one third of the coil from the rear. This must be done with high voltage OFF.

4. The output coil is then rotated counterclockwise rapidly until a dip in plate current is shown. Off resonance operation for more than a few seconds will damage the output tube. The plate clip may then be moved toward the rear a turn or two at a time until the loading is proper as shown by a current of 150 milliamperes at the dip.

Low frequency coils are adjusted in the antenna tuning unit by rotating the tap switch L35 after connecting the flexible lead to an antenna coil tap. If maximum antenna current is not reached with a final amplifier plate current of 150 milliamperes, the tap should be moved and adjustment repeated.

The meter patch cord is connected between jacks J5 and J3 to read grid current, using the tap switch for the various grids and between jacks J5 and J1 to read the modulator plate current.

The antenna ammeter is not used for frequencies below 600 kilocycles.

Schematics are shown in Figs. 96, 97 and 98.

Currents and voltages in typical operation are shown in the following tabulation. Where two readings are given the first applies to the low frequency band between 300 and 600 kilocycles and the last reading for the high frequencies.

Oscillator grid current	.4 to .8 milliampere
Intermediate amplifier grid current	3 to 4 milliamperes
Power amplifier grid current	25 to 21 milliamperes
Power amplifier plate current	143 to 130 milliamperes
Static modulator plate current	15 milliamperes
Maximum signal modulator current	150 milliamperes

TUBE	PLATE VOLTAGE	SCREEN VOLTAGE	SUPPRESSOR VOLTAGE
Oscillator	250	200	..
Intermediate amplifier	350	300	..
Power amplifier	1000	380	40
Speech amplifier	500	...	..
Modulator	1000	...	..
Dynamotor supply	1050	...	..

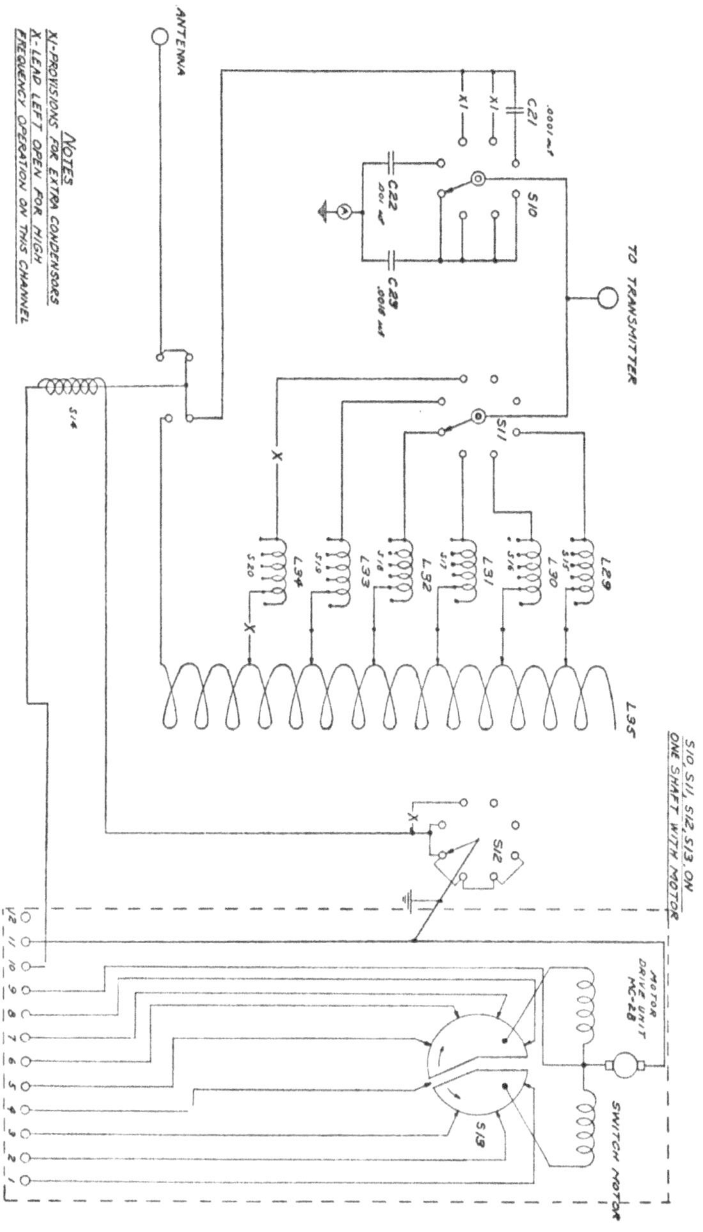

Courtesy Bendix Radio Corp.

FIG. 97—BENDIX MT-36B ANTENNA TUNING UNIT FOR TA-2D TRANSMITTER

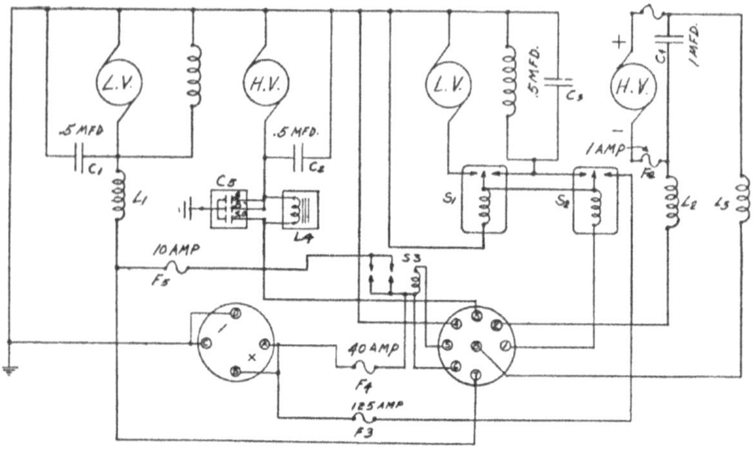

Fig. 98

The following parts will be found in the transmitter:

R1	.25 megohm	C2, C3, C5, C6, C7, C9, C10	.01 mf
R2, R5, R6	500 ohms	C4, C8	.0002 mf
R3, R20, R24	20,000 ohms	C11, C13	.002 mf
R4	50,000 ohms	C17, C19	1 mf
R7	1,500 ohms	C15, C20	1 mf
R9	.2 megohm		
R10	10,000 ohms		
R11	2 ohms		
R12, R13, R14	1,000 ohms		
R15, R21	15,000 ohms		
R18	100 ohms		
R22	90 ohms		
R23	1 ohm		

QUESTIONS ON CHAPTER VII

1. In the Western Electric 25A transmitter V3 and V4 are:

( ) a. 6L6 modulators. ( ) b. Oscillator and doubler respectively. ( ) c. Radio output tubes.

2. In the 25A transmitter, F3 is a fuse in the:

( ) a. Dynamotor primary (12 volts). ( ) b. High voltage supply circuit. ( ) c. Filament leads.

3. In the 25A transmitter M1 reads the:

( ) a. Final amplifier cathode current. ( ) b. Final amplifier plate current. ( ) c. Final amplifier screen grid current.

4. In the 25A transmitter, Y1 is:

( ) a. The oscillator condenser. ( ) b. The oscillator switch. ( ) c. A double crystal unit.

5. In the 25A transmitter, R11.1 is:

( ) a. A resistor to reduce the oscillator voltage. ( ) b. A resistor to reduce the audio voltage. ( ) c. A resistor to reduce the grid voltage.

6. The resistor R10, in the 25A, is:

( ) a. A dummy antenna resistor. ( ) b. A static leak resistor. ( ) c. A voltage dropping resistor for the final tube.

7. In the RCA AVT 7B transmitter, the 807 is the:

( ) a. Final amplifier. ( ) b. The oscillator. ( ) c. The audio amplifier.

8. In the AVT 7B transmitter, relay 30:

( ) a. Closes the dynamotor circuit. ( ) b. Short-circuits the transmitter antenna. ( ) c. Transfers the antenna.

9. In the AVT 7B, the cathode of the 6A6 goes to:

( ) a. The cathode resistor and to a condenser. ( ) b. Ground directly. ( ) c. The bias in the vibrator power unit.

10. The oscillator grid current in the Western Electric 8A transmitter will usually be between:

( ) a. .4 and 2.0 amperes. ( ) b. .4 and 2.0 milliamperes. ( ) c. 1 and 5 milliamperes.

11. The United Air Lines ES271 transmitter has a single oscillator tube because:

( ) a. This gives greater efficiency. ( ) b. There are two, the statement is incorrect. ( ) c. A single tube simplifies switching.

12. Night frequency should be adjusted last in the ES271 transmitter because:

( ) a. The oscillator can be better adjusted. ( ) b. Night is easier to adjust. ( ) c. Condenser switching makes this necessary.

13. The RCA AVT 12B transmitter uses how many relays to shift frequency?

( ) a. 3. ( ) b. 2. ( ) c. 4.

14. The AVT 12B uses choke 70 near the 42 tube for:

( ) a. Regulating bias voltage on the 42. ( ) b. Microphone current filtering. ( ) c. Bias filtering.

15. In the AVT 12B the potentiometer arm on the 42 tube bias resistor is for:

( ) a. Regulating the microphone voltage. ( ) b. Regulating the bias. ( ) c. Filtering the bias.

16. In the W4A transmitter, R1 is used for:

( ) a. Oscillator plate voltage reduction. ( ) b. Static drain purposes. ( ) c. To furnish grid return and bias.

17. The plate voltage cutoff switch in the W4A is useful for:

( ) a. Neutralizing. ( ) b. Turning the plate voltage on and off the transmitter. ( ) c. Overcurrent protection of the final amplifier.

18. The 13C Western Electric transmitter shows resistors in each crystal unit for the purpose of:

( ) a. Providing oscillator grid bias. ( ) b. Keeping crystals warm. ( ) c. Providing a steady drain on the battery.

19. The 13C transmitter shows a lamp in series with each tube:

( ) a. To hold the filament voltage constant. ( ) b. To show the tube has not burned out. ( ) c. To furnish light for servicing.

20. The switch D10, in the 13C transmitter, serves to:

( ) a. Accomplish frequency change. ( ) b. Regulate plate voltage. ( ) c. Regulate screen voltage.

21. The relay S5 in the 13B transmitter is:

( ) a. A glass enclosed, magnetically operated, plate voltage switch. ( ) b. A dynamotor primary relay. ( ) c. In reality just another fuse.

22. In the 13C transmitter, the tube V5 not only modulates the plates of V3 and V4 but also V2:

( ) a. To accomplish better modulation. ( ) b. It modulates the three screens and not the plates. ( ) c. V5 is not a modulator tube, but an oscillator.

23. The 13CB transmitter extends the frequency range of the 13C to the:

( ) a. Upper radio range frequencies. ( ) b. Ultra high frequencies. ( ) c. 12 megacycles range.

24. The Collins 17D transmitter uses the C100D tube for:

( ) a. Regulation of voltage. ( ) b. Controlling the bias on the first 807. ( ) c. A negative resistance type crystal oscillator.

25. The 17D uses the drum drive motor to:

( ) a. Return the drums home before selection. ( ) b. Move the stops for selection. ( ) c. Drive the lugs hard against the stops.

26. Switch 3 in the 17D always turns in the shortest direction to the properly numbered relay wire because:

( ) a. This does not require as many wires. ( ) b. Ring sectors 5A and 5B determine motor rotation direction. ( ) c. Open sector 6 is resting on a contact.

27. The 27A Western Electric transmitter low voltage dynamotor:

( ) a. Runs a fan. ( ) b. Does not select frequency. ( ) c. Supplies voltage for associated receivers only.

28. The Bendix TA-2D transmitter utilizes the following number of radio coils when set up for eight frequencies:

( ) a. 3. ( ) b. 16. ( ) c. 24.

29. The TA2D transmitter utilizes a split field in the frequency selector motor:

( ) a. To control direction of motor travel. ( ) b. To supply a spare winding in case of failure. ( ) c. So as to operate on two phase.

Chapter VIII

RECEIVER FUNDAMENTALS

Receivers are very much like transmitters in their principle of operation. Due to the minute signals which are received, the radio frequency currents cause inappreciable changes in plate and grid current in the tubes employed for radio amplification. Receivers are generally more difficult to maintain and service than transmitters due to their greater complexity.

Class A amplifiers are used almost exclusively in receivers with the exception of receivers with a Class B audio output stage. The usual receiver of the superheterodyne type, which is employed almost exclusively today, contains most of the following elements:

1. Radio frequency amplifier
2. Frequency converter (including oscillator)
3. Intermediate frequency amplifier
4. Detector
5. Audio amplifier
6. Automatic volume control
7. Squelch circuit
8. CW beat oscillator

The frequency converter and intermediate amplifier are not used in simple tuned radio frequency receivers, few of which are used in aircraft service.

Radio Frequency Amplification—The antenna entering the receiver and on which the signals are intercepted is matched to the receiver in much the same way that the transmitter output stage was matched to its antenna. Two-way, or company communication, receivers are connected through a send-receive relay so that the receiver is attached to the antenna during periods of reception, while the transmitter utilizes the same antenna, with the receiver disconnected, for transmission.

Modern radio range receivers are often arranged to be operated at the same time that the transmitter is in use, with no interruption to low frequency reception. Thus the co-pilot may use the transmitter and company receiver for communication while the pilot uses the range receiver.

The tuned circuit used in conjunction with the first amplifier tube is adjusted to give a maximum signal in the receiver when attached to the antenna. This is usually determined by the use of a signal generator. A signal generator is in reality a small transmitter which is well shielded so that none of the energy leaks out. At one point a shielded transmission line leaves the signal generator for attachment to the receiver antenna and ground connections.

The signal generator is capable of delivering voltages in the order of a few microvolts to the receiver and thus simulates the condition of the receiver operating at variable distances from a transmitter. Between the signal generator and the receiver is connected a dummy antenna which is the exact electrical equivalent of the antenna with which the receiver is to be used, Fig. 99. The receiver

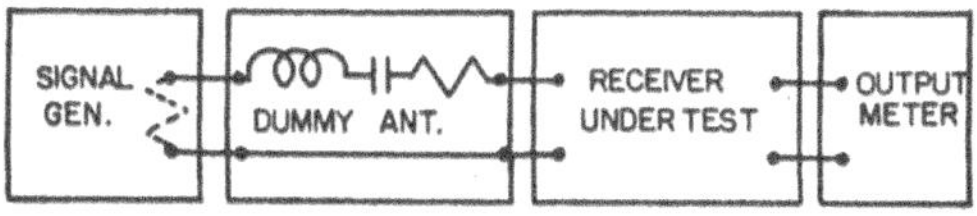

Fig. 99—Use of Dummy Antenna

may be attached later to the antenna without retuning its first tuned circuit, which is affected by any change of antenna impedance.

With the usual range receiver, the dummy antenna consists of a small mica condenser of about 100 micromicrofarads capacity. Company receivers require a dummy antenna whose impedance will be different for the day and night frequencies and will depend upon the particular antenna installation.

The radio frequency tube in the receiver does not draw grid current under any condition. If this were so the tube would present a low resistance in parallel with the tuned circuit and this would broaden the tuning of the circuit and cause a loss of signal voltage as well.

The antenna transformer may have a primary having a greater, equal or less number of turns than those of the tuned secondary. However, in a tuned circuit which is lightly loaded, as in the case of the tube that draws no current, the voltage gain of the transformer may still be great. In one radio range receiver, the primary has 500 turns and the secondary has but 200 turns. The step up to the grid, even with its fewer turns, is about five to ten times in voltage. Some receivers used for reception in the low frequency band on aircraft do not have any primary and the antenna is brought through condensers to the grid of the tube at the top of the tuned circuit. There may still be some increase of signal voltage as the current in the antenna flows through the antenna coil, which has a high reactance. The current through the coil determines the voltage to the grid. Thus by tuning an antenna the voltage delivered to the grid will be much higher than the original antenna voltage if attached to the grid directly with no tuned circuit. Several antenna circuits are shown in Figs. 57G to 57L.

Amplifier tube gain depends on bias. If the bias is of the normal value, the gain will be about thirty times in the low frequency band. If tubes used for radio frequency amplifiers were reduced in gain rapidly with small change of grid voltage there would exist a condition of partial rectification so that unwanted signals might modulate wanted ones. Further the control of the tube would be critical with small changes of bias voltage. For this reason the grid structure is wound so that the grid wires are close together except for one or two that have been omitted, leaving a wider space for electrons to travel through to the plate. Thus the tube has the characteristics of both a low and a high amplification factor tube. At high negative biases it has a low amplification factor but at less negative voltages it has a high amplification factor. Thus the tube may require a bias change from negative three volts to negative forty volts to reduce the gain of an amplification stage from thirty to a value of one. Such tubes are known as exponential, or remote cutoff, tubes.

The plate tuned circuit of the radio frequency tube, or the grid circuit to which the plate primary is coupled, is also tuned to maximum efficiency by using the signal generator. A tuned radio frequency receiver of simple design is shown in Fig. 56.

Frequency Conversion—All superheterodynes are equipped with a frequency converter for changing the incoming frequency to one which is more easily amplified and where the selectivity can be better governed. Thus if a signal is received at 5000 kilocycles, or on day frequency, it is better to convert this signal to a low frequency of 385 kilocycles for example. This frequency is called the intermediate frequency of the receiver.

When a signal enters the receiver at 5000 kilocycles and it is desired to change it to 385 kilocycles, it will require a local, or first oscillator, in the receiver to accomplish this change. The process is like modulation. The oscillator in the receiver combines with the incoming signal, just as the voice combines with the carrier in a transmitter. The incoming carrier could be combined with an oscillator frequency at either 4615 or 5385 kilocycles. In either case the plate circuit of the converter tube would be tuned to 385 kilocycles. The frequency sum or difference, without regard to algebraic sign, will be present.

In this case, assume the signal combined with a 5385 kilocycle from the oscillator and the difference between this and 5000 kilocycles gives 385 kilocycles, or the intermediate frequency signal. This is amplified in the intermediate amplifier which has its tuning fixed for this frequency. If it is desired to tune to another frequency, the oscillator and the radio frequency tuning can be adjusted to any new frequency so that the difference between the signal and the oscillator frequencies will always be 385 kilocycles. If it is not 385 kilocycles, there will be no signal amplification and the signal will be rejected. Of course the sidebands and the carrier together are converted so they bear the same relationship at intermediate frequencies that they bear at the higher frequencies.

It will be noted that if the oscillator is at 5385 kilocycles and a signal arrives at the receiver on 5770 kilocycles, this signal should be received. It would be, except for the fact that the high frequency tuned circuits are selective enough to eliminate a signal so far removed from the tuned frequency of 5000 kilocycles. The signal existing on the other side of the oscillator frequency by the intermediate frequency amount (at 5770 kc) is called the image and it is the main purpose of the tuned high frequency circuits to eliminate this image.

Tracking—When ganged condensers are used to tune a superheterodyne receiver the oscillator should track with the radio frequency tuned circuits, maintaining the exact intermediate frequency difference. Practically, the usual oscillator approximates this quite closely. The oscillator covers a smaller frequency ratio than that of the tuned circuits. It covers the same kilocycle range but at a higher frequency than the tuned circuits. It is necessary, therefore, to provide two adjustments for the oscillator circuit. The adjustment at the high frequency end is conventional. The receiver is set to the calibrated frequency desired and the oscillator trimmer is adjusted for maximum signal. The oscillator trimmer for this adjustment is electrically like the radio frequency trimmer adjustments, which are tuned for maximum also.

Oscillator Adjustment—At the low frequency end of the dial, the oscillator must be retrimmed with variation of a variable trimmer condenser in series with the oscillator tuning section of the ganged condenser. To accomplish the low frequency tuning, the receiver is set to some frequency about ten per cent higher than the lowest frequency. The generator is also tuned to this frequency and the receiver is tuned slowly back and forth ("rocked") through the frequency and the maximum output reading noted. Then the series padding condenser is slightly moved to a new value and the receiver "rocked" through the generator frequency noting the new maximum. The "rocking" for maximum is repeated until the maximum output is obtained, at which time the oscillator is tracking perfectly with the other tuned circuits at the low frequency end. High frequency end adjustment (at a frequency about ten per cent lower than the highest frequency) must be repeated after the low frequency adjustment described.

Intermediate Amplification—The intermediate amplifier in the receiver serves to raise the signal to a point where it is great enough to be detected and turned into an audio or voice frequency.

It has been said that the gain and selectivity are better at these frequencies. With good coils and condensers, the percentage selectivity at almost all frequencies is about the same. Thus, a tuned circuit at 5000 kilocycles reduces an unwanted signal at 25 kilocycles difference by three times. A tuned circuit at 385 kilocycles will reduce an unwanted signal twenty-five kilocycles different by forty times. The arithmetic or absolute frequency difference selectivity thus increases rapidly as the frequency is lowered, even though the percentage selectivity remains the same. A circuit at 1000 kilocycles is but one tenth as effective for a given interference kilocycle separation as one at 100 kilocycles.

Further, with two tuned circuits, each of which might reduce the signal twenty times, both combined will result in a reduction of unwanted signal by 400 times (their product). In most intermediate amplifiers there are about six tuned circuits and they will reduce a signal ten kilocycles removed from the desired one by a factor of about 1000 times. Fig. 100 shows average selectivities of one to six tuned circuits.

The need for such a selective intermediate amplifier is evident as stations are spaced about every fifteen kilocycles in the high frequency spectrum and every three kilocycles in the radio range band.

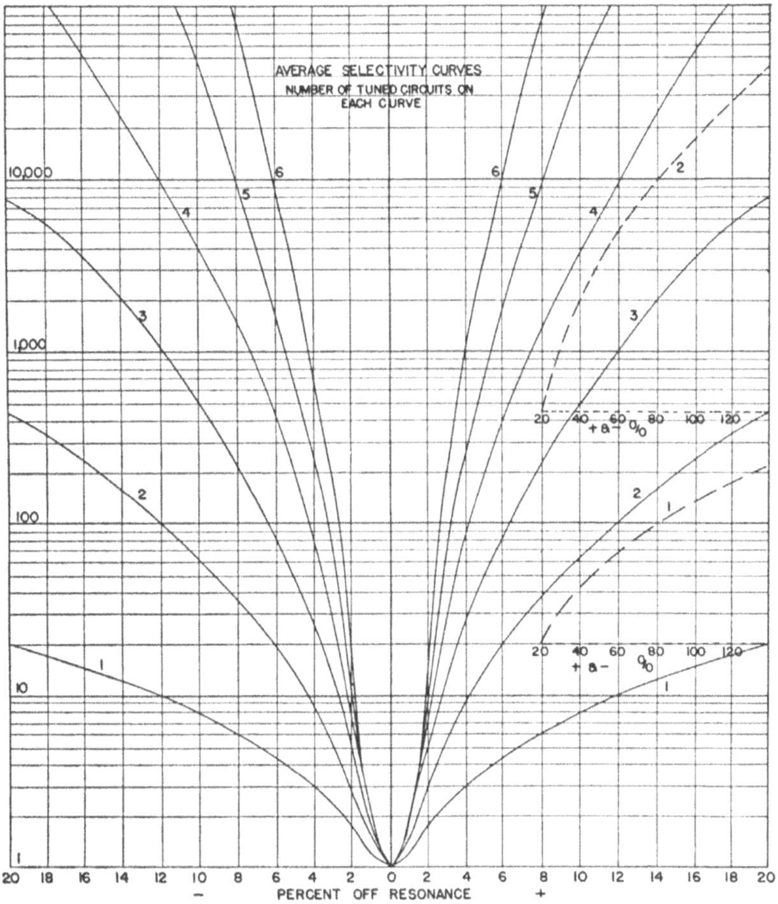

Fig. 100

The receiver will often be very near the station sending the unwanted signal.

Detection—After the signal level has been raised by the gain in the radio, converter and intermediate amplifiers it must be detected and turned into an audible frequency. This is done in the last or second detector (the converter is called the first detector sometimes). It has been shown that there are sidebands created which exist on both sides of the carrier and are spaced at the voice frequency difference. Thus if a 1000 cycle audio signal is transmitted at 5000 kilocycles there will be three frequencies emitted by the transmitter, 4999 and 5001 kilocycles (which are sidebands) and 5000 kilocycles (the carrier). When this signal enters the receiver it will be converted to an intermediate frequency by combination with 5385 kilocycles, for example. The intermediate frequency signal will consist of three frequencies; 386, 384 and 385 kilocycles respectively. The two sidebands still differ by 1000 cycles from the 385 kilocycle intermediate frequency carrier.

The three signals (386, 384 and 385) are combined and give, in their various combinations, a number of frequencies which are in the medium high frequency range such as 771, 769 and 770 kilocycles. These are filtered out with the 385 kilocycle order signals by capacitors in the circuits following the detector. The only frequencies which are not removed are the two differences which both give 1000 cycles and the frequency difference between the two sidebands, or 2000 cycles. This last frequency represents distortion.

Detector Types—There are two methods of detection and these are plate and grid detection. In plate detection the tube is operated at such a bias that the negative half of the radio frequency wave is removed. This is shown in Fig. 62. Plate detection is very similar to grid modulation as a comparison between the two will show. Grid detection, on the other hand, utilizes the characteristics of the grid circuit as a rectifier. The main function of the plate circuit of the tube is to amplify the rectified or audio frequency. The operation may best be seen by reference to Fig. 62. Diodes operate in the same manner but no audio amplification results.

It will also be noticed that the tube does not operate in the same manner for weak and strong signals. A strong signal is any signal at the detector which is above a value of about a volt. The difference in the operation is shown in Fig. 62.

Square Law Detectors—A weak signal detector is a square law detector in that the audio output varies as the square of the input voltage. Thus when the received signal becomes twice as great, the output will be four times as great in voltage. The square law detector adds considerable distortion at higher percentage modulations. It has its use in flying radio ranges in that the difference between the "A" and "N" signals will appear to be much greater than the difference in the two radio voltages would indicate, due to the square law operation of the detector (except on simultaneous ranges).

Linear detectors require a strong signal at the second detector of at least three or four volts. Such detectors are seldom used in aircraft and the audio amplifier would have small gain or be unnecessary. This type of design is not as satisfactory, as more intermediate amplification is required to obtain sufficient sensitivity.

Audio Amplification—The audio amplifier serves to raise the signal to a suitable level in the headphones. Most audio amplifiers will deliver half a watt of signal to the headphones although a level of over about ten milliwatts becomes uncomfortable when the headphones are worn close to the head.

One reason for having the high signal level available is that there may be more than one pair of headphones operated from the receiver. When two sets are operated only one half the energy will be fed to each. Other losses may cut down the maximum amount of signal to about fifty to seventy-five milliwatts maximum.

The signal will increase in the headphones as the radio sensitivity is raised up to a point where the signal overloads the audio frequency stages. From then on the signal will be distorted and range signals ("A" and "N") may actually reverse.

The reason for the reversal of the range signals is as follows. Let it be supposed that the "A" signal was just twice as strong as the "N" signal. The sensitivity control of the receiver is now advanced up to the point of overload and the "A" will still be twice as strong as the "N.". After overload the "A" signal producing overload will decrease in amplitude as the sensitivity control is advanced. Therefore, the further advancement of the sensitivity control will cause the "A" signal to decrease but as the "N" signal is not overloading as yet it will continue to increase in value with sensitivity increase. Thus a point will be reached where the "A" has been reduced by overload and the "N" has become stronger due to the advance of the sensitivity control. Thus the "N" will be louder than the "A" and the signals will be reversed, giving an incorrect indication to the pilot of his position with respect to the range leg.

For this reason it is very important that range receivers have sufficient audio output so that overload will produce an uncomfortable signal or distortion and show its presence. Receivers in which overload causes a very rapid decrease of the overloading signal are dangerous as a comfortable signal may be received if the sensitivity is advanced far enough into the overload region to reduce signals sufficiently. Receivers should be designed to give a highly distorted and loud signal beyond overload.

Automatic Volume Control—Three automatic volume control circuits are shown in Figs. 101, 102 and 103. The principle of operation of the volume control circuit is this: When the signal at the second detector is rectified it will give rise to a certain amount

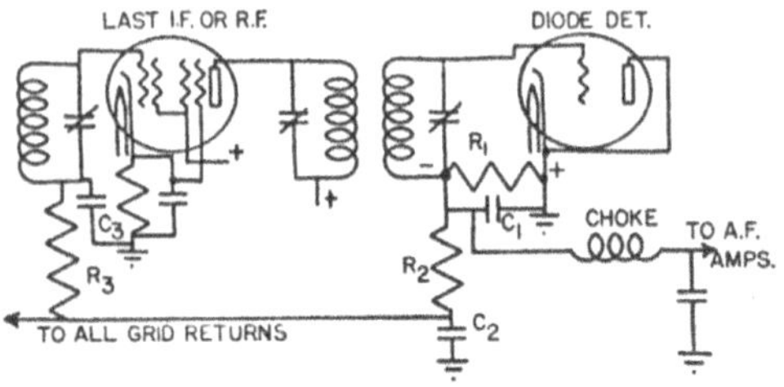

FIG. 101—AUTOMATIC VOLUME CONTROL

of direct current which will flow in the resistor in series with the diode and across which the audio voltage is produced. This voltage will be proportional to the incoming signal. If the grid return leads of one or more of the amplifying stages ahead of the second detector are connected to the voltage, the grids will become more negative

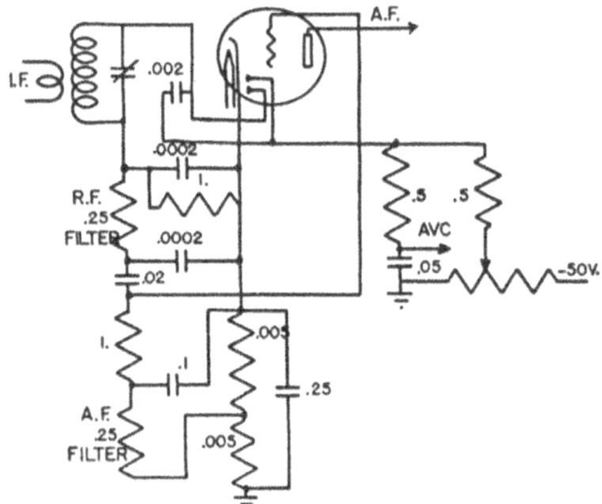

FIG. 102—AUTOMATIC VOLUME CONTROL

as the signal increases. Thus the gain of the receiver will be lowered as the signal becomes greater. The overall result is that the receiver output increases very slowly with a large ratio increase of input signal, as in Fig. 104.

The automatic volume control cannot be applied easily to range receivers for it tends to iron out the differences between the "A" and "N" signals, particularly when close to a range leg. It also makes it impossible to determine readily the increase or decrease in signal when orientation problems are being solved. It has application

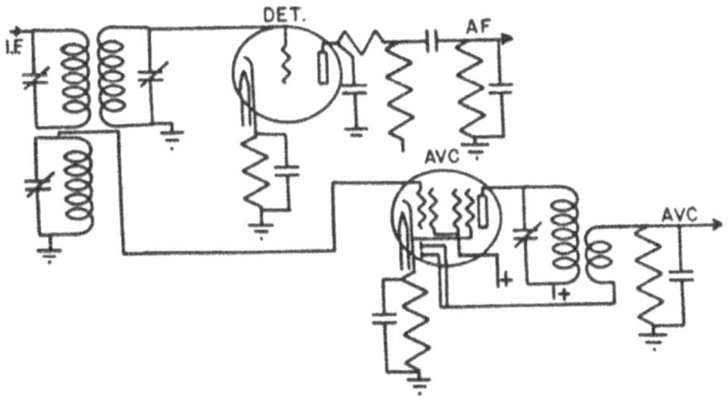

FIG. 103—AMPLIFIED AUTOMATIC VOLUME CONTROL

in both the ultra high frequency and the high frequency communication receivers.

The sensitivity control on receivers with AVC usually regulates the cathode voltage. It will be recalled that the more the cathode voltage is made positive, the more the grid is effectively made negative. The automatic volume control circuits usually control only the grid return circuits while the sensitivity control acts upon the cathodes. Both have the same result but the circuits are usually isolated in this manner.

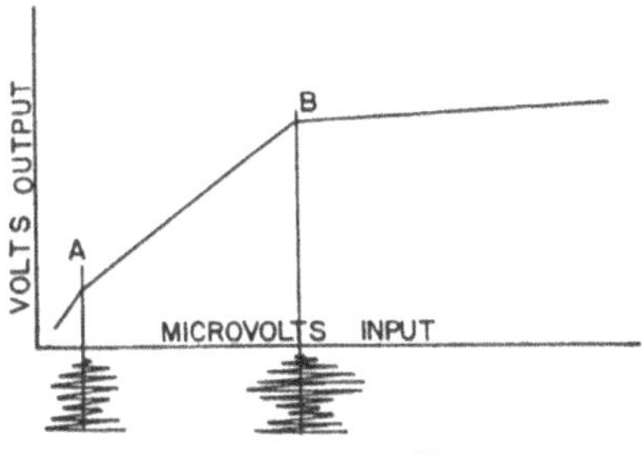

FIG. 104—AUTOMATIC VOLUME CONTROL ACTION

Squelch Circuits—It will be seen by the curves of the automatic volume control circuit (Fig. 104) that below the point where the automatic volume control circuit takes hold the signal decreases linearly with the input. As noise will be heard when the transmitter carrier of the ground station is not being radiated, there will be a certain amount of noise that will be disturbing to the pilot. If a circuit is employed that will greatly reduce the audio gain when there is no carrier, it is possible to remove the noise when the receiver is not actuated by an incoming signal. There is some danger of the pilot missing a call from the ground station if the receiver sensitivity is just low enough so that the squelch circuit will not operate. Generally the improvement is marked and especially so at the ultra high frequencies. The receiver sensitivity on the ultra high frequencies can be made ample with the squelch circuit in operation due to lack of heavy static. When no carrier is on the air the receiver will then pick up none of the constant residual noise background from ignition systems and the like.

Fig. 105 illustrates one type of squelch circuit. The 57 tube is normally supposed to be biased to three volts for operation. If the 55 is drawing plate current, the grid of the 57 tube will be biased by

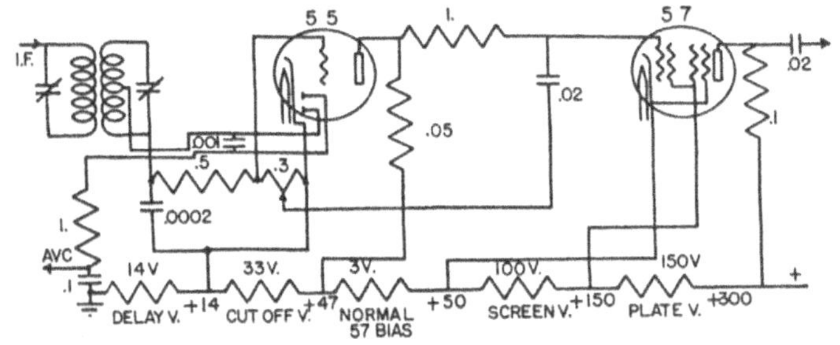

Fig. 105—Automatic Noise Control—Squelch Circuit

the three and thirty-three bias voltages. As signal into the 55 tube is raised it will cause the tube to draw less plate current due to increased bias. This will cause the voltage drop in the .05 megohm resistor to decrease and finally the 55 tube will draw no plate current. When this occurs the normal three volt bias will appear on the 57 tube and it will amplify the audio signal. Audio voltage is fed to the 57 through the .02 microfarad condenser from the audio voltage rectified at the .3 megohm diode resistor.

Range Antennas—The antenna used with a range receiver is quite important in the proper operation of the receiver. There are several types of antenna used for such reception at the present time and each has a slightly different characteristic.

One of the older types of antenna, which is being revived again, is the vertical self-supporting, or guyed, mast. The newer version of this antenna is a whip similar to those used on automobiles. Whips of either forty-two or sixty inches in length will be found in service. The whip may be mounted in one of two places, either on top or below the plane. The characteristics are different for the two locations when the plane passes over the cone of silence of a radio range but otherwise there is little or no choice between the two positions.

With the whip below the plane the cone of silence of the station will be very sharp and clear and the silent point will actually give a period, depending upon altitude, when no signal at all can be heard. The exact location along the lower surface of the plane will make some difference as to whether the cone occurs exactly over the station or slightly before or after passing the station. The location with the shortest possible leadin to the receiver is usually to be preferred. As these antennas represent a rather low capacity in themselves (20 to 30 mmf.) it is important that the leadin capacity in the plane be kept very low. Running the leadin parallel to the skin and near it for long lengths will greatly reduce the signal. In fact it will reduce signals more than with any other types of antennas. If reasonable care is taken the antenna will be better than others as far as actual received signal is concerned.

When the whip is mounted above the plane the results will be quite different. As the plane approaches the station the signal level will not build up as rapidly as before because the wing and fuselage will be coming between the station and the plane, due to the plane having considerable altitude above the range station. At the point of passage over the station there will be some signal resulting which comes over the edges of the wing and approaches the antenna from a direction other than "end on." (When located below the plane, the projected area is nearly zero when over the station.) As the plane passes the cone of silence over the station the wing will interfere still more with the top whip's signal received from the station and a point will be reached where the signal over the leading and trailing edges of the wing will arrive at the antenna out of phase and a cone of silence will be observed which may be some fifteen degrees beyond the vertical line over the station. Thus the cone of silence will be displaced from its true position. The rise of signal beyond this point will be fairly rapid. Thus the antenna is not as good as when in a position below the plane. With the advent of cone of silence ultra high frequency markers the aural cone over the station possibly need not be located as accurately as before.

The fin antenna extends from a point some twelve feet forward of the vertical fin to a point at the top leading edge of the fin. Unluckily this antenna cannot be used as efficiently on installations of radio apparatus forward in the plane as there is a loss in the leadin which may have to be as much as fifty feet in length. Transformers with transmission lines have been developed to be used on such leadins and aid greatly. The antenna is not applicable to transports with short or double vertical fins because of lack of vertical height. The antenna gives a good cone of silence at a point approximately over the station. The stabilizer does not interfere noticeably with its operation nor does the fuselage (which is usually fairly small at this point).

The V antenna is mounted below the fuselage and well forward. The leading edge of the V is usually attached to the air speed pitot head and insulated from it. The two other ends are situated on supports to the rear and to the side below the leading edge of the wing. By regulating the angle which these wires make at their apex it is possible to control the cone of silence characteristic so that the cone of silence will occur over the station. This quality is very useful although the pickup qualities of the antenna are not quite as great as those of the whip antenna and the antenna is more apt to pick up ignition noise.

Another common location of antennas is under the nose of the plane and parallel to the fuselage, either as an inverted T or L antenna. These antennas are not as subject to icing but their pickup and cone of silence qualities are inferior to the others. The cone of silence will usually lead the station by five degrees or so and the cone is not as distinct and sharp, nor are the surges associated with the cone as pronounced. The antenna is a fairly good all purpose antenna however.

Itinerant aircraft have T antennas which are almost parallel to the fuselage and have short leadins to the receiver. The cone of silence characteristics can be greatly improved on many such installations by changing the amount of flat top fore and aft of the leadin. Usually the whole leadin should be as vertical as possible through to the receiver and the fore and aft flat top length should be the same. The nearer the leadin end of the wire comes to the engine, the more ignition interference will be encountered.

In some cases the leadin is run through a shielded lead after entering the plane. The usual result is to severely reduce the amount of pickup of the antenna and the shielding improvement from noise in the ignition and battery system is of little consequence. A few trial flights with changes in the antenna system will serve to improve conditions markedly in most cases.

QUESTIONS ON CHAPTER VIII

1. The radio frequency amplifier tube in a superheterodyne is used for:
( ) a. Reducing image interference due to amplification. ( ) b. Providing amplification, principally. ( ) c. Providing better volume control. ( ) d. Reducing distortion.

2. The gain of an amplifier tube will:
( ) a. Increase as the bias is made more negative. ( ) b. Vary in accordance with the incoming signal. ( ) c. Increase as the cathode is made more positive. ( ) d. Decrease with more negative grid bias.

3. Frequency conversion requires:
( ) a. Frequency multiplication. ( ) b. Presence of sidebands. ( ) c. A local oscillator. ( ) d. A tube of linear characteristics.

4. A radio frequency receiver separates stations 10 kilocycles apart at 300 KC. Presumably it will separate stations under like conditions when tuned to 3000 KC, which have a separation of:
( ) a. 100 kilocycles. ( ) b. 1000 kilocycles. ( ) c. 10 megacycles. ( ) d. 10 kilocycles.

5. A carrier of 1000 kilocycles has sidebands at 990 and 1010 kilocycles. The signals pass through a converter so that the carrier is reduced to 100 kilocycles. The new sidebands will be:
( ) a. 99 and 101 kilocycles. ( ) b. 0.11 and 0.09 megacycles. ( ) c. 90 and 110 cycles. ( ) d. 90 and 101 kilocycles.

6. The intermediate frequency of a superheterodyne is 100 kilocycles and the oscillator is operating at 250 kilocycles. What radio signals could be received best with no R.F. selectivity preceding the converter?
( ) a. 350 and 450 kilocycles. ( ) b. 50 and 150 kilocycles. ( ) c. 200 and 300 kilocycles. ( ) d. 350 and 150 kilocycles.

7. The oscillator in a tunable superheterodyne is usually operated at a frequency that is:
( ) a. The same as the radio frequency. ( ) b. Lower than the radio frequency. ( ) c. Higher than the radio frequency. ( ) d. Lower than the intermediate frequency.

8. The "rocking for maximum" is used for:
( ) a. Adjusting the oscillator at the low frequency end of a band. ( ) b. Transmitter adjustments. ( ) c. Adjusting the antenna coupling. ( ) d. Tuning stages in the intermediate amplifier.

9. Intermediate amplifiers are usually:

( ) a. More selective than lower frequency amplifiers. ( ) b. More selective than the higher radio frequency amplifiers. ( ) c. Less selective with decreasing frequency. ( ) d. Used to provide buffers between oscillator and converter.

10. Detection is usually "square law" at:

( ) a. Higher frequencies. ( ) b. Lower frequencies. ( ) c. High voltages. ( ) d. Low voltages.

11. Square law detectors will give four times as much voltage output when the signal voltage into them is:

( ) a. Squared. ( ) b. Raised four times. ( ) c. Doubled. ( ) d. Halved.

12. Higher voltage detectors are usually:

( ) a. Linear. ( ) b. Square law. ( ) c. Found in aircraft. ( ) d. Used for radio range receivers.

13. Audio amplifiers when overloaded will produce:

( ) a. Absolutely uniform output. ( ) b. Range course "reversal." ( ) c. Desired harmonics. ( ) d. Clearer reception.

14. Automatic volume control:

( ) a. Always improves radio range reception. ( ) b. Allows the pilot to hear a fade or increase of signal. ( ) c. Makes "A" and "N" signals more distinct. ( ) d. Holds the output fairly constant.

15. Squelch circuits are used to:

( ) a. Make signals clearer during transmissions. ( ) b. Eliminate the carrier and secure better detection. ( ) Reduce noise when no carrier is received. ( ) d. Regulate the "tone" of a receiver.

16. Range antennas should be:

( ) a. As nearly vertical as possible. ( ) b. As nearly horizontal as possible. ( ) c. As close to the fuselage as possible. ( ) d. Brought in through high capacity leadins.

17. T antennas are used because:

( ) a. They have less wind resistance. ( ) b. They stimulate a good vertical antenna. ( ) c. Horizontal wires give the antenna directivity. ( ) d. The horizontal portion reduces the signal.

18. A dummy antenna is used:

( ) a. To replace a lost antenna in flight. ( ) b. To load the receiver output during measurement. ( ) c. To reduce signal from a signal generator. ( ) d. To aid in adjustment of the antenna circuit.

CHAPTER IX

COMMERCIAL AIRCRAFT RECEIVERS

A great deal of information on commercial aircraft receivers has been assembled in this chapter. The practical application of those principles set forth in the previous chapter will be seen in each receiver. The material has been graduated from the simple receivers of a few tubes up to the receivers whose circuits are relatively complicated. A study of the material, in the order given, with a careful tracing of each circuit will familiarize the reader with almost every modification of circuit to be found in present receivers and in most receivers that will be produced for many years to come.

Some of the receivers have had their tube complements revised to take advantage of newer or less expensive tube types but the circuits in such cases have been changed but little. Most receivers in use will be found to agree with the diagrams given.

Tables showing the most important information are placed at the beginning of each section. Dimensions are overall and approximate. All dimensions show the control face as height and width.

When separate dynamotors are shown their battery drain is not given, as one dynamotor may supply two or three receivers. Most 200 volt dynamotors draw about four amperes at twelve volts.

9B Western Electric Receiver—

	Frequency Range	*Alignment Frequencies*
With L coils	1500 to 2450 KC	2350 KC
With M coils	2400 to 3900 KC	3000 KC
With H coils	3900 to 6100 KC	5500 KC
Intermediate	None used—tuned radio frequency receiver	

Power unit: Dynamotor, external. 200 volts, 35 ma (approx.)
Current consumption: Receiver 2.5 amps (approx.) at 12 volts
Power output: Approximately 200 milliwatts at 500 ohms
Tube complement: R. F. Amplifiers (3) 245A, Detector 245A, Output 244A, Ballast lamp 6B

	HEIGHT	WIDTH	DEPTH	WEIGHT
Receiver	$8^5/_8$″	11″	$13^3/_8$″	16.6 lbs.

This receiver is a companion unit for the 8A transmitter and was the first standard receiver in aviation. It is of the tuned radio frequency type and some modifications of this receiver have replaced the gang condenser with two adjustable condensers per section with a switch to select between them. One is then used for night or low frequency and the other is adjusted for the day or high frequency.

Control of gain is obtained by varying the screen grid voltage as shown on the schematic. This voltage is limited to a maximum of 90 volts by means of 50,000 ohm series resistor placed from the 200 volt plate supply source to the top of the 50,000 ohm controlling potentiometer (Installation, Fig. 106).

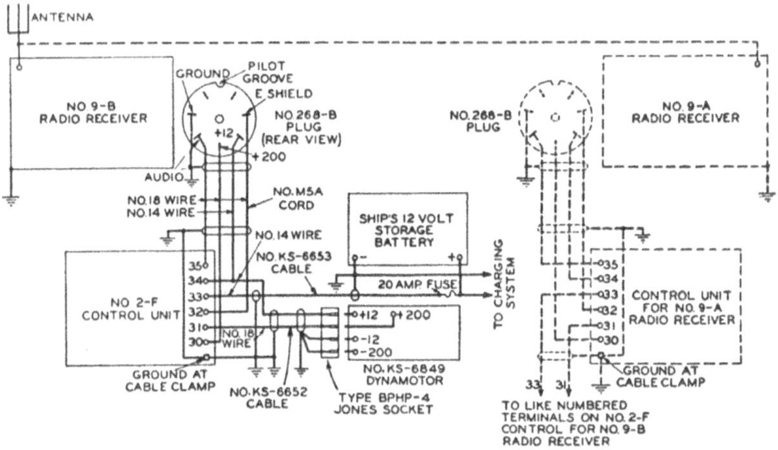

Courtesy Western Electric Co.

FIG. 106—AIRCRAFT INSTALLATION, 9 TYPE RECEIVERS

The schematic (Fig. 107) shows values of parts. A ballast lamp is used to hold filament voltages constant.

The antenna taps are provided for different capacity antennas. Those with the highest capacity are used with the smallest internal antenna capacitor.

9D Western Electric Receiver—

	Frequency Range	*Alignment Frequency*
Radio range band	218 to 515 KC	400 KC
Intermediate frequency	None Used—tuned radio frequency receiver	
Power unit	Dynamotor, external	200 volts 35 milliamperes (approx.)
Current consumption	2.5 amperes (approx.) at 12 volts	
Power output	Approximately 200 milliwatts at 500 ohms	
Tube complement	RF amplifiers (2) 245A, Detector 245A, Audio amplifier 244A, Output 244A, Ballast lamp 6B	

	HEIGHT	WIDTH	DEPTH	WEIGHT
Receiver	8⅝″	11″	13⅜″	16.4 lbs.

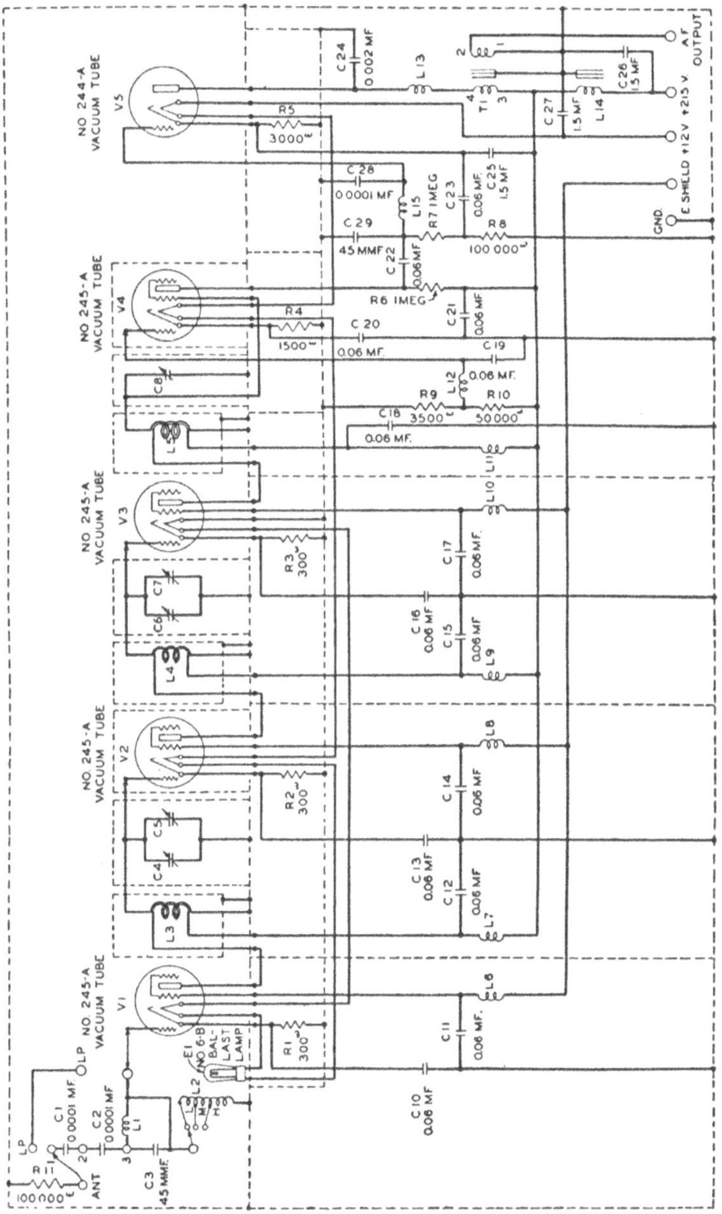

Courtesy Western Electric Co.

FIG. 107—SCHEMATIC DIAGRAM, 9B RECEIVER

This receiver has been in use for as long a period as the 9B. It has the same weight as the 9B and is of the same size and construction. Control of gain is obtained in the same manner and except for frequency range and use is much like the former receiver.

The schematic diagram is shown in Fig. 108 which also shows the circuit constants.

AVR 10 and AVR 10A Receivers—

	Frequency Range	*Alignment Frequency*
Range band	200 to 400 KC	400 KC
Intermediate	Tuned radio frequency receiver	

Power unit: Vibrator in power unit
Power consumption: 0.7 amperes at 12 volts. 1.4 amperes at 6 volts
Power output: 35 milliwatts at 600 ohms
Tube complement: RF amplifier-detector 6F7, Audio and output 6F7

	HEIGHT	WIDTH	DEPTH	WEIGHT
Receiver unit	5 1/2″	6 1/2″	3 1/2″	3 lbs.
Power unit	5 1/2″	6 1/2″	3 1/2″	4.5 lbs.

One of the simplest aircraft receivers in use today is the AVR 10 which uses but two tubes.

The AVR 10 operates from a 12 volt battery while the AVR 10A operates from a 6 volt battery.

The receiver (schematic Fig. 109) has one stage of radio frequency in the pentode section of the first 6F7 tube. After amplification the signal passes to the triode section of the same tube which is a detector. The signal is then amplified at audio frequency first by the triode section of the second 6F7 and is finally amplified and fed to the headphones by the pentode section of the second tube.

Alignment is accomplished by one adjustment at 400 kilocycles. A signal is applied through a dummy antenna consisting of a 100 mmf condenser to the antenna post and C30 is adjusted by rocking the tuning back and forth through the signal while adjusting C30 to a value giving the maximum signal as the receiver passes through the signal. Strong signal inputs require suitable volume control regulation as it is relatively easy to overload this receiver.

17A, 17AX and 17AXR Western Electric Receivers—

	Frequency Range	*Alignment Frequency*
Range band	190 to 450 KC	380 KC
Broadcast	550 to 1500 KC	1350 KC
High frequency	2800 to 3500 KC	3100 KC
Intermediate	96 KC	96 KC

Either broadcast or high frequency, not both, in addition to range frequencies.

Power unit: External dynamotor, 200 volt, 30 milliamperes or 135 volts, 20 milliamperes
Power consumption: 17A—1.1 amperes whether at 6 or 12 volts
17AX, 17AXR—1.4 amperes at 12 volts
Power output: 700 milliwatts, 600 ohms (175 milliwatts at 135 volts plate)
Tube complement: Converter-Oscillator 6A7
Detector-audio 6B7
Output 41
RF 78 (17AX only)

	HEIGHT	WIDTH	DEPTH	WEIGHT
17A Receiver......	7 5/8″	8 1/4″	7 7/16″	9.1 lbs. with shocks
17AX Receiver....	7 5/8″	12 3/4″	7 7/16″	12 lbs. with shocks
KS-7543 Dynamotor	6 7/8″	8	4 7/16″	10.25 lbs. with shocks

The wiring diagram for receivers for serial numbers 101 to 399 are shown in Fig. 110 while the wiring diagram for serial numbers over 400 is shown in Fig. 111. There are some differences between these schematics.

Both types of receivers are built to cover the ranges 190 to 450 kilocycles and 550 to 1500 kilocycles although a number of receivers were especially built to cover the band from about 2800 to 3500 kilocycles instead of the broadcast band. Some receivers in the field have undergone a similar conversion, the change being in the radio frequency transformers. A number of receivers (17AX) have had a loop tuning arrangement added in which a stage of amplification is incorporated, as described later.

A glance at the wiring diagrams will show that the first tube is used as a combined oscillator and converter. The second tube is an intermediate frequency amplifier operating at 96 kilocycles and a detector. The audio frequency, after detection, is passed again through the amplifier portion of this tube so that both intermediate and audio amplification take place together. The third and last tube is an audio power amplifier. The sensitivity is about fifteen microvolts for an audio output of fifty milliwatts.

The "varistor" element in the output circuit is beneficial in reduction of static surges in the headphones, particularly when anti-static phones are not employed. Fig. 112 shows how high impedance phones may be connected to the output. With non-antistatic high impedance phones, the varistor should be left in the circuit at all times for best results.

The following table gives the approximate voltages in the receiver with the volume control at maximum (10) and the antenna disconnected.

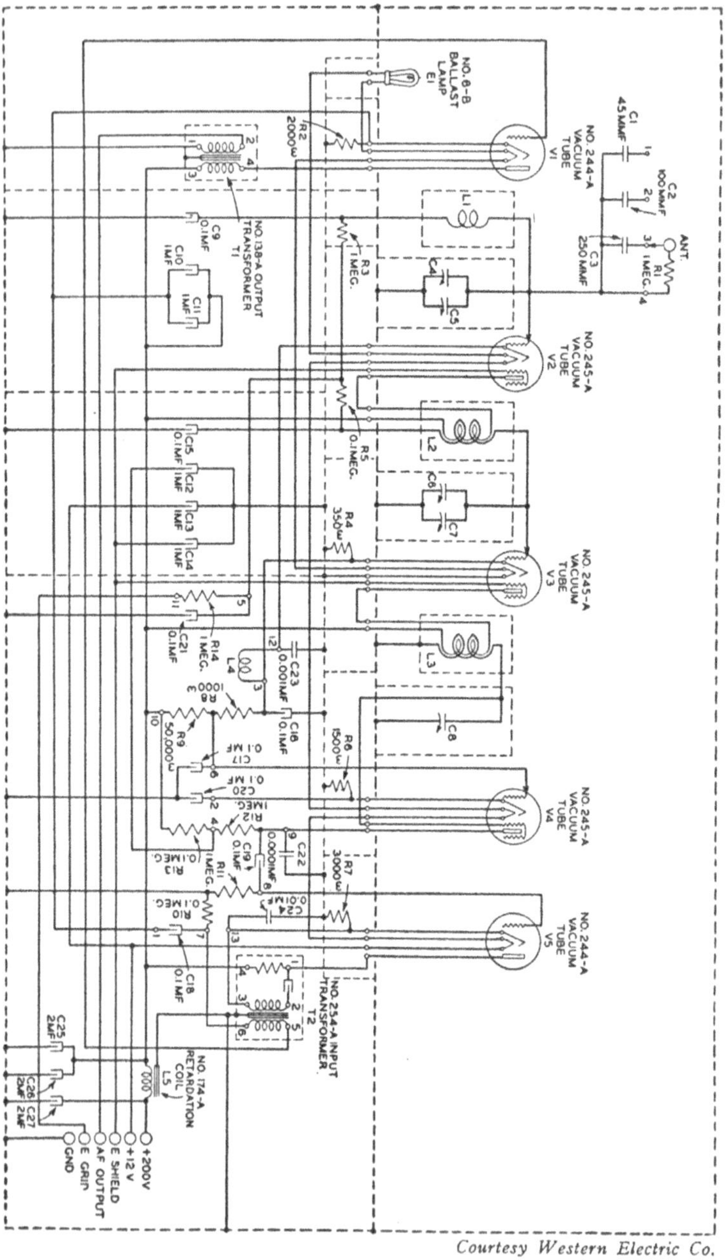

Courtesy Western Electric Co.

FIG. 108—SCHEMATIC DIAGRAM—9D RECEIVER

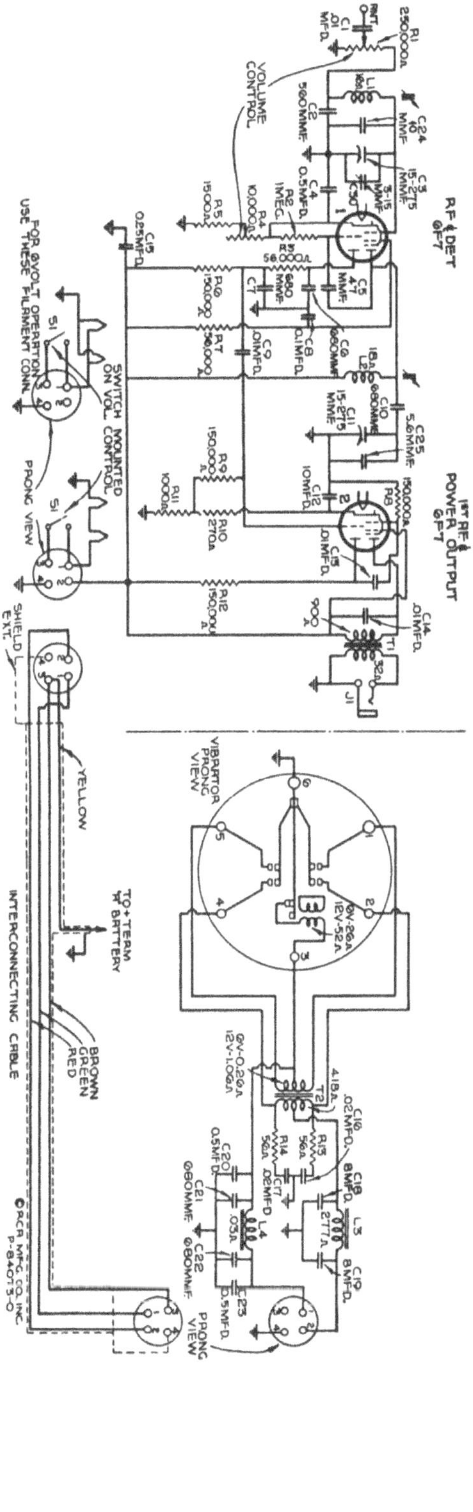

Fig. 109—AVR 10 and AVR 10A Receivers

Courtesy RCA Mfg. Co.

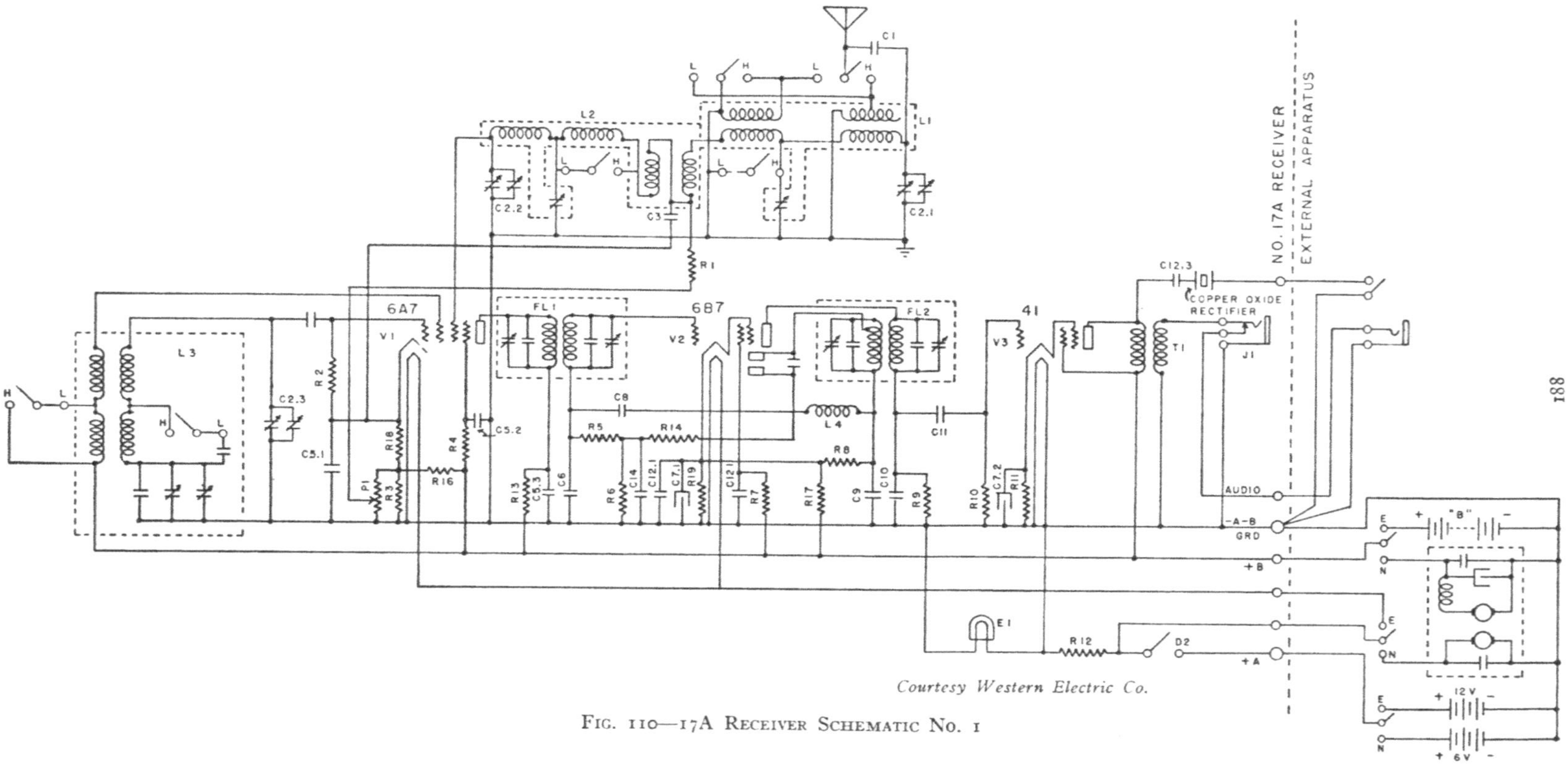

Courtesy Western Electric Co.

FIG. 110—17A RECEIVER SCHEMATIC NO. 1

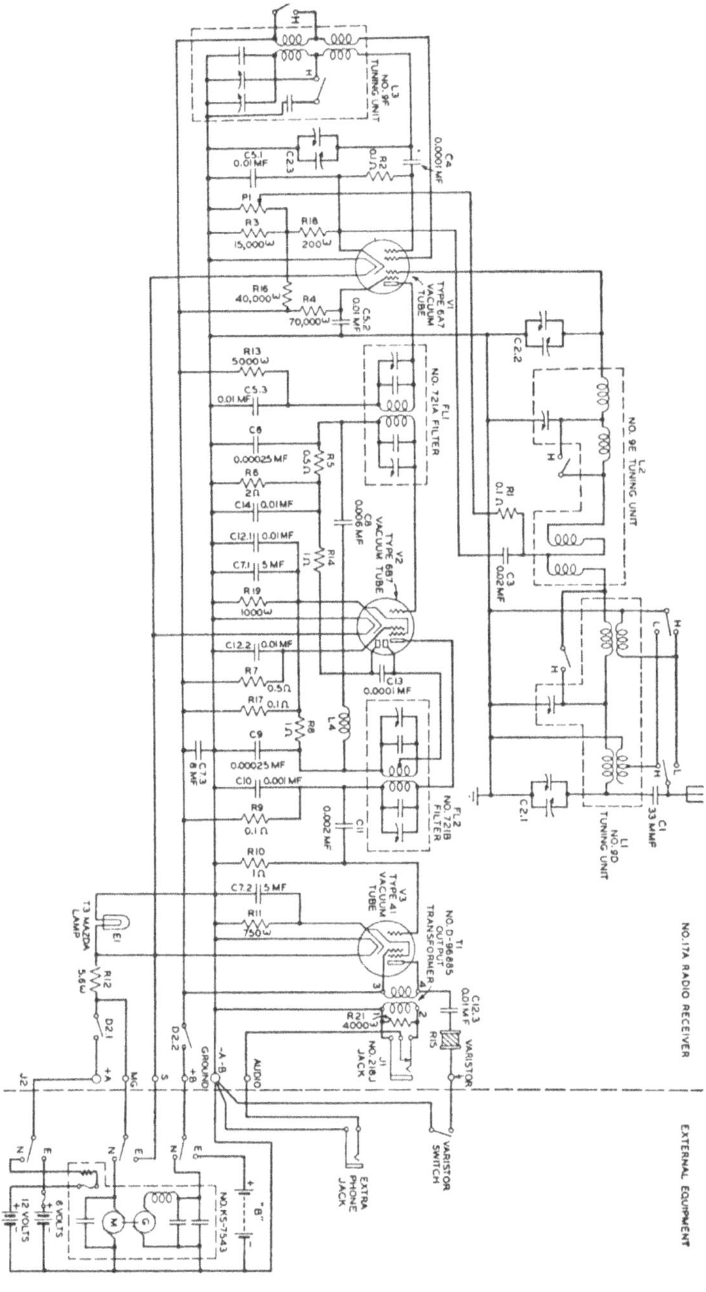

FIG. 111—17A RECEIVER SCHEMATIC NO. 2

Courtesy Western Electric Co.

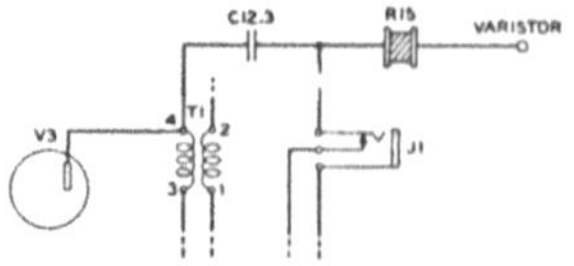

Courtesy Western Electric Co.

FIG. 112—USE OF HIGH IMPEDANCE HEADPHONES WITH 17A RECEIVER

CIRCUIT	VOLTAGE
Total plate supply	170
Plate of 6A7	161
Grids 3 and 5 of 6A7	95
Grid 2 of 6A7	168
Cathode of 6A7	63
Plate of 6B7	59
Screen of 6B7	36
Cathode of 6B7	2.7
Plate of 41	142
Screen of 41	170
Cathode of 41	10

The following is the parts list for 17A receivers, serial numbers 101 to 399 inclusive.

C1	33 mmf
C3	.02 mf
C4, C13	100 mmf
C5.1, C5.2, C5.3, C12.1, C12.2, C12.3, C14	.01 mf
C6, C9	250 mmf
C7.1, C7.2	5 mf
C8	.006 mf
C10	.001 mf
C11	.002 mf

R1, R2, R9, R17	.1 megohm
R3	15,000 ohms
R4	70,000 ohms
R5, R7	.5 megohm
R6	2 megohms
R8, R10, R14	1 megohm
R11	375 ohms
R12	5.6 ohms
R13	5,000 ohms
R15	variable (varistor)
R16	40,000 ohms
R18	200 ohms
R19	1,000 ohms
Volume control	50,000 ohms

The part values for the 17A with serial numbers over 400 are slightly different and their values are shown on the schematic of that receiver, Fig. 111.

Alignment is accomplished by either tuning in a station or by using a calibrated signal generator at 1400 kilocycles (or at the high end of the high frequency band). The trimmer on the top of the gang condenser at the rear is tuned for maximum signal and then the other two trimmers on the gang condenser are tuned for maximum and the process repeated once.

To adjust the range band the dial is set to a known frequency at the high end of the dial, say 380 kilocycles. The trimmer at the left top of the rear can is then trimmed for maximum signal. The trimmers on the other two cans forward of this can are also adjusted for maximum signal and the process repeated once. The unused trimmer at the top right of the back can is the low frequency trimmer and can be adjusted for better alignment of the receiver at the low frequency end of the beacon band by the "rocking for maximum" method. After this is done the receiver must again be readjusted at the high frequency end of the beacon band.

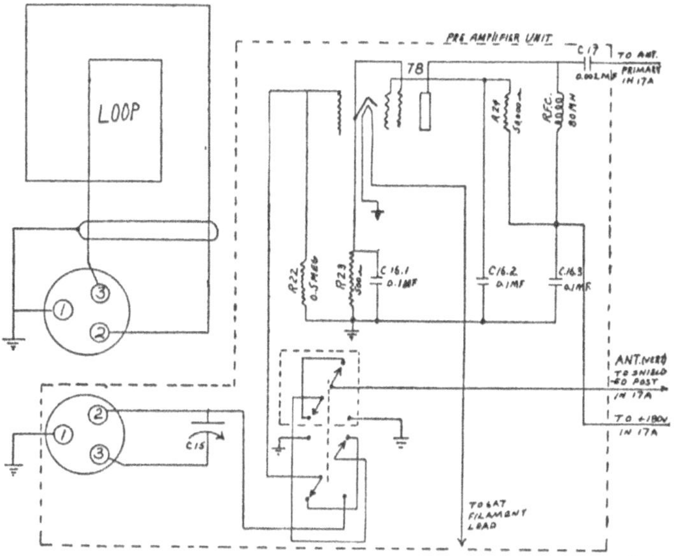

Courtesy TWA

FIG. 113—LOOP AMPLIFIER FOR WESTERN ELECTRIC 17AX RECEIVER

Intermediate frequency alignment is accomplished with the small trimmers on the intermediate frequency (96 kilocycles) transformers.

Fig. 113 shows the airline loop amplifier addition to the 17A receiver which converts the receiver to a 17AX (17AXR if remote controlled in addition). Condenser C15 is another gang condenser exactly like the one in the 17A. It is ganged to the 17A condenser drive with a chain and sprockets. A small compensating condenser at 75 mmf is used to trim the loop at the high frequency end when installed in a plane. Figs. 114, 115, 116, 117 show certain installa tion plans.

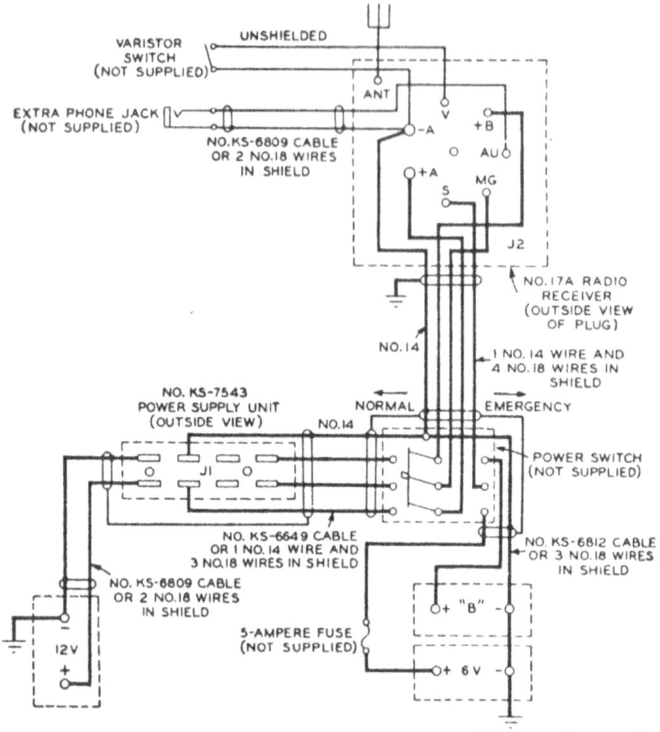

Courtesy Western Electric Co.

FIG. 114—17A RECEIVER WITH NORMAL 12 VOLT SUPPLY AND EMERGENCY 6 VOLT SUPPLY

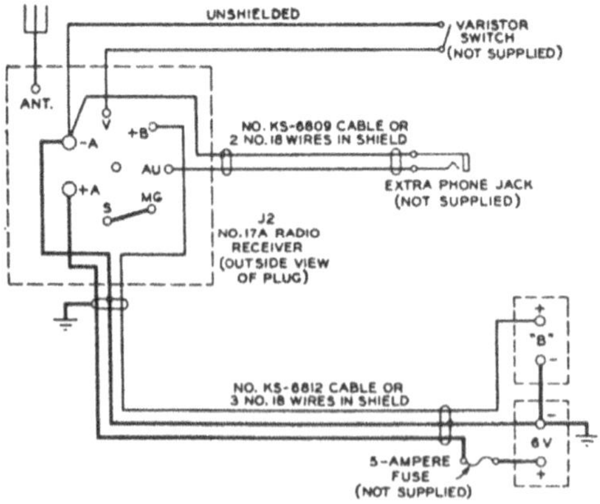

Courtesy Western Electric Co.

FIG. 115—17A RECEIVER FOR 6 VOLT SUPPLY

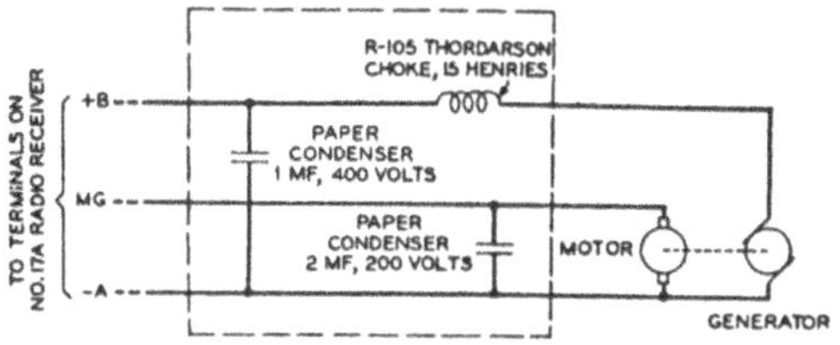

Courtesy Western Electric Co.

FIG. 116—17A RECEIVER, FILTER FOR USE WITH UNFILTERED DYNAMOTOR SUPPLY

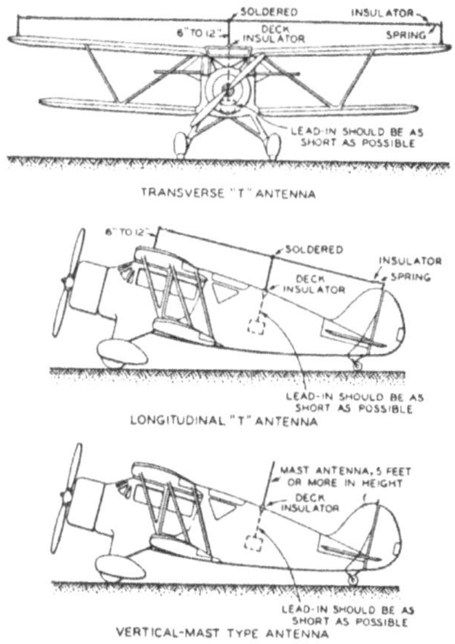

Courtesy Western Electric Co.

FIG. 117—TYPICAL ANTENNAS FOR USE WITH SMALL AIRCRAFT

AVR 7 and AVR 7A RCA Receivers—

	Frequency Range	*Alignment Frequency*
Range band (7 and 7A)	200 to 400 KC	400 and 220 KC
Broadcast (7 only)	550 to 1500 KC	1400 and 600 KC
High frequency (7A only)	2300 to 6800 KC	6150 KC only
Intermediate	175 KC	175 KC

Power unit: Internal vibrator in receiver
Power consumption: AVR 7 1.3 amps., AVR 7A 1.4 amps at 12 volts
Power Output: 600 milliwatts at 12 volts

Tube complement: RF amplifier—6D6 Detector-oscillator—6A7
I.F. Amplifier and detector—6B7
Audio and output—6F7

	HEIGHT	WIDTH	DEPTH	WEIGHT
Receiver	8 3/4"	6 3/4"	9 1/8"	16.7 lbs. including mount
Other equipment except antenna to complete installation				5.3 lbs.

The receiver sensitivity is about 10 microvolts for 50 milliwatts output in the beacon band and somewhat less for the high frequency bands.

It will be found that the variable condenser is exactly in mesh at 197, 535 or 2205 kilocycles for the respective bands and this can be used for setting calibration where required.

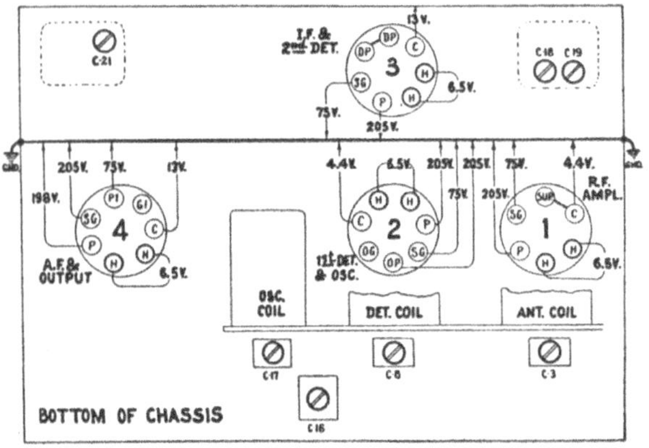

Courtesy RCA Mfg. Co.

FIG. 118—AVR 7 SOCKET VOLTAGES—13.2 VOLTS BATTERY
See text for AVR 7A

Alignment of the intermediate frequency amplifier is made by adjusting trimmers C18, C19 and C21 for maximum output (Fig. 118) preferably using a signal generator, set to 175 kilocycles, attached to the grid of the second tube (6A7).

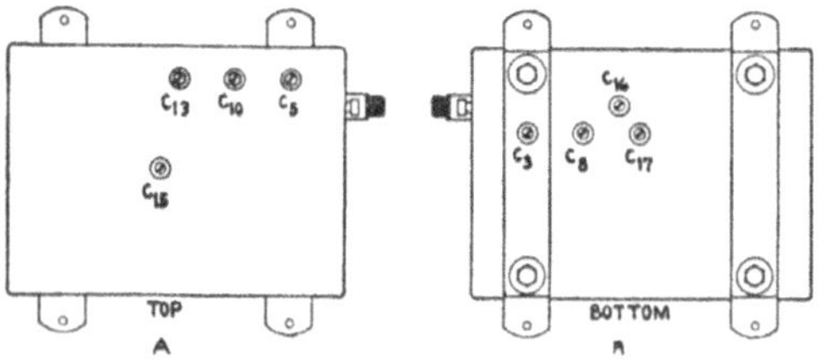

Courtesy RCA Mfg. Co.

FIG. 119—AVR 7 TRIMMER LOCATIONS
Beacon adjustments on bottom and broadcast and communication on bottom.

Trimmers for the radio range band are located on the underside of the chassis while those for the high frequency band are located on the top side of the chassis. The higher frequency band must be aligned first. Fig. 119A shows the location of the high

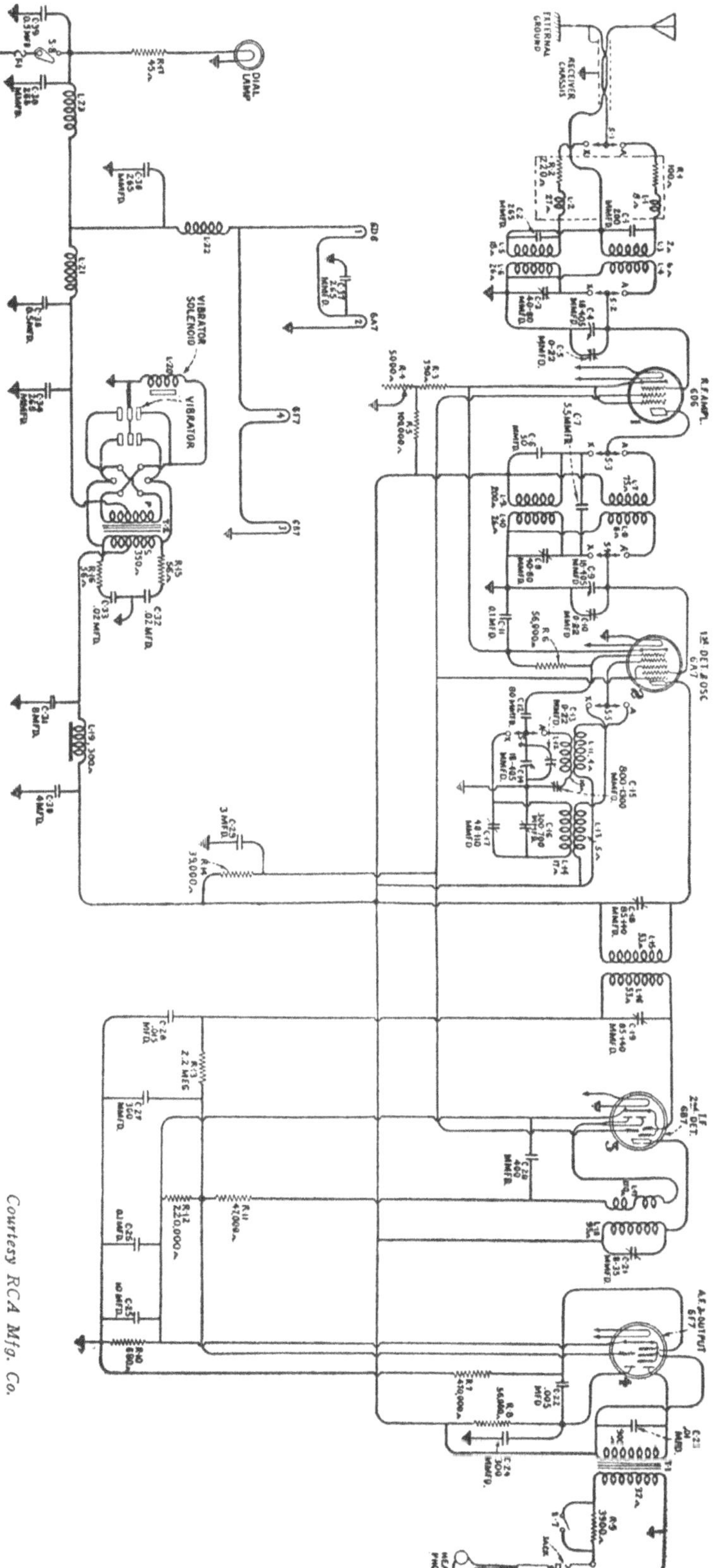

FIG. 120—AVR 7 SCHEMATIC

Courtesy RCA Mfg. Co.

frequency trimmers. In the AVR 7 the signal generator connected to the antenna post is tuned to 1400 kilocycles; in the 7A it is tuned to 6150 kilocycles. The band switch is pulled out for the high frequency band.

Trimmers C5, C10 and C13 are adjusted for maximum signal. The test oscillator is then moved to 600 kilocycles, for the AVR 7, and the 600 kilocycle trimmer is adjusted so that as the tuning is rocked through 600 kilocycles the maximum signal is found. The adjustment at the high end is again repeated. The AVR 7A does not have this adjustment.

The range band is aligned after the high frequency bands. A signal at 400 kilocycles is fed to the receiver after the band change has been pushed in. Then C3, C8 and C17 are adjusted for maximum signal (Fig. 119B). The receiver is now tuned to a signal at 220 kilocycles and with the same rocking adjustment described, C16 is adjusted for maximum output. The adjustment is repeated at the 400 kilocycle point finally.

Fig. 118 shows the voltages found testing the AVR 7 receiver with a high resistance voltmeter, with the volume control at maximum, at 13.2 volts battery and with no signal. The screen voltages of the AVR 7A are 93 instead of 75 volts on sockets 1, 2 and 3 and cathode voltages are 4 volts on sockets 1 and 2 (Figs. 120 and 121).

25A Western Electric Receiver—

	Frequency Range	*Alignment Frequency*
Range band	195 to 415 KC	400 and 200 KC
Broadcast and marine	495 to 1200 KC	1100 and 510 KC
High frequency band	2800 to 6800 KC	6500 KC
Crystal control frequency	Two in the high frequency band	
Intermediate frequency	455 KC	455 KC
Beat oscillator	450 to 460 KC	454 or 456 KC

Power unit: Vibrator supply in power unit
Power consumption: 3 amperes (approx.) at 12 volts.
Power output: 500 milliwatts at 500 ohms (approx.)
Tube complement: Radio frequency amp.—6K7 Converter-osc.—6K8
Intermediate amp. —6K7 Det.-AVC-audio—6Q7
Output audio —6F6 Beat oscillator —6C5

	HEIGHT	WIDTH	DEPTH	WEIGHT
Receiver	6″	7 5/8″	4 7/8″	4.2 lbs.
Power unit	5 3/4″	9 3/4″	6″	9.3 lbs.
Cables (10′)				2.8 lbs.

The 25A receiver is a companion unit for the 25A transmitter. The tuning unit contains the first two tubes and may be separated over a distance up to 10 feet from the intermediate and power unit. The schematic is shown in Fig. 122.

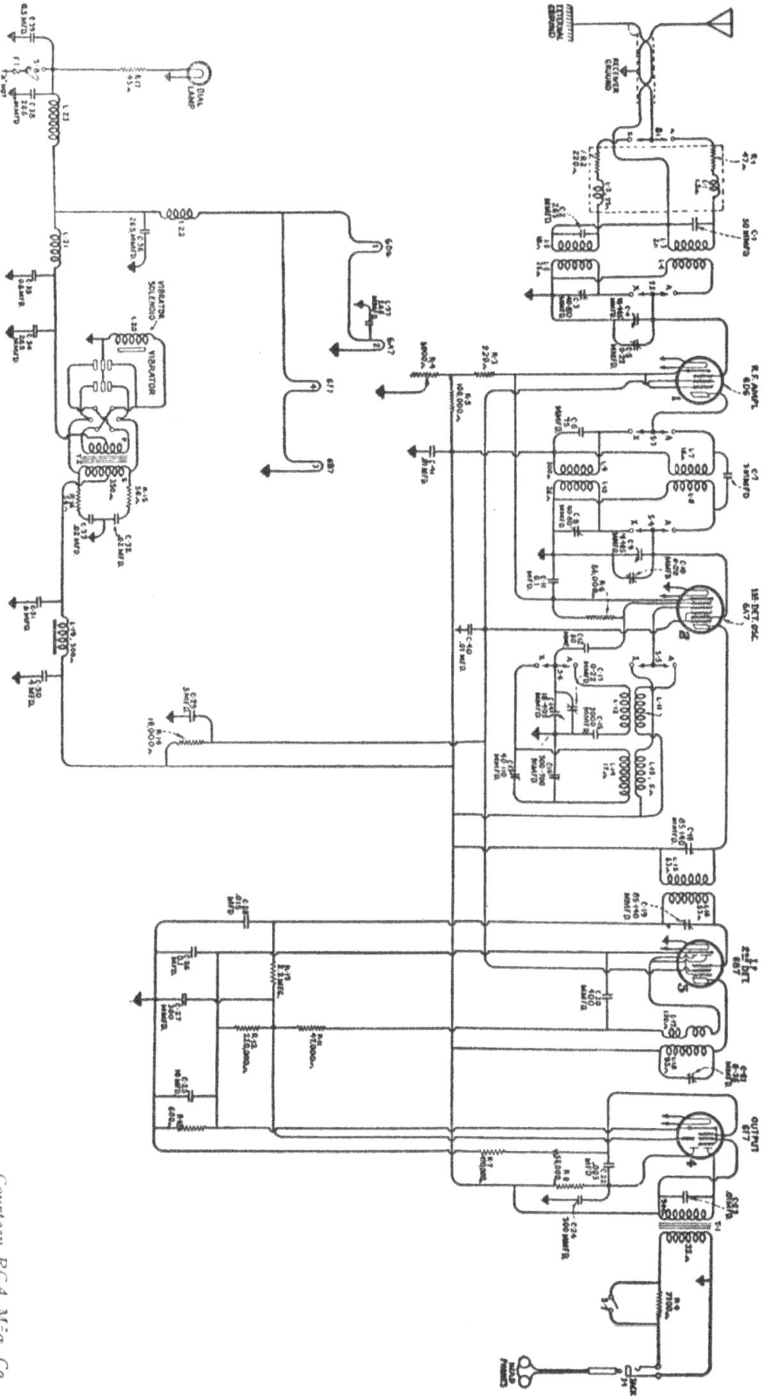

Fig. 121—AVR 7A Schematic

Courtesy RCA Mfg. Co.

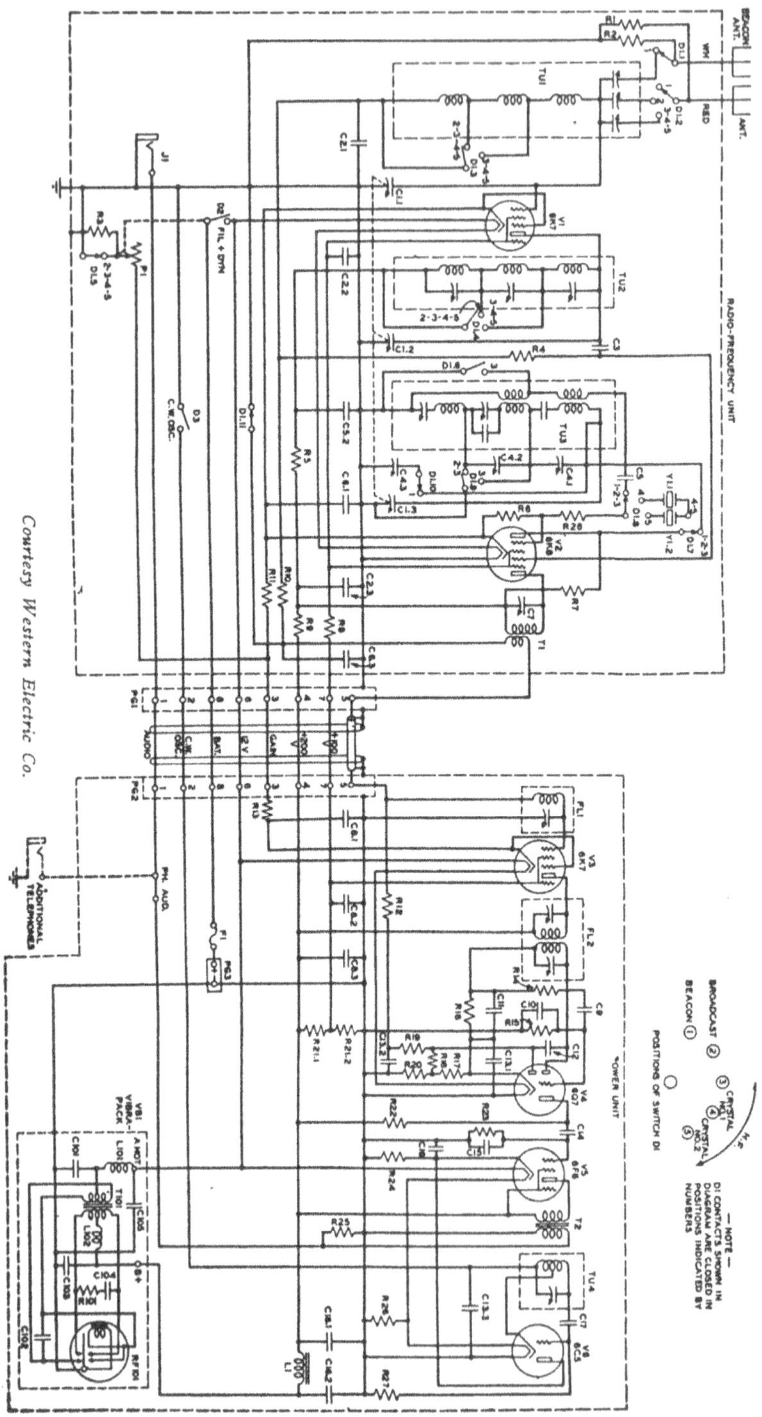

Courtesy Western Electric Co.

Fig. 122—25A Western Electric Receiver Schematic

Adjustment of C7 at the right rear of the radio frequency unit and the condensers in cans FL1 and FL2 will serve to align the intermediate frequency.

The alignment screws for the radio frequency unit are located on that unit and adjustment should be made with the cover can in place. The high frequency is the top adjustment and the low frequency the lower adjustment. Alignment should be made on the high frequency band first and then the next two lower bands in order.

The values, in Fig. 122, for the receiver components are:

R1, R2, R4, R15, R18, R19	1 megohm
R3	300 ohms
R5, R8, R9, R28	1,000 ohms
R6, R7	20,000 ohms
R11	250 ohms
R10, R12, R14, R16, R22	.1 megohm
R13	100 ohms
R17, R20, R25	2,000 ohms
R21	6,500 and 10,000 ohms
R23	.5 megohm
R24	750 ohms
R26	16 ohms
R27	50,000 ohms
C1	ES-613347-1 special
C2, C6	each a triple .1 mf
C3, C5	500 mmf
C4	ES-613420-8 special
C9, C14	.006 mf
C10, C11, C15, C17	250 mmf
C12	100 mmf
C16	10 mf
C18	two 8 mf

R-3-AB Lear Receiver—

	Frequency Range	*Alignment Frequency*
Radio range 1	200 to 400 KC	380 KC
Broadcast 2	670 to 1,750 KC	1,650 KC
Communication 3	2900 to 6,500 KC	6,000 KC

Or with the same radio range band and either of the following pairs:

Communication 2	1200 to 2,800 KC	2,600 KC
Communication 3	2800 to 6,500 KC	6,000 KC
Communication 2	2800 to 6,500 KC	6,000 KC
Communication 3	6500 to 14,500 KC	13,500 KC
Crystal control	On any band when switched in (one freq.)	
Intermediate	455 KC	455 (see notes)

Power unit: Internal dynamotor
Power consumption: 2.9 amperes at 12 volts
Power output: 500 to 1,000 milliwatts at 300 ohms

Tube complement: RF—6K7
Det. Osc.—6K8
1st If—6K7
2nd IF and AVC—6B8
Det. and Audio 6C8G
Output and beat oscillator—6P7G

	HEIGHT	WIDTH	DEPTH	WEIGHT
Receiver	4 11/16″	6 15/16″	5 3/8″	3.3 lbs.
Power and IF Unit (G-3-AB)	6 3/4″	6 5/8″	11 15/16″	9.0 lbs.
Accessories				2.6 lbs

The unit may also be used in conjunction with the G-30-AB power supply unit, Fig. 123 and the T-30-AB transmitter, Fig. 67. The schematic of the tuner end of the R-3-AB receiver is shown in Fig. 124.

The receiver is divided into two units. The tuner contains the radio frequency and detector-oscillator unit which feeds the intermediate and audio amplifier in the G-30-AB or GT-30-AB units.

Alignment is made at the high frequency end of each of the bands at the approximate frequencies shown in the table of receiver characteristics. The trimmers are marked in the respective units.

Care must be taken in the alignment of the intermediate transformer T-5. The primary and secondary circuits should be aligned for maximum output feeding the generator to the second intermediate tube (6B8) grid. After this is done the generator should be tuned to one side or the other and the two peaks found should be adjusted by manipulation of the trimmers so that they are centered equally by about 2.5 kilocycles either side of 455 KC and so that the two peaks are equal in rise of signal. After this process, the remainder of the intermediate amplifier may be aligned in the normal manner, feeding the signal generator to the desired grids until the complete intermediate amplifier has been aligned.

The parts in the R-3-AB receiver and G-30-AB power unit have the following values.

R10	10,000 ohms
R11, R15, R32	40,000 ohms
R12	100 ohms
R13	500 ohms
R14, R18, R21, R22, R28, R17	.25 megohm
R31, R48	5,000 ohms
R19, R25	1,000 ohms
R23, R26, R27, R42, R47, R45, R49	100,000 ohms
R33, R34, R41, R44	2,500 ohms
R24, R29, R30, R40, R43	1 megohm
R36	1.7 ohms
R46	20,000 ohms

C21, C20, C23	8 mf
C26, C34	2 mf
C22, C29, C30, C31	.1 mf
C36	.5 mf
C24, C25, C33, C32, C35, C47, C48, C49, C58, C59, C62	.01 mf
C27, C28, C60, C61, C76	250 mmf
C38	100 mmf
C73	.5 mf
C40, C44, C50, C72	10 mmf
C45, C56	.001 mf
C51	5 mmf
C64	50 mmf
C65	190 mmf
C66	45 mmf
C70	1,250 mmf
C71	2,500 mmf
C57, C73, C74	15 mmf
C75	20 mmf

AVR 7D, 7E, 7F, 7G RCA Receivers—

	Frequency Range	*Alignment Frequencies*
Range band (x)	195 to 420 KC	400 and 200 KC
Broadcast (A)	495 to 1400 KC	1200 and 550 KC
High frequency (B)	2300 to 6700 KC	6000 and 2700 KC
Intermediate	460 KC	460 KC

Power unit: Vibrator in receiver
Power consumption: 2.1 amperes at 12 volts
Power output: 700 milliwatts at 600 ohms
Tube complement: R. F. amplifier 6K7
Converter 6A8
I. F. amplifier 6K7
Detector and audio 6R7 (new models 6Q7)
Output and CW oscillator 6F7

	HEIGHT	WIDTH	DEPTH	WEIGHT
Receiver	9 1/8″	8 7/16″	15 5/16″	See below
Type 7D	9 1/8″	8 7/16″	15 5/16″	30.9
Type 7E	9 1/8″	8 7/16″	15 5/16″	31.9
Type 7F	9 1/8″	8 7/16″	15 5/16″	31.1
Type 7G	9 1/8″	8 7/16″	15 5/16″	32.1
Control unit	6 5/8″	5 3/4″	1 7/8″	4.0

The AVR 7D is the basic receiver to which is added a crystal control of the high frequency oscillator in the 7E or a continuous wave beat frequency oscillator in the 7F and both the crystal control and beat oscillator in the 7G. The schematic wiring diagram shows the 7G receiver, Fig. 125, with appropriate markings for the parts omitted in other receivers.

A change has been made in these receivers which is shown in the schematic between the point E on the last intermediate transformer and the grid of the detector-first audio tube grid in the addi-

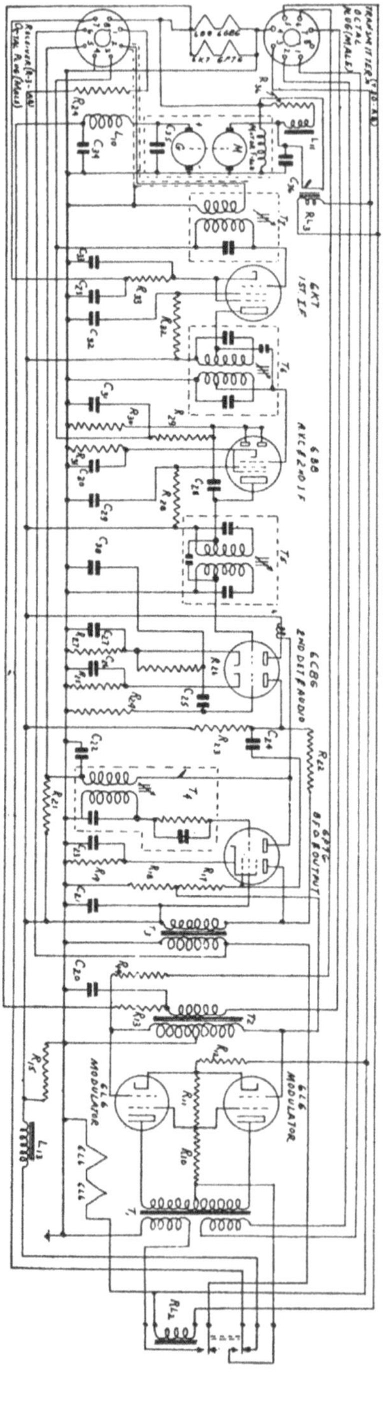

Courtesy Western Electric Co.

FIG. 123—LEAR G-30-AB TRANSMITTER AND RECEIVER POWER SUPPLY INCLUDING INTERMEDIATE AND AUDIO AMPLIFIERS

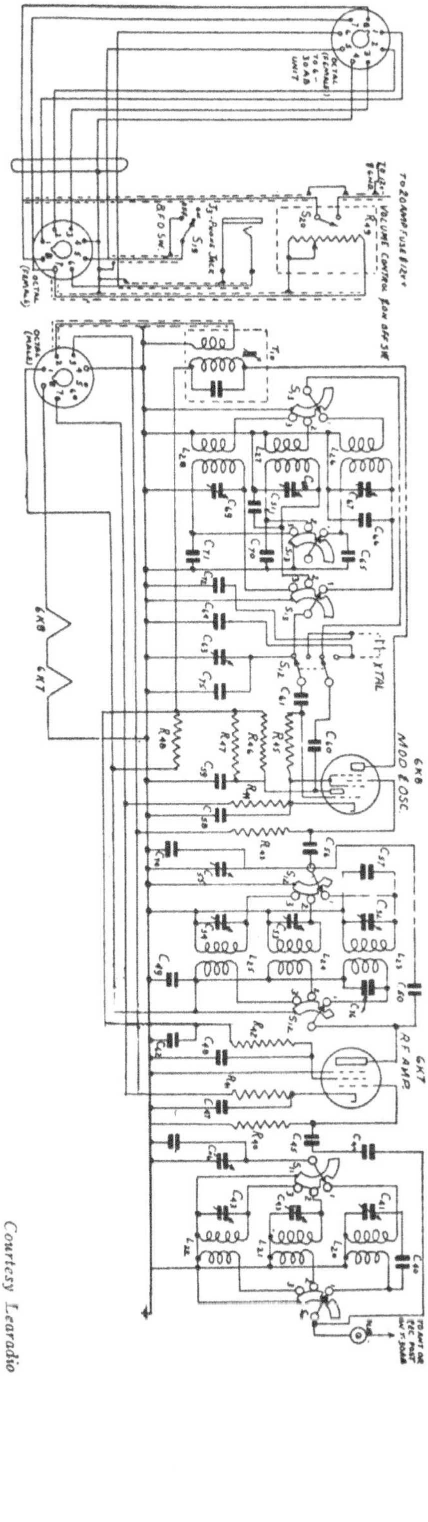

FIG. 124—LEAR G-3-AB RECEIVER SCHEMATIC

Courtesy Learadio

tion of a coupling condenser and resistor from grid to ground. The 6R7 tube is then changed to a 6Q7.

Fig. 126 shows the location of all trimmers that are required. A 150 mmf condenser should be placed in series with the signal generator lead in the following. Output is adjusted for maximum in each case and alignment should take place in the order given. A

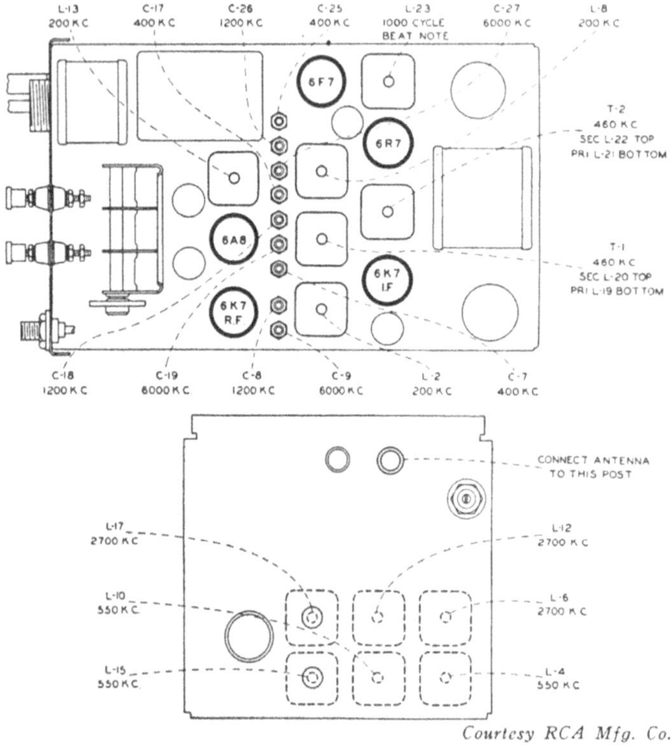

Courtesy RCA Mfg. Co.

FIG. 126—ARRANGEMENT OF PART AND TRIMMERS—AVR 7 D TO G RECEIVERS

radial line on the dial at the low frequency end shows the pointer position at which the condenser plates should be in full mesh.

The intermediate frequency is adjusted first with a generator at 460 kilocycles and the dial at 2300 kilocycles in the "B" band (shift control "out"). The generator is connected through the capacitor to the grid caps without removing the grid wire in each case. The signal generator is attached to the grid of the 6K7 intermediate frequency tube and L22 and L21 are aligned. The generator is then attached to the 6A8 converter grid cap and L20 and L19 are aligned. If the receiver is equipped with a continuous wave oscillator it is turned on and adjustment of L23 is made for a beat note of about 1000 cycles by ear (which is nearly the same frequency that is used on radio ranges). After this adjustment the oscillator is turned off again.

The radio frequency circuits are adjusted by connecting the generator through the condenser to the antenna post. In the first and fourth adjustments at 6000 kilocycles in the table following, the oscillator trimmer is adjusted to the peak of least capacity if two peaks can be found. This peak is the one with the trimmer pulled out farthest. To check this alignment the receiver (leaving generator set) is then tuned to 5080 kilocycles and the generator should be heard again as 6000 kilocycles is now at the image. If the image is not heard the oscillator is tuned to the low frequency side of the signal and a point on the trimmer should be found which is correct. The following table is used for further alignments.

GENERATOR FREQUENCY	BAND SELECTOR	CIRCUIT	SYMBOL
6000	"B" out	"B" Osc.	C–27
6000	" "	"B" Det. and R.F.	C–19, C–9
2700	" "	"B" Osc., Det., R.F.	L–17, L–12, L–6
6000	" "	"B" Osc.	C–27
6000	" "	"B" Det., R.F.	C–19, C–9
1200	"A" center	"A" Osc., Det., R.F.	C–26, C–18, C–8
550	" "	" " " "	L–15, L–10, L–4
1200	" "	" " " "	C–26, C–18, C–8
400	"X" in	"X" Osc., Det., R.F.	C–25, C–17, C–7
200	" "	" " " "	L–13, L–8, L–2
400	" "	" " " "	C–25, C–17, C–7

Voltages to be found in normal operation as measured to ground are taken with the volume control at minimum rather than the usual maximum volume setting, with the receiver tuned to 550 kilocycles. The CW switch is in "phone" position except when measuring the 6F7 triode voltages.

CIRCUIT VOLTAGE	6K7 R.F.	6A8 DET.-OSC.	6K7 I.F.	6R7 DET.-AUDIO	6F7 OUTPUT
Heater	0, 6	0, 6	0, 6	6, 12	6, 12
Cathode	46	46	46	0	14
Grid 1	0	11	0		
Grid 2	157	190	157		
Grid 3	46	0	46		
Grid 4	.....	157			
Grid pentode #2, volts	.....	.....	.....	....	250
Plate	250	250	250		
Plate triode	.....	.....	.....	43	95
Plate pentode	.....	.....	.....	....	240
Other grids and diodes	.....	.....	.....	0	0

Addition of the crystal oscillator is shown in Fig. 127.

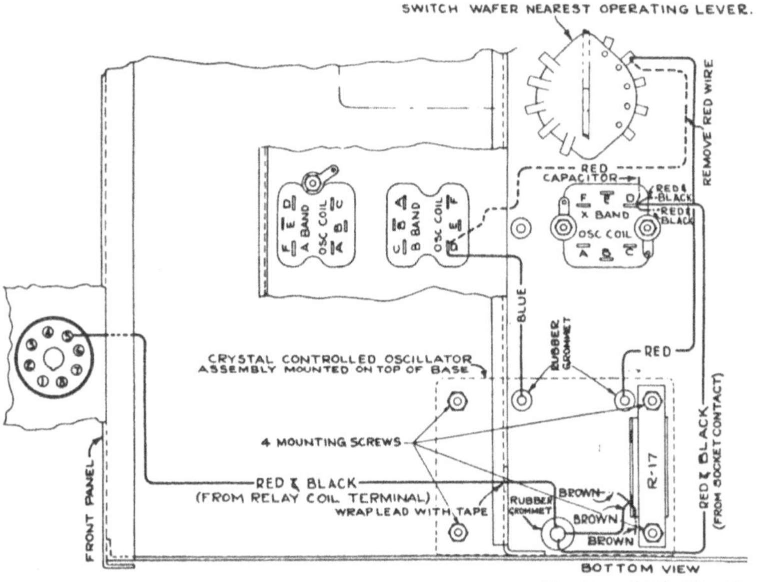

Courtesy RCA Mfg. Co.

FIG. 127—AVR 7 TYPE RECEIVER MODIFICATION

20A, 20B, 20AA and 20BA Western Electric Receivers—

	Frequency Range	*Alignment Frequency*
Range band (all types)	188 to 420 KC	380 and 200 KC
Broadcast (20A and 20B)	550 to 1,500 KC	1350 and 600 KC
Broadcast-Marine (20AA, BA)	485 to 1,200 KC	1050 and 500 KC
Medium high (all types)	1500 to 4,000 KC	3800 and 1600 KC
High frequency	4000 to 10,000 KC	9500 and 3500 KC
Intermediate	96 KC	96 KC

Trimming is seldom needed on the low frequency end of each band.

Power unit: Separate dynamotor, 55 milliamperes at 200 volts
Power consumption: 0.9 amperes at 12 volts
Power output: 700 milliwatts at 500 ohms
Tube complement: RF and IF amplifier 6F7
Converter 6A7
2nd IF and 2nd Det. and audio 6B7
Output 41

	HEIGHT	WIDTH	DEPTH	WEIGHT
Local control receiver	8 3/4"	13"	8 1/4"	15.25 lbs. with shocks
CW oscillator attachment	....	...	.....	1.25 lbs.
KS 7543 dynamotor	6 7/8"	8"	4 7/16"	10.25 lbs.

The 20 type is a four band superheterodyne receiver. The 20A and 20AA receivers are direct controlled on the face of the receiver while the 20 B and the 20BA have complete remote control. Crystal control of two high frequency channels of the high frequency bands may be added to the receiver. Wiring diagrams of the 20 types are shown in Fig. 128 and Fig. 129.

Fig. 130 shows the interconnections while Fig. 131 shows the crystal conversion diagram. Fig. 132 shows the 27A control unit schematic.

Alignment is accomplished by supplying signals from a generator to the receiver near the highest frequency end of each band and adjusting the trimmers for that band. Fig. 133 shows the arrangement of antenna trimmers which are adjusted in the plane, unless a suitable dummy can be used. The remaining trimmers to be adjusted for each band are:

BAND	TRIMMERS
200 to 400 kilocycles	C6, C10
Low frequency end only of above	C14
550 to 1500 kilocycles	C8, C12
1500 to 4000 kilocycles	C7, C11
4000 to 10,000 kilocycles	C9, C13

Individual inductance trimmers are provided on each coil which serve to adjust the inductance of each coil for best tracking near the low frequency end of each band but normally their adjustment is not required. If adjustment must be made the complete high frequency adjustment is repeated for the band after such inductance adjustment.

Intermediate frequency trimmers are provided on the transformers which are adjusted at 96 kilocycles.

Voltages to be found at the tube sockets are shown in the following table and are obtained with varistor off and the automatic volume control switch on "MVC" (manual volume control). The antenna terminal should be grounded during voltage measurements.

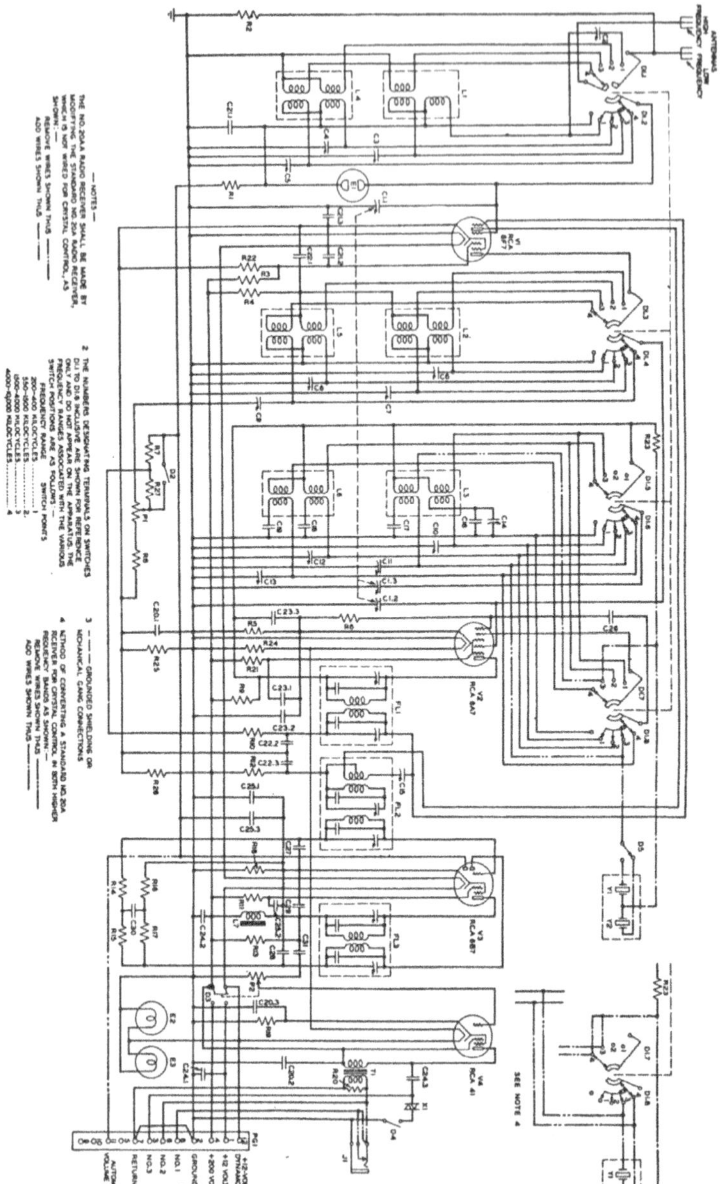

Courtesy Western Electric Co.

FIG. 128—SCHEMATIC OF 20A AND 20AA RECEIVERS

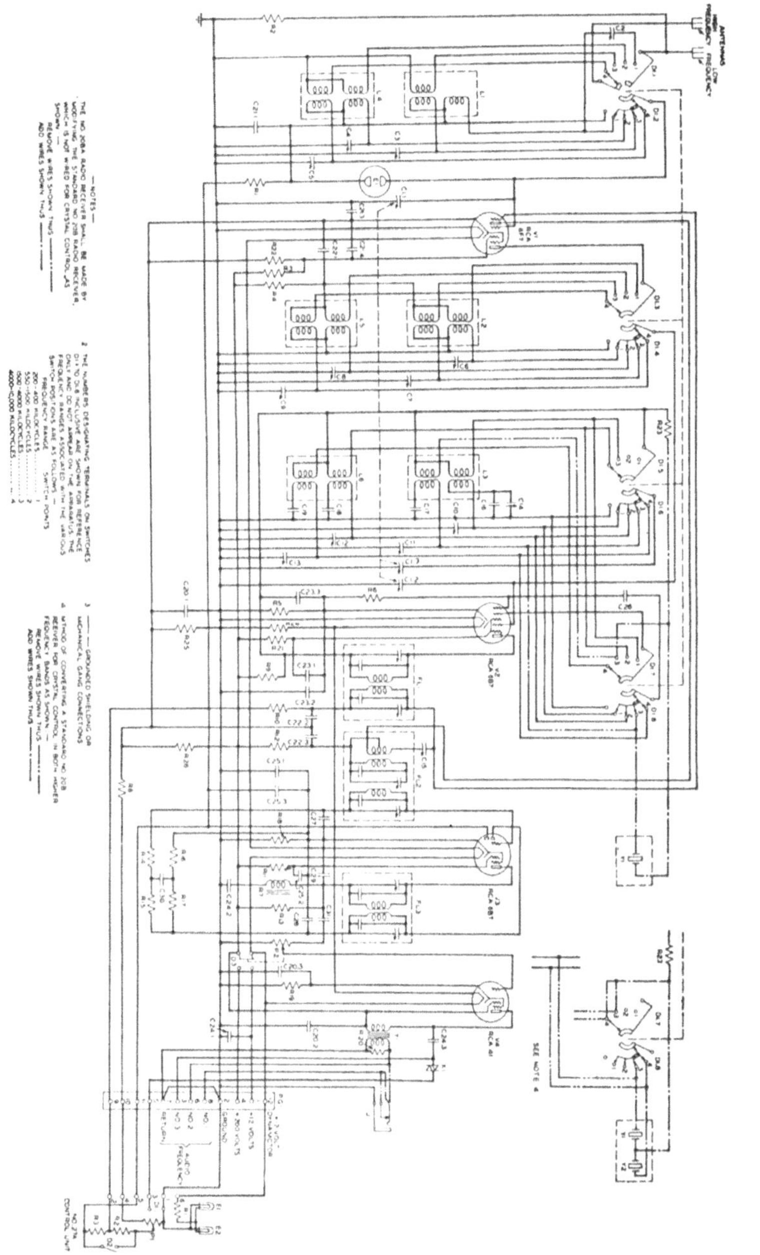

Courtesy Western Electric Co.

FIG. 129—SCHEMATIC OF 20B AND 20BA RECEIVERS WITH 27A CONTROL UNIT

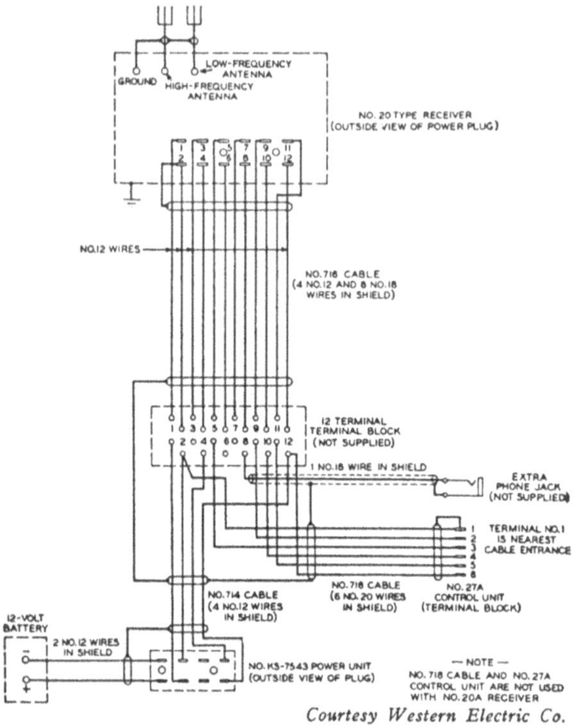

Courtesy Western Electric Co.

FIG. 130—EXTERNAL CONNECTIONS—20 TYPE RECEIVERS

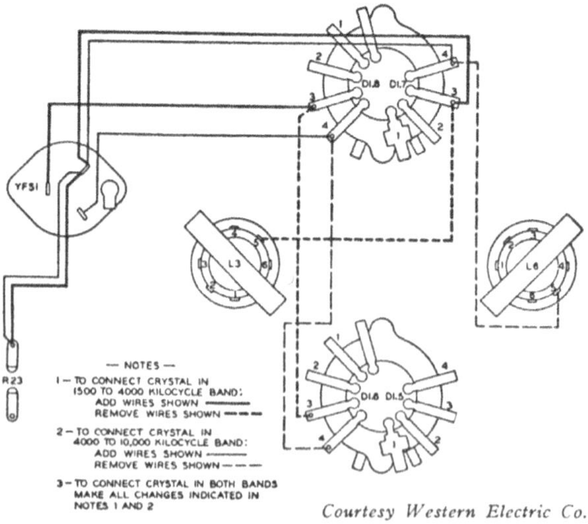

Courtesy Western Electric Co.

FIG. 131—CONVERSION DIAGRAM FOR 20A OR 20B RECEIVERS—CRYSTAL CONTROL

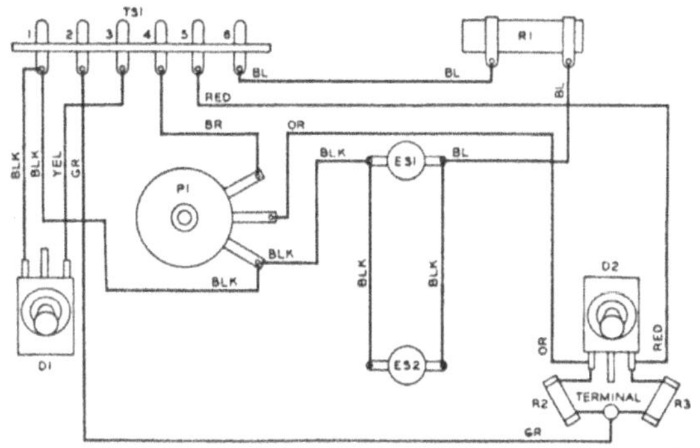

Courtesy Western Electric Co

FIG. 132—27A CONTROL UNIT SCHEMATIC

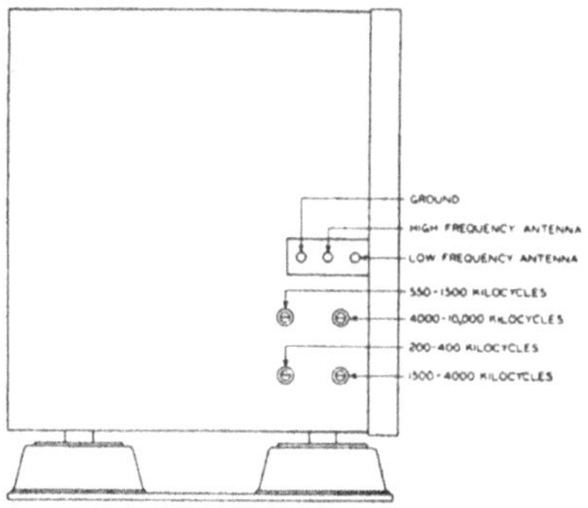

Courtesy Western Electric Co.

FIG. 133—20 TYPE RECEIVER ALIGNMENT AND ANTENNA LOCATIONS

TUBE	POINT	VOLTAGE
Power plug	Terminal 1 and 12	11.75–12.25
Power plug	Terminal 4	190–210
V1	Cathode	75–85
V1	Pentode plate	185–205
V1	Screen grid	150–170
V1	Triode grid	35–45
V1	Triode plate	150–170
V2	Cathode	6.5–7.5
V2	Oscillator plate	185–205
V2	Modulator plate	185–205
V2	Screen grid	70–90
V3	Cathode	70–80
V3	Diode 1	70–80
V3	Diode 2	10–15
V3	Plate	160–180
V3	Screen grid	135–155
V4	Cathode	12–15
V4	Plate	165–185
V4	Screen grid	190–210

All voltages are measured with a 1000 ohm per volt meter on the 300 volt scale.

The following parts are identified with the schematic.

C16	170 mmf
C17	1050 mmf
C18	500 mmf
C19	3150 mmf
C20.1	8 mf
C20.2	8 mf
C20.3	5 mf
C21, C22, C23, C24, C25, Three	.1 mf
C26	.001 mf
C27	150 mmf
C28, C29	250 mmf
C30, C31	.006 mf
E1	Neon overload lamp

Pilot lamps shown are T3, 14 volt Mazda

P1	50,000 ohms, R.F. gain
P2	500,000 ohms, audio gain
R1, R10, R13, R14, R16	100,000 ohms
R2	2 megohms
R3	6,000 ohms
R4, R5, R9	1,000 ohms
R6, R21	50,000 ohms
R7, R27	250,000 ohms
R8	2,400 ohms
R11	30,000 ohms
R12, R26	15,000 ohms
R15, R17	1 megohm
R18, R23	10,000 ohms
R19	750 ohms
R20, R25	4,000 ohms
R22	20,000 ohms
R24	62 ohms

In control unit:

R1	100 ohms max.
R2, R3	250,000 ohms

12D and D-98647 Western Electric Receivers—

	Frequency Range	*Alignment Frequency*
Low fixed frequency	2750 to 4300 KC	At crystal frequency
High fixed frequency	4300 to 6500 KC	At crystal frequency
CW oscillator	383 to 387 KC	385 KC
Intermediate	385 KC	385 KC

Power unit: Separate dynamotor, 40 milliamperes at 200 volts
Power consumption: 3.2 amperes at 12 volts
Power output: 500 milliwatts at 500 ohms (approx.)
Tube complement: RF amplifier 283A
Converter 283A
Crystal Osc. 283A
I. F. amplifiers (2) 283A
Detector 303A
Output 285A
CW Osc. 283A

	HEIGHT	WIDTH	DEPTH	WEIGHT
Receiver	8 5/8″	10″	14 1/2″	17.5 lbs. with mtg.

The 12D receiver is a two frequency crystal controlled receiver for the high frequency service. The D-98647 receiver is similar but has no CW oscillator. The receiver with tubes and two crystals weighs 15 pounds 9 ounces without mounting. The receiver may be adjusted to any one frequency in the band from 2750 to 4300 kilocycles and any other frequency in the band from 4300 to 6500 kilocycles. A crystal controlled beating oscillator for continuous wave reception can be provided. The schematic is shown in Fig. 134 and a modification by one airline in Fig. 135.

A test bench setup for testing the receiver is shown in Fig. 136. A beat oscillator control potentiometer of 2000 ohms resistance may be used to control oscillator beat note and is shown attached to terminal 10. The main heterodyne oscillator in the receiver may be non-crystal controlled and controlled remotely by means of a trimmer attached to terminals 11 and 12. Normally, this type of oscillator will not be found in airplane installations and the crystal oscillator is used instead.

The following change must be made if the receiver is to be fed by a low (40 to 60 ohm) impedance transmission line. The slate lead replaces the red lead and the black lead replaces the yellow lead on switch D1. (The slate and black leads are taped up near the switch in new receivers.) The red and yellow leads are then grounded on a nearby lug.

The first step in tuning the receiver is to place the links on the antenna transformers at the correct settings. For low impedance lines the following connections are made: H to 2 and L to 2. The H and L refer to the high and low frequency bands. For other antennas which resonate at higher frequency than the frequency used connections are:

L to 1 for 2750 to 3300 kilocycles
L to 2 for 3300 to 4300 kilocycles
H to 1 for 4300 to 5200 kilocycles
H to 2 for 5200 to 6500 kilocycles

The antenna condensers may be reached through the front of the receiver and may require readjustment with the antenna to be used. The oscillator and radio frequency trimmer are adjusted for maximum signal in the usual way. The oscillator condenser is set at a point well removed from the point where oscillation suddenly starts or stops.

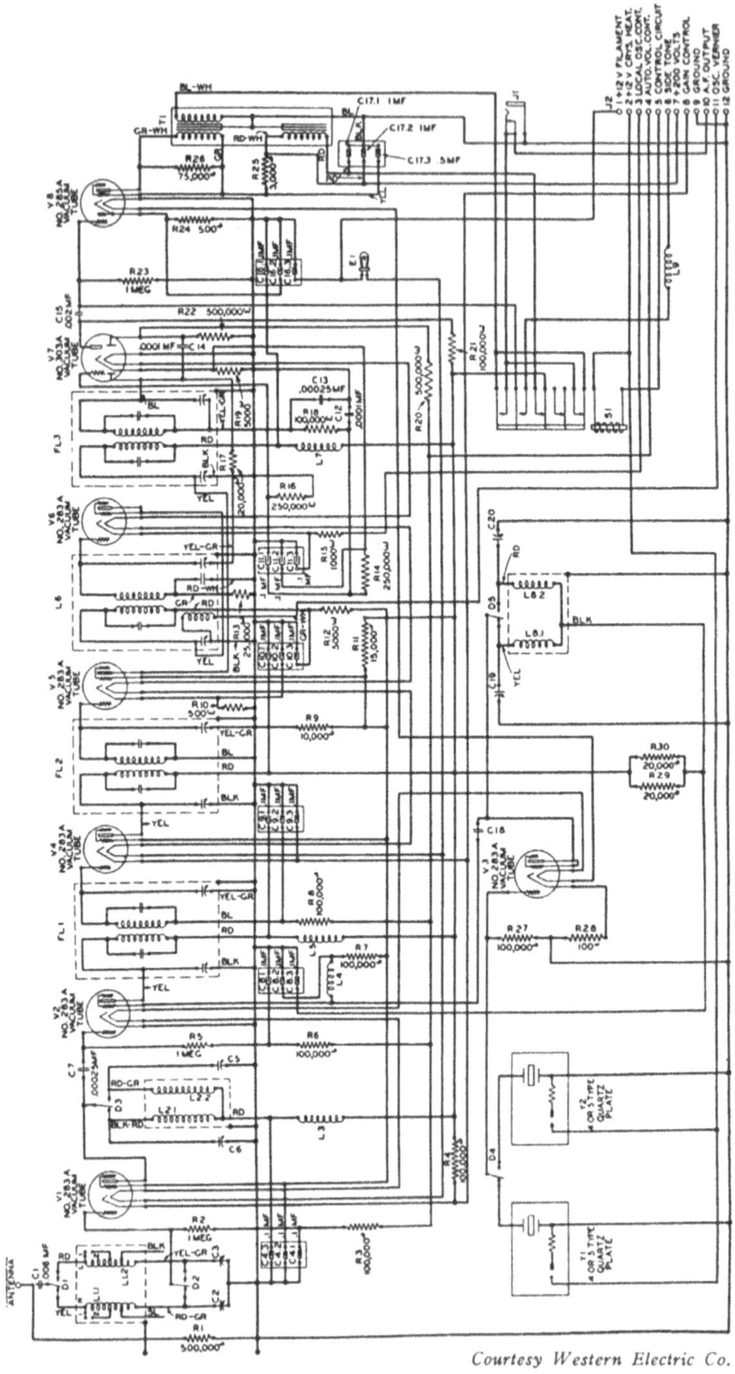

Courtesy Western Electric Co.

FIG. 134—SCHEMATIC 12D RECEIVER

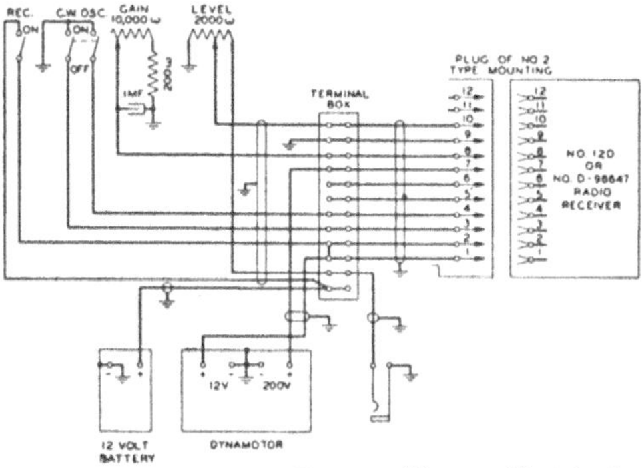

Courtesy Western Electric Co.

FIG. 136—TEST BENCH SETUP FOR 12 TYPE RECEIVERS

The following voltages are normal in the receiver as measured to the chassis with the antenna post grounded and the gain control at maximum.

Radio frequency plate V1	185	Volts
Radio frequency screen V1	60	"
Radio frequency cathode V1	0.8	"
Converter plate V2	185	"
Converter screen V2	20	"
Oscillator plate V3	95	"
First intermediate plate V4	185	"
First intermediate screen, V4	60	"
First intermediate cathode V4	0.8	"
Second intermediate plate V5	185	"
Second intermediate screen V5	60	"
Second intermediate cathode V5	1.0	"
CW oscillator plate and screen V6	50	"
Detector cathode V7	3.5	"
First audio plate V7	80	"
Output audio plate V8	140	"
Output audio screen V8	150	"
Output audio cathode V8	7.2	"
Plate supply voltage	200	"

In some receivers it will be found that the tube types have been changed from those shown in the schematics and the voltages may differ from those in the table above.

The parts used in the receiver have the following values:

C1	.006 mf	E1	Western Electric Ballast 114A
C2, C3, C5, C6, C19, C20	100 mmf (var.)	R1, R20, R22	.5 megohm
C4.1, C4.2, C4.3, C8.1, C8.2, C8.3, C9.1, C9.2, C10.1, C10.2, C10.3, C11.1, C11.2, C11.3, C16.1, C16.2, C16.3	.1 mf	R2, R5, R23	1 megohm
C7, C13	250 mmf	R3, R4, R6, R7, R8, R18, R21	0.1 megohm
C12, C14	100 mmf	R9	10,000 ohms
C15	.002 mmf	R10, R24	500 ohms
C17.1, C17.2	1 mf	R11	15,000 ohms
C173	.5 mf	R12, R19	5,000 ohms
C18	1.5 mf	R13	25,000 ohms
		R14, R16	.25 megohm
		R15	1,000 ohms
		R17, R29, R30	20,000 ohms
		R25	3,000 ohms
		R26	75,000 ohms
		R28	100 ohms

14B Western Electric Receiver—

	Frequency Range	*Alignment Frequency*
Range band	200 to 400 KC	380 and 200 KC
Fixed band	about 50 KC	278 KC (airport)
Intermediate	96 KC	96 KC

Power unit: Separate dynamotor, 40 milliamperes at 200 volts
Power consumption: 0.7 ampere at 12 volts
Power output: approximately 500 milliwatts at 500 ohms
Tube complement: R. F. amplifier 78
Converter 6A7
I. F. amplifier (2) 78
Detector-audio 85
Output 41

	HEIGHT	WIDTH	DEPTH	WEIGHT
Receiver	8 15/16″	10 5/16″	16″	17 lbs. with shocks

The 14B radio receiver is a companion receiver to the 12D and covers the radio range band from 200 to 400 kilocycles continuously. It weighs 17 pounds with tubes but without the plug and mounting base. An electrically operated switch tunes the receiver instantly to 278 kilocycles for airport traffic control use.

The schematic is shown in Fig. 137. The intermediate amplifier operates at 96 kilocycles. An automatic volume control is incorporated which will only function normally on high percentage modulation stations. With the new simultaneous radio ranges it is well not to operate the receiver at high outputs as fades and increases

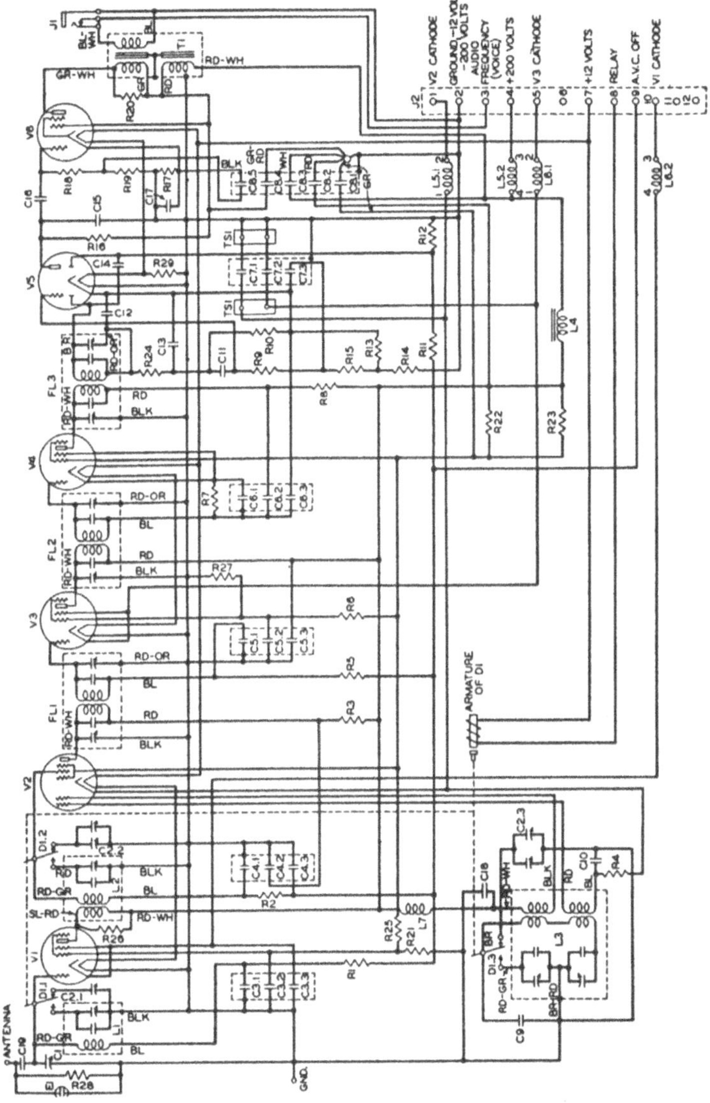

Courtesy Western Electric Co.

FIG. 137—SCHEMATIC OF 14B RADIO RECEIVER

of signal will be less apparent due to the AVC. In normal use the sensitivity control is adjusted for reasonable signal and decreased as the range station is approached.

After alignment of the receiver it will show a slight double peak in the output when tuned through a signal near the high frequency end and this is entirely normal. In fact, tuning should be adjusted so that both peaks, which are very close together, have about the same maximums. The antenna trimmer at the front of the receiver should be adjusted to the antenna used or with an identical dummy antenna. Alignment of the radio frequency circuits as well as those

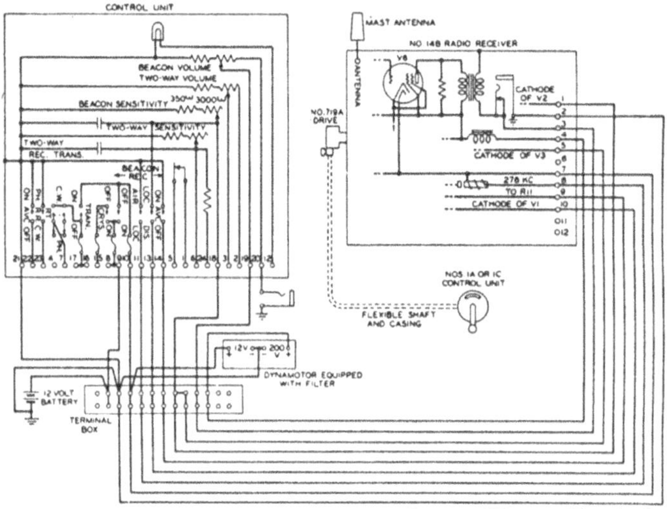

Courtesy Western Electric Co.

FIG. 138—RECEIVER SYSTEM WIRING DIAGRAM—14B

associated with the 278 kilocycle quick shift are readily made with the case removed from the receiver.

A schematic of the attachment of the 14B to a control unit type 8C is shown in Fig. 138. It will be found that the control unit has been simplified in transport installations and that the knobs for the audio gain control (range and two-way volume) have been removed. The reason for removal of these is that overload may accidentally be encountered with the 14B if the sensitivity is left at maximum and the pilot uses the audio gain control only. This will cause erroneous course signals. The two-way 12D receiver audio volume knob is often omitted as the gain control is used primarily in conjunction with the automatic volume control which is usually on at all times.

Approximate voltages on the 14B receiver measured at maximum sensitivity and with the antenna grounded are (some differences will exist where other tubes have been substituted):

TUBE	CATHODE	SCREEN	PLATE
V1	2.8	47	170
V2	3.2	85	150
V2 (osc.)	....	...	170
V3	2.8	50	170
V4	5.8	85	150
V5	16	...	95
V6	17.5	190	180

Supply voltage—200 volts.

The following are the values of the parts in the receiver (colors refer to leads from condenser cans):

C1, C3, C4, C5, C6, C7	(3) .1 mf each	
C8 (can)	(2) 2 mf "	(red-red & green)
	(2) 1 mf "	(green-white)
	(1) .5 mf "	(black-black)
C9	15 mmf	
C10, C15	.002 mf	
C11, C16	.03 mf	
C12, C13	200 mmf	
C14, C19	100 mmf	
C17	2 mf	
C18	.1 mf	

V1, V3, V4	RCA 78 tube
V2	RCA 6A7 tube
V5	RCA 85 tube
V6	RCA 41 tube
E1 and R28	Overload lamp
R1, R2, R5, R15, R20, R26, R19	0.1 megohm
R3, R7, R8, R13, R14	5,000 ohms
R4, R16	50,000 ohms
R6, R25	10,000 ohms
R9, R11, R10, R24	.5 megohm
R12, R28	1 megohm
R17	1,400 ohms
R18	.25 megohm
R21, R27	15,000 ohms
R22, R23	16,000 ohms
R29	63 ohms

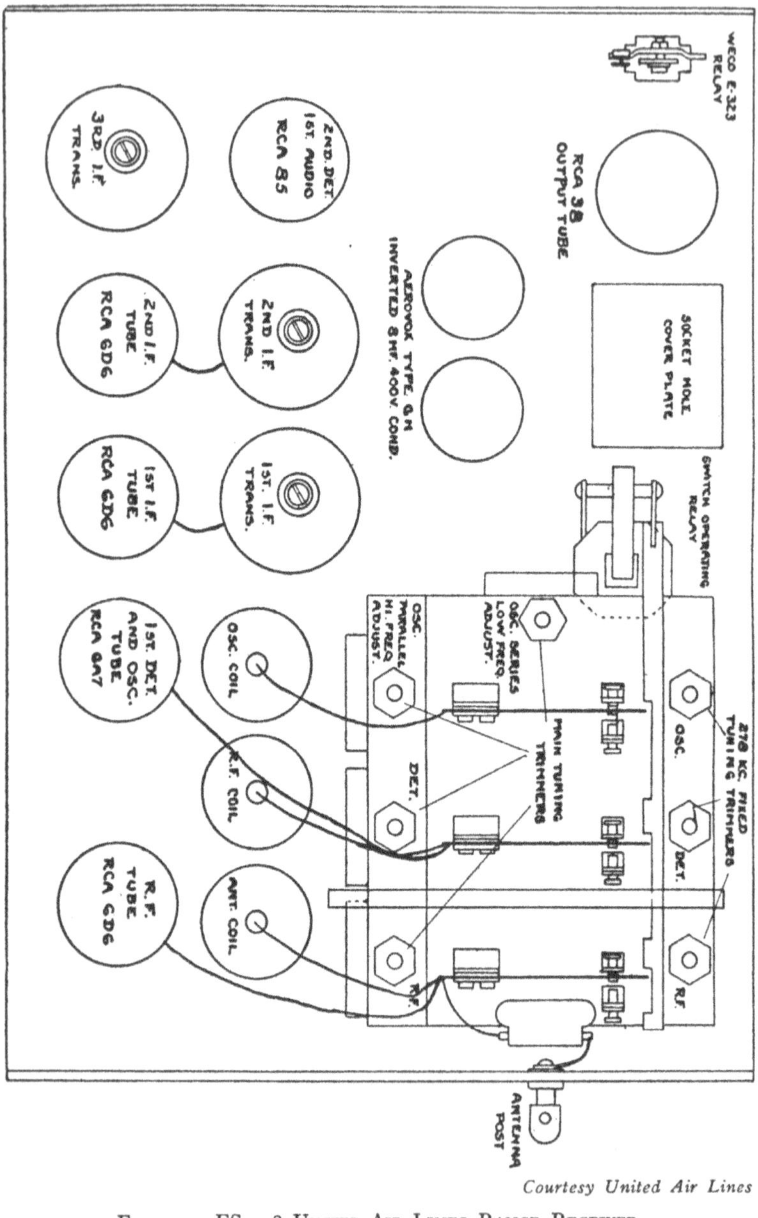

Courtesy United Air Lines

FIG. 139—ES 198 UNITED AIR LINES RANGE RECEIVER

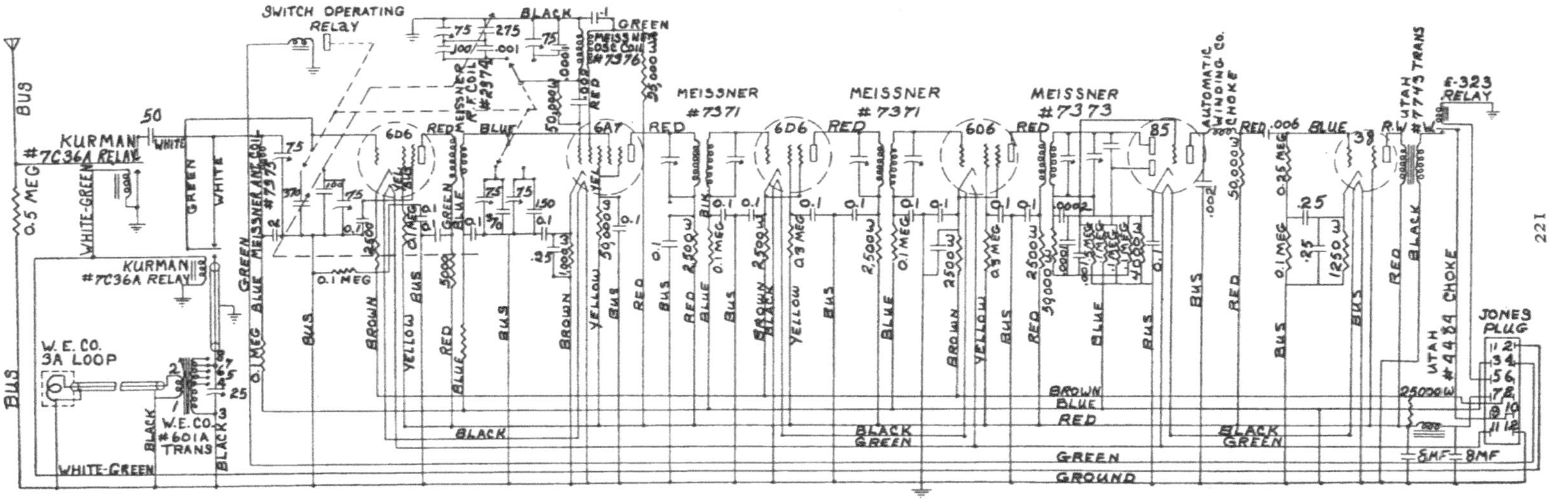

Courtesy United Air Lines

FIG. 140—SCHEMATIC OF UNITED AIR LINES ES 198 RECEIVER

ES 198 United Airlines Receiver—

	Frequency Range	Alignment Frequency
Range frequencies	200 to 400 KC	380 and 200 KC
Fixed frequency	250 to 325 (approximately)	278 KC (airport)
Intermediate	96 KC	96 KC

Power unit: External dynamotor, approximately 200 volts, 40 milliamperes
Power consumption (filaments): approximately 1 ampere at 12 volts
Power output: 600 milliwatts at 400 ohms
Tube complement: RF amplifier 6D6
Converter-Osc. 6A7
IF amplifier (2) 6D6
Det.-audio 85
Output 38

This receiver is a conventional superheterodyne radio range receiver and is equipped for quick shift to 278 kilocycles for airport traffic control use.

The alignment is made in the conventional manner for any such superheterodyne. The location of trimmers is shown in Fig. 139. There are two intermediate frequency adjustments on the under chassis side of the intermediate transformers which are not shown. A 100 mmf condenser should be used for alignment and in series with the signal generator. If the exact antenna capacity is known, its value should be used instead, for otherwise the antenna trimmer of both the 278 kilocycle quick shift and first tuned circuit should be adjusted in the airplane.

A schematic of the receiver with part values may be found in Fig. 140.

W4A Receiver, Western Electric or RCA—

	Frequency Range	Alignment Frequency
Range receiver	200 to 400 KC	380 and 220 KC
Medium high freq.	2800 to 4000 KC	4 frequencies used in group
High frequency	4000 to 6000 KC	4 frequencies used in group
Intermediate	515 KC	515 KC

Power unit: External dynamotor, 200 volts, 60 milliamperes
Power consumption: 1.8 amperes at 12 volts
Power output: 500 milliwatts, each receiver, at 500 ohms
Tube complement: Each half contains the following tubes:
RF amplifier 78 Converter-Osc. 6F7
IF amplifiers (2) 78 Detector-audio 85
Output 38

	HEIGHT	WIDTH	DEPTH	WEIGHT
Complete receiver	7″	20 1/2″	15″	33 lbs.
Control unit	5 3/4″	11 11/16″	7″	6.6 lbs
Emergency tuning dial				.7 lb.
Dynamotor unit. See W4A transmitter description				

This is a companion unit to the W4A transmitter described in the previous chapter. Control units and power supplies were shown in that description.

The unit consists of two very similar complete receivers. One is used for radio range reception from 200 to 400 kilocycles and the other for reception of high frequencies. There are two bands in the high frequency receiver, one between 2800 and 4000 kilocycles, the other between 4000 and 6000 kilocycles. In each band there is a choice of four crystal frequencies, all of which must lie in a total band width about two per cent of the average radio frequency. Thus eight channels are provided, four in each group.

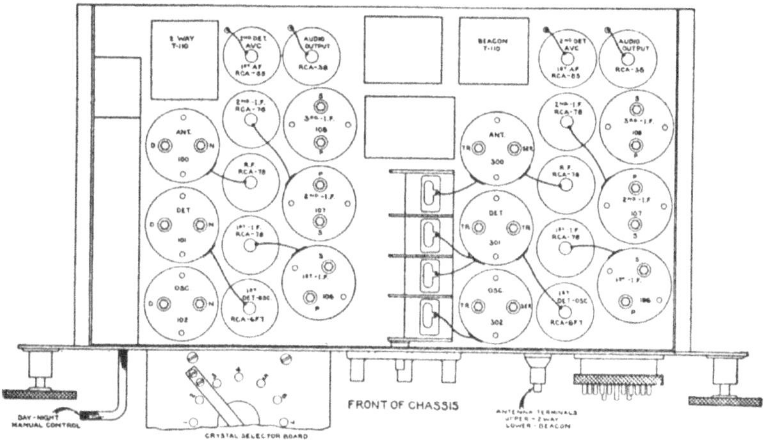

Courtesy RCA Mfg. Co.

FIG. 142—PLAN VIEW OF W4A RECEIVER

An electric range frequency drive tunes the receiver remotely. An emergency dial to replace this electric drive is also supplied.

Remote tuning is provided by control of the tuning condenser rotation through a small direct current motor and a switch which grounds either of the field coil terminals, depending on the rotation desired. This is shown in the receiver schematic Fig. 141 at the upper right hand portion of the diagram. An autosyn unit attached directly to the dial drive shaft shows the frequency to which the receiver has been tuned, by the remote indicator in the cockpit.

The high frequency remote control is accomplished by a notching switch actuating the crystal selector. The operation is described in the W4A transmitter section and the unit is identical with the transmitter crystal unit except that the day and night cam-operated contacts operate the day-night solenoids under the cover at the left edge of the receiver. The switch solenoids are of the impulse type and break the current to the coil as they reach their operated position through two special contacts and an arm on the switch board. A manual control to operate this switch, in case of failure, is provided and extends through the front of the receiver and side of the crystal selector cover where appropriate markings are provided.

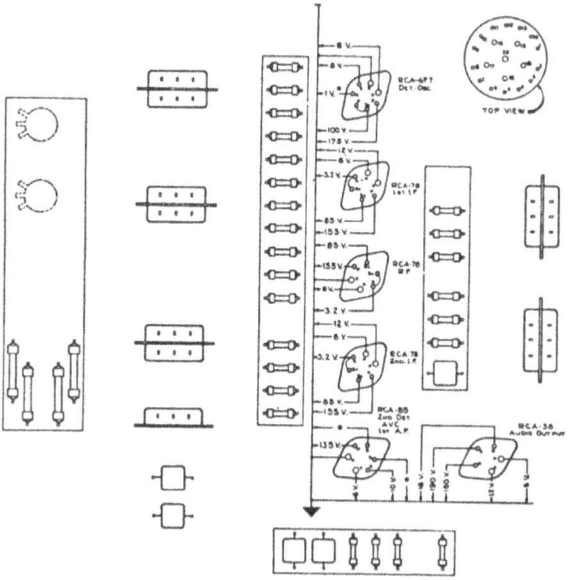

Courtesy RCA Mfg. Co.

FIG. 143—SOCKET VOLTAGES—W4A RECEIVER

Intermediate frequency alignment is accomplished by attaching the signal generator to the grid cap of the proper 6F7 and aligning the six intermediate transformers for each receiver, as shown in Fig. 142, to 515 kilocycles.

Range frequency alignment is made by adjusting the trimmers on the three cans to the right of, and beside, the ganged condensers. The alignment should be made at about 360 kilocycles, using a dummy antenna of 100 mmf in series with the antenna post. There are two trimmers shown in Fig. 142 which should not be touched ordinarily. These are marked "SER" and the one on the rear can adjusts the capacitive coupling between the primary and secondary of the antenna transformer. In case the trimmer is disturbed it should be reset to 17 mmf with a capacity meter. The front series trimmer is the low frequency adjustment for the range oscillator and should be adjusted by "rocking for maximum" at about 220 kilocycles. The oscillator operates on the high frequency side of the radio range band, 515 kilocycles higher than the radio range frequency being received.

High frequency alignment is made after selecting the proper frequency by adjusting either the "D" or "N" trimmers. The "D"

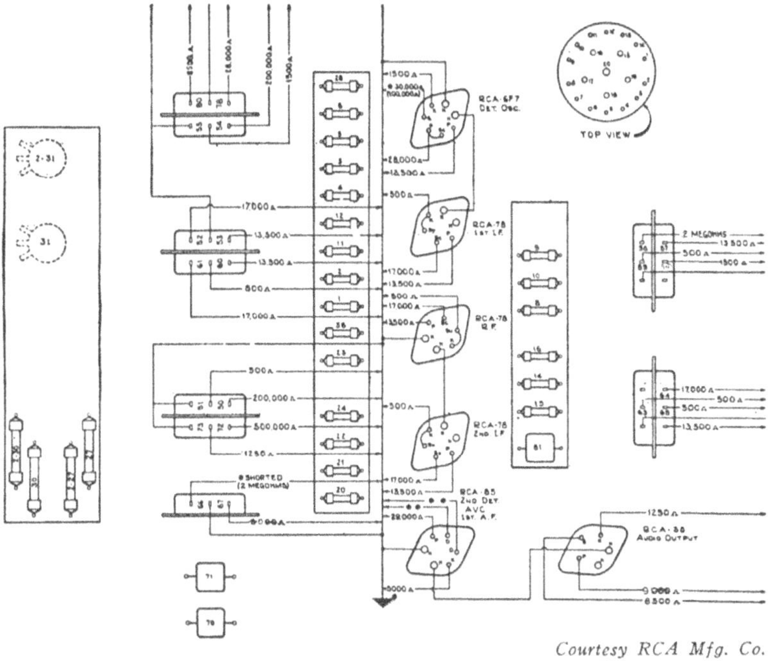

Courtesy RCA Mfg. Co.

FIG. 144—CIRCUIT RESISTANCE VALUES—W4A RECEIVER

or Day frequency trimmers are for the highest frequency and the "N" or Night frequency trimmers are for the lowest frequencies. Where several frequencies are used in a band the tuning should be made on the mean frequency or one nearest the center. In adjusting the "D" trimmer for the detector there may be two points of maximum output. The crystals normally operate on the low frequency side of the frequency desired so that the two points correspond to the crystal frequency and the radio frequency to be received. The trimmer should be set to the least capacity point of the two which is the most counterclockwise of the two points.

Voltage and resistor value diagrams are shown in Figs. 143 and 144. Note that only half of the receiver is shown in each diagram as the base wiring of the receiver for the individual receiver units is identical. There is one change that will be found in some receivers. The resistor R22 in the plate circuit of the 85 is usually changed to give reasonable audio output as judged by pilot tests. In one group of receivers this value has been lowered from 20,000 ohms to 2500 ohms for this reason. Resistor R222 is not changed in the range receiver as high audio output is desirable to prevent accidental overload which might lead to course reversal.

The two potentiometers on the central baseboard control the sensitivity of the two receivers at maximum volume control setting. Normally they are adjusted so that receiver sensitivities are about fifteen microvolts absolute, limiting sensitivity to one fifth of the maximum possible.

It will be noted that the receiver plate supply is not disconnected during transmission as the radio frequency output of the transmitter furnishes sidetone by leakage to the receiver with its antenna disconnected at the transmitter relay. Enough stray coupling exists in the transmitter for sidetone to be heard. The antenna relay in the transmitter grounds the receiver antenna lead in some receivers but the grounding contact may be removed giving better operation.

RA-2B Bendix Receiver—

	Frequency Range	*Alignment Frequency*
Medium high frequencies......	2700 to 4500 KC	Any four, crystal control
High frequencies.............	4400 to 7000 KC	Any four, crystal control
Intermediate frequency.......	385 KC	385 KC

Power unit: External dynamotor, 250 volts, 65 milliamperes
Internal 800 cycle, 115 volts can be added
Power consumption: (Rec. alone) 1.6 amperes at 12 volts
Crystal heaters 1 amp each when operating
Power output: 500 milliwatts at 500 ohms, each channel
Tube complement: RF amplifier 6K7G
Converter 6L7G
Crystal Osc. 6K5G
IF amplifiers (2) 6K7G
AVC amplifier 6K7G
Detector 6H6G
Audio 6C5G
Output (dual) (2) 6K6G

	HEIGHT	WIDTH	DEPTH	WEIGHT
Receiver with tubes and crystals..	8 1/2″	10 1/2″	17 3/4″	23.8 lbs.
MR-14A control unit...........	5 1/2″	7 1/2″	2 1/2″	1.8 lbs.

This is the first receiver described which has a dual output channel, as shown in schematic, Fig. 145. The advantage is that one output channel can be used by the pilot and the other by the co-pilot. Either may switch his output circuit to his phones and mix it with any other receiver output without affecting the other pilot using the same receiver. In conventional single output circuits the two pilots may not use the same receiver together and then individually listen to another receiver.

Alignment of the intermediate amplifier is performed by means of the adjustments on the top of the intermediate frequency cans. There are eight such adjustments. The top adjusters on the three cans, starting from the front panel and extending back along the edge, should be adjusted for maximum output while keeping the signal being fed into the converter tube grid at a low level. The can beside the third transformer is the automatic volume control amplifier and should be adjusted for minimum output when supplying enough signal to the receiver to secure automatic volume control operation.

Oscillator adjustment is made in the proper band by inserting a Bendix crystal in the proper holder. The crystal frequency should be greater than the frequency received by 385 kilocycles. A 300 volt meter may be connected between the junction of L7 and L8 for determining the correct crystal oscillator tank adjustment. The voltage rises when the crystal starts oscillating and the adjustment condenser should be adjusted to a slightly higher capacity than that required to just start oscillation. Touching the oscillator grid (V3) should stop oscillation only until the finger is removed. The crystal oscillator and radio frequency tuning adjustments are located under cover strips on the side of the tuning can and on the front panel (antenna stage). The trimmer condensers are labelled to agree with the schematic numbering.

The input stage is equipped with tapped antenna coils which are adjustable by moving the switch sector located on top of the tuning can forward and under the can cover. Adjustment for the tap giving best sensitivity with the antenna employed is correct.

A relay D1 may be used for interphone or sidetone and it disconnects the voltage from portions of the radio amplifiers so that no radio signal will be received during such use.

With 200 volts plate supply the screen voltages on all but the output tube will be about 85 volts. Biases will be about 15 volts except for V3 at zero, V8 at two volts and the output tubes at 21 volts. Plate voltages will be about 180 volts or more except for 70 volts on V8.

The complete electrical selecting mechanism is removable with the switch shaft by loosing its four mounting screws on the front panel and pulling it out. A plug and jack serve to open the connections automatically when this is done.

The electrical frequency shift requires eight wires between the selecting sector switch and the control unit. When the motor starts to run, a solenoid engages a sectional gear tooth in the revolving motor gear and the switch turns until, at the proper position of release, the gear sector is released and the motor coasts to a stop. The control unit schematic is shown in Fig. 146.

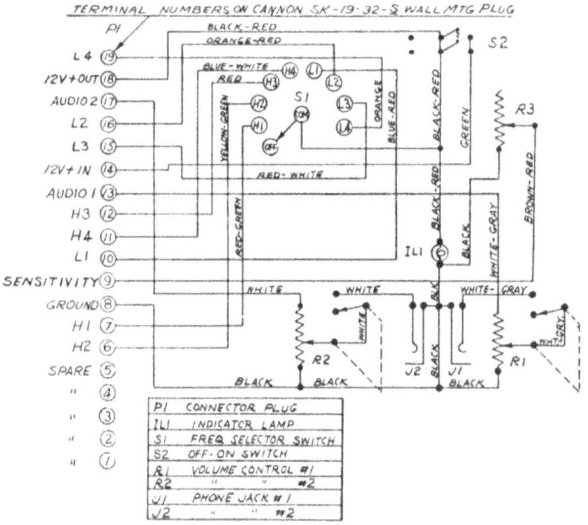

Courtesy Bendix Radio Corp.

FIG. 146—MR-14A REMOTE CONTROL UNIT FOR RA-2B RECEIVER

RA-4C Bendix Receiver—

	Frequency Range	*Alignment Frequencies*
Range band	195 to 415 KC	400 and 200 KC
Fixed frequency	250 to 300 KC	278 KC (airport)
Intermediate frequency	175 KC	175 KC

Power unit: External dynamotor, 220 volts, 50 milliamperes or by conversion to 800 cycles, 115 volts

Power consumption: (filament only): 1.2 amperes at 12 volts (plus .5 ampere with 278 relay on)

Power output: 500 milliwatts at 250 ohms for each output tube
Tube complement: RF amplifier 6K7G
Converter 6L7G
IF amplifiers (2) 6K7G
Det. Audio 6R7G
Oscillator 6C5G
Output tubes (2) 6K6G

	HEIGHT	WIDTH	DEPTH	WEIGHT
Receiver	8 1/8″	10″	11 9/16″	14.6 lbs.
Remote drive	3″	6″	4 1/8″	1.3 lbs.
Control unit	8 1/4″	6″	2 1/2″	2.25 lbs.
Power supply unit	6 3/16″	5 3/16″	3 7/8″	5.75 lbs.
MR22B antenna coupling unit				.25 lb.
Shock mounting for receiver (MR46A)				2.8 lbs.

This radio range receiver is a companion unit for the RA-2B communication receiver.

In adjusting the remote control, the receiver dial is set at zero with the remote control dial at 195 kilocycles.

To align the intermediate amplifier a signal generator is attached through a .1 mfd condenser to the grid of the converter tube V3, which is then grounded through a half megohm resistor to the chassis. Trimmers (Fig. 147) C43, C33, C42, C29, C27 and C26 are adjusted for maximum output in the order named, then returning over the adjustments in the reverse of the order.

The radio frequency alignment is made by feeding a 400 KC signal into the MR-22B coupling unit terminal one, through a 150 mmf coupling condenser with terminal two of the unit grounded to the generator. (Terminals two and five of the antenna unit should be connected to the receiver number two plug lead. Antenna unit terminal six should be connected to the number one plug terminal. A .005 mfd condenser should be placed between terminals five and six of the antenna unit.) With a dial reading of 88.5 the trimmers C20, C16 and C2 are peaked for maximum output (C2 is reached through the plate covered hole on the front panel).

Alignment at a dial reading of 5.9 with a 200 kilocycle signal is then performed by means of C44 and the adjustments repeated at 400 KC.

The 278 KC channel is aligned by adjusting C10, C12 and C4 for maximum output with a 278 kilocycle signal input. (A temporary strap between plug terminals six and seven will operate the electric shift.) The adjustment must be made before the 200 and 400 KC adjustment or incorrect alignment will result.

The final alignment procedure is to feed a 175 KC signal into the antenna unit as above and adjust L2 (located on the chassis deck between VI and the panel) for minimum signal. This trap circuit

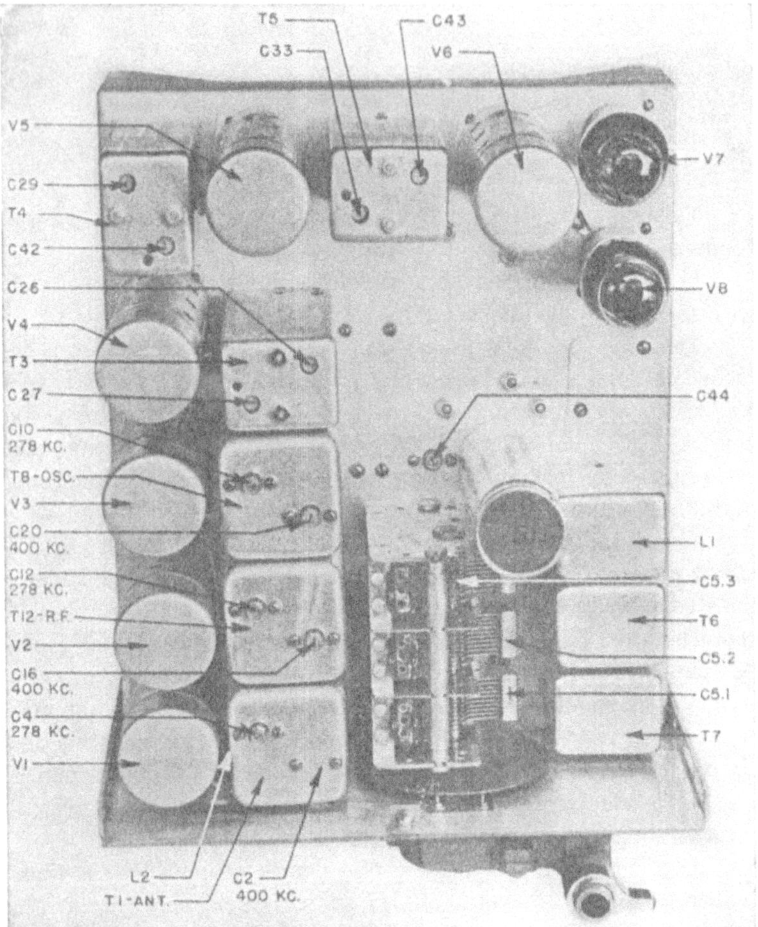

Courtesy Bendix Radio Corp.

FIG. 147—RA-4C RECEIVER, TOP VIEW OF CHASSIS

prevents intermediate frequencies appearing on the antenna from being amplified efficiently. Figs. 148, 149, and 150 show schematics and interwiring.

The following voltages will be found with voltages of 14 and 200 supplying the receiver. Maximum sensitivity is used except for readings in the last column.

TUBE	PLATE VOLTAGE	SCREEN VOLTAGE	CATHODE V MAX. SENS.	CATHODE V MIN. SENS.
V1 RF	160	130	5	55
V2 Oscillator	100	...	0	0
V3 Mixer	195	130	12	14
V4 IF	170	80	3	55
V5 IF	200	75	3	55
V6 Detector	55	−3	3	3.5
V7, V8 Output	180	200	20	23

RA-6A and RA-6B Bendix Receivers and MN-2A Bendix Direction Finders—

	Frequency Range	*Alignment Frequencies*
Range Band (I)	195 to 415 KC	400 and 200 KC
Broadcast (II)	545 to 1350 KC	1300 and 550 KC
Medium high (III)	2600 to 4500 KC	4450 and 2650 KC
High (IV)	4400 to 7000 KC	6900 and 4450 KC
Bands III and IV may be equipped with one crystal each.		
Intermediate	465 KC	465 KC
CW Oscillator	450 to 480 KC approximately	466 KC

Power Unit: Separate dynamotor 50 milliamperes at 230 volts or, on batteries, 25 milliamperes at 135 volts (reduced output).

Power consumption: 1.5 amperes at 12 volts normal or 1 ampere at 12 volts for emergency battery operation with one output stage and dial light cut out.

Power output: 500 milliwatts at 250 ohms, each output.

Tube Compliment: RF amplifier—6K7G
Converter—6L7G
IF amplifier—6K7G
Det. Audio—6B8G
Oscillator—6C5G
Output—(2)—6K6G
CW oscillator—(if used) 6C5G

	HEIGHT	WIDTH	DEPTH	WEIGHT
Receiver	7 7/8″	9 15/16″	12″	15.9 lbs. complete
MR-36B mounting base (dimensions included above)				1.0 lb.
MP-12A dynamotor				5.8 lbs.
MR-20A antenna transformer				
MR-26A remote control unit				4.5 lbs.
Emergency batteries 12–135 volts (Burgess 3T10X90)				9.4 lbs.
MN2A D/F complete with call control units, cables, etc.				15 lbs. approx.

When the receiver is equipped with crystal control in each of the high frequency bands the selection of that band automatically selects the crystal controlled frequency with no approximate tuning, as in most other receivers of similar type.

The RA-6B is similar to the 6A except that the 6B is equipped with remote mechanical control shafts.

The MR-26A remote control with two mechanical drive units MR-34B or MR-34A may be used.

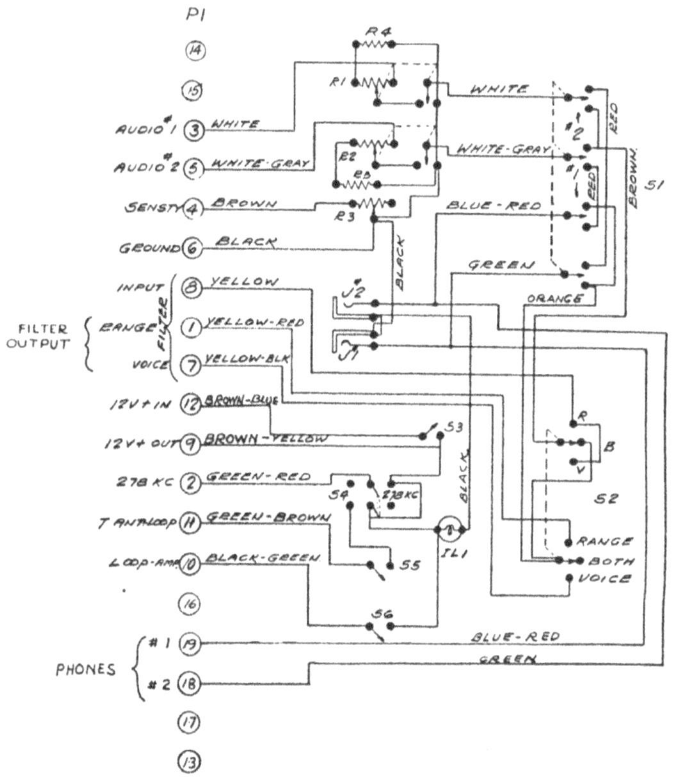

Courtesy Bendix Radio Corp.

FIG. 149—MR-18A REMOTE CONTROL FOR RA-4C RECEIVER

The CW oscillator cannot be used with high frequency crystal control as well because of mechanical space limitation.

The power switch on the front panel has four positions:

1. Off.
2. Normal "on" using dynamotor.
3. Filaments on plane's battery and "B" battery (135 V) for plates.
4. Auxiliary 12 and 135 volt batteries.

In the last two positions only one output tube operates. When using the remote receiver the volume control turns the receiver off in its most counterclockwise position.

The intermediate amplifier is aligned by attaching the signal generator to V2, the converter, through a .01 condenser to the tube grid with cap removed and grounding the grid through a half megohm resistor. With the receiver tuned to about 300 kilocycles a 465 modulated signal is used and C17, C18, C19 and C20 are adjusted for peak output (See Fig. 151). The beat oscillator is turned on

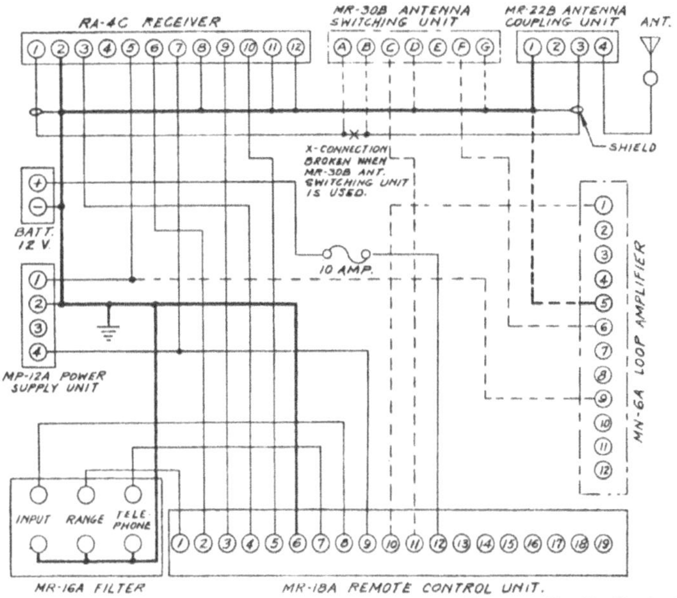

Courtesy Bendix Radio Corp.

FIG. 150—INTERCONNECTIONS RA-4C RECEIVER

and C40 adjusted for a 1000 cycle beat note after which the CW switch is turned off.

The oscillator adjustments in this receiver are made with the signal generator still attached to V2, as in the above paragraph. After tuning the receiver to each of the following frequencies the adjustment trimmer number following the frequency is adjusted for maximum output (using a small signal input). 400 KC—C9, 200 KC—C13, 1300 KC—C10, 550 KC—C14, 4450 KC—C11, 2650 KC—C15, 6900 KC—C12 and 4450 KC—C16.

A 100 mmf condenser is attached to the antenna post of the receiver (or through the MR-22A antenna transformer if used for Band I) and the adjustments are made for the antenna and radio

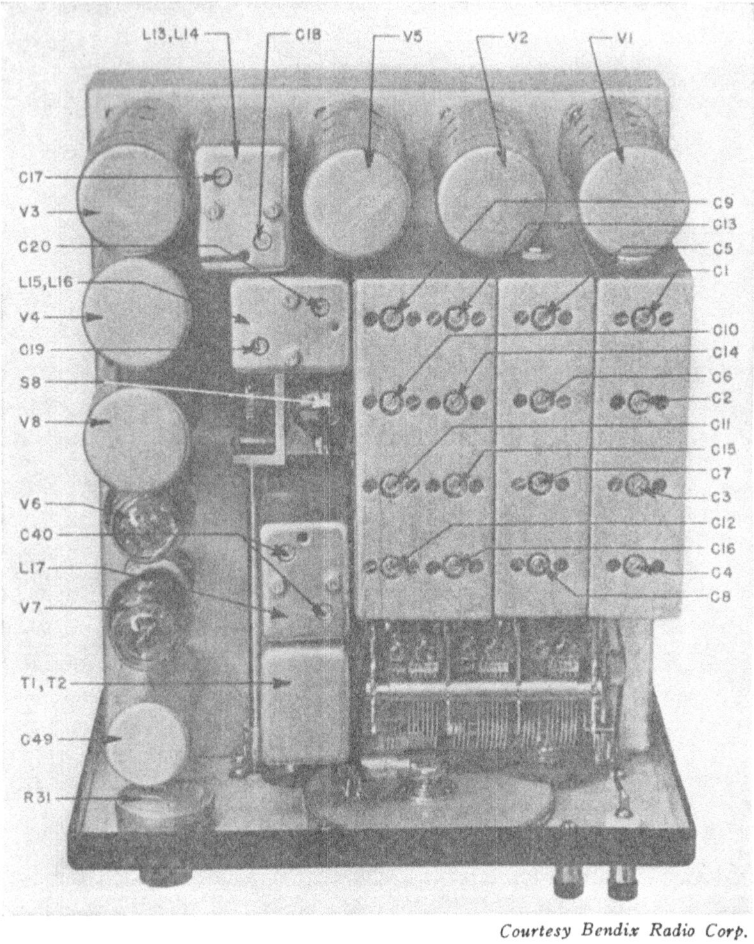

Courtesy Bendix Radio Corp.

FIG. 151—RA-6 RECEIVER TRIMMER LOCATIONS

stages in the following order as before: 400 KC—C1, 1300 KC—C2 and C6, 4450 KC—C3 and C7, 6900 KC—C4 and C8. Various schematics and pertinent data are shown in Figs. 152 to 156.

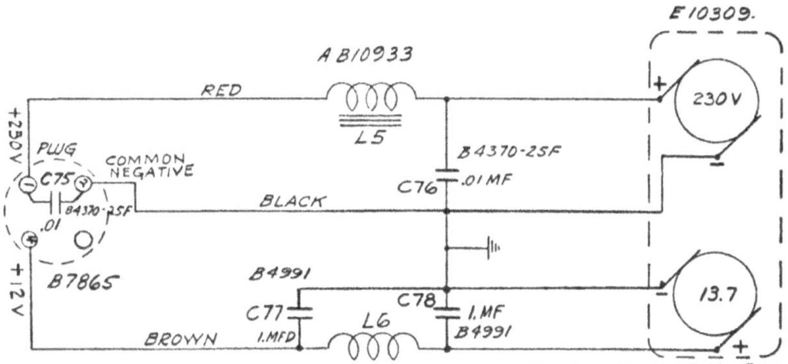

Courtesy Bendix Radio Corp.

FIG. 153—MP-12A SCHEMATIC FOR RA-6 RECEIVERS

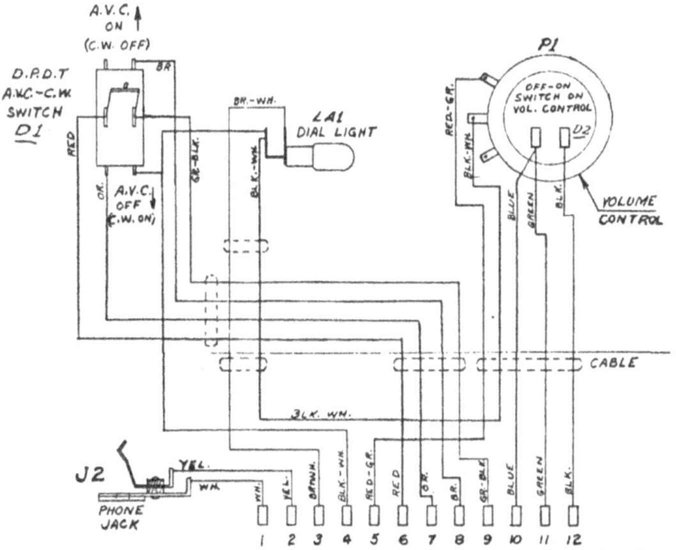

Courtesy Bendix Radio Corp.

FIG. 154—MR-26A REMOTE CONTROL

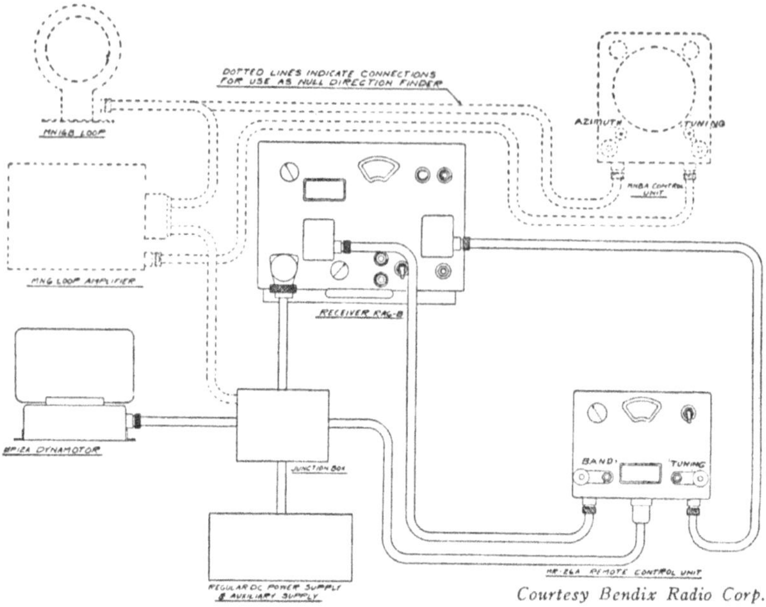

Courtesy Bendix Radio Corp.

FIG. 155—BENDIX LOOP EXTERNAL CONNECTIONS

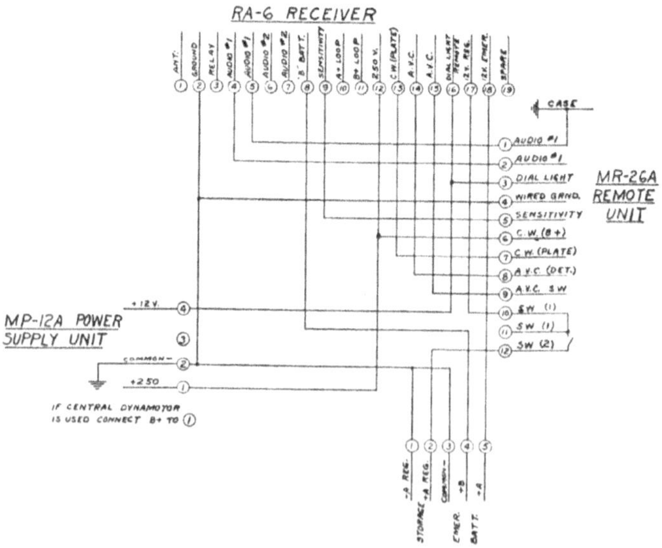

Courtesy Bendix Radio Corp.

FIG. 156—RA-6 JUNCTION BOX CONNECTIONS

Voltages to be found between the socket terminals to ground are as follows with the sensitivity control at maximum and the AVC "on" except for V8 readings:

	SOCKET TERMINAL NUMBER							
	1	*2*	*3*	*4*	*5*	*6*	*7*	*8*
V1 RF amplifier...	0	0	230	90	2.8	0	6	2.8
V2 Converter......	0	6	230	140	...	0	12	6
V3 IF amplifier....	0	12	230	95	2.8	0	6	2.8
V4 Detector.......	0	0	82	...	...	82	6	6.5
V5 Oscillator......	0	0	120	...	0	0	6	0
V6-V7 Output.....	0	12	225	230	0	0	6	23
V8 Beat Osc.......	0	0	55	0	0	0	6	0

A 300 volt meter of 1000 ohms per volt is used for all voltages over 20 volts.

MN-2A Aural Null Direction Finder, Bendix—This unit may be used with this receiver or any radio range receiver on the market having reasonable sensitivity. A loop amplifier tube serves to raise the signal received and to match the ten inch separately tuned rotatable loop to the receiver. Schematics of the various parts of the system are shown in Figs. 157 and 158 and need no explanation.

MN-12A Bendix Radio Compass—

	Frequency Range	*Alignment Frequencies*
Range band.....................	200 to 400 KC	400 and 200 KC
Broadcast band..................	550 to 1500 KC	1500 and 600 KC
Intermediate frequency...........	465 KC	465 KC

Power Unit: External dynamotor, 250 volts, approx. 75 milliamperes
Power consumption: 5 amperes at 12 volts including dynamotor
Power output: 500 milliwatts at 500 ohms, approx.
Tube complement: Antenna amplifier 6K7
Converter 6L7
Oscillator 6C5
IF amplifier 6K7
Detector 6R7
Audio 6K7
Audio Osc. 6J7
Loop amplifier 6K7
Balanced modulator 6N7
Meter audio 6K7
Output 6K6G

	HEIGHT	WIDTH	DEPTH	WEIGHT
MN-12A with tubes and base..	9 1/4″	9 13/16″	16 3/4″	22.25 lbs.
MN-14A loop and mounting...	12 5/8″	6 7/8″	20 1/4″	4.2 lbs.
MN-10A Azimuth control.....	2 5/32″	4 1/8″	5 3/16″	.6 lb.
MN-18A Control unit........	3″	6 1/4″	8 3/4″	3.0 lbs.
IN-2A Left-Right indicator....	2 1/2″	2 1/2″	3 1/2″	1.4 lbs.
MP-12A power supply.......	5 5/8″	4″	6″	5.8 lbs.
Junction Box (MS-20A) and cables..........	(approximately)			14.0 lbs.

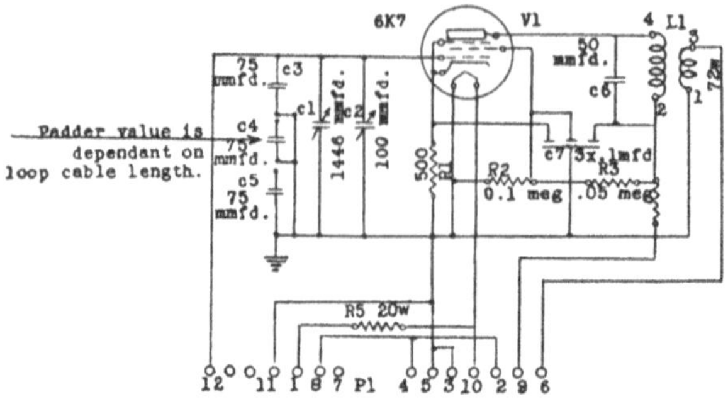

Courtesy Bendix Radio Corp.

FIG. 157—MN-6A LOOP AMPLIFIER

The loop is mounted in a plastic housing which has a drag of approximately four pounds at 250 miles per hour. The MN-10B control is used, instead of the MN-10A, when the azimuth dial is mounted upside down.

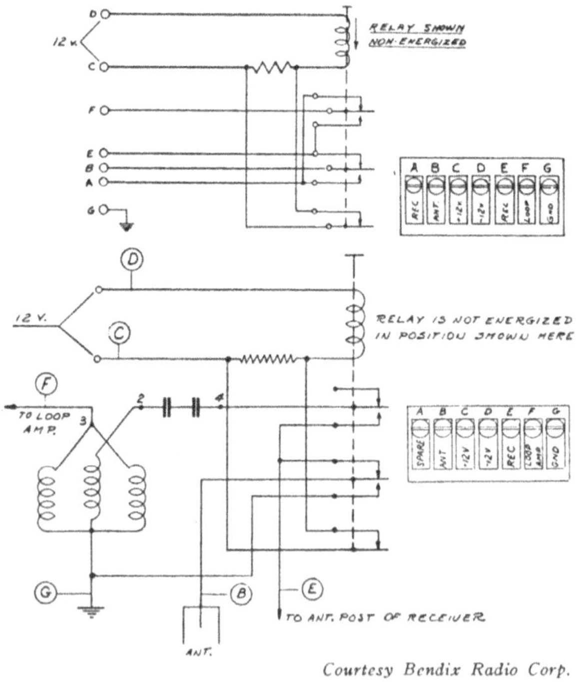

Courtesy Bendix Radio Corp.

FIG. 158—MR-30A AND MR-30B ANTENNA SWITCHING UNITS

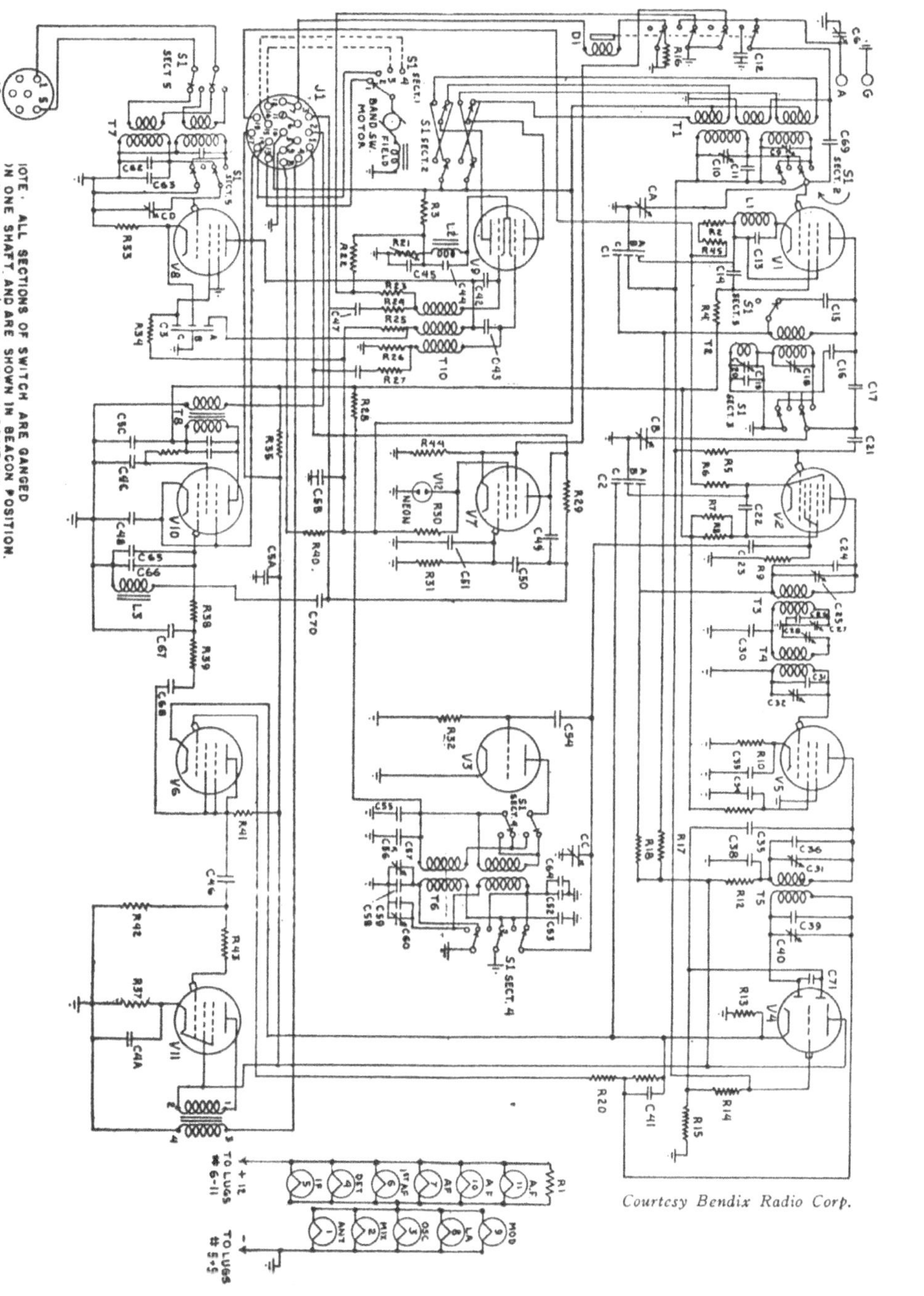

Courtesy Bendix Radio Corp.

FIG. 159—MN-12A RADIO COMPASS

The control switch S3 allows the compass (Fig. 159) to be operated (1) as a standard receiver (plate supply removed from

Courtesy Bendix Radio Corp.

FIG. 160—MN-12A RADIO COMPASS ALIGNMENT SCREWS

V7, V8 and V9) in the ANT position, (2) as a loop receiver in the LOOP position (Stops V7, grounds the vertical antenna and unbalances the modulator V9) and (3) as a radio compass (COMP).

It is important that the slate battery wire be connected to the positive battery terminal and the black wire to the negative terminal.

Alignment of the compass unit is accomplished by moving the plane well away from buildings and tuning in a signal between 200 and 250 kilocycles. Condenser C6 reached through the front panel should be adjusted for maximum deflection of the tuning meter indicator (not the right-left indicator).

Control R 21 should be adjusted for full scale deflection (needle

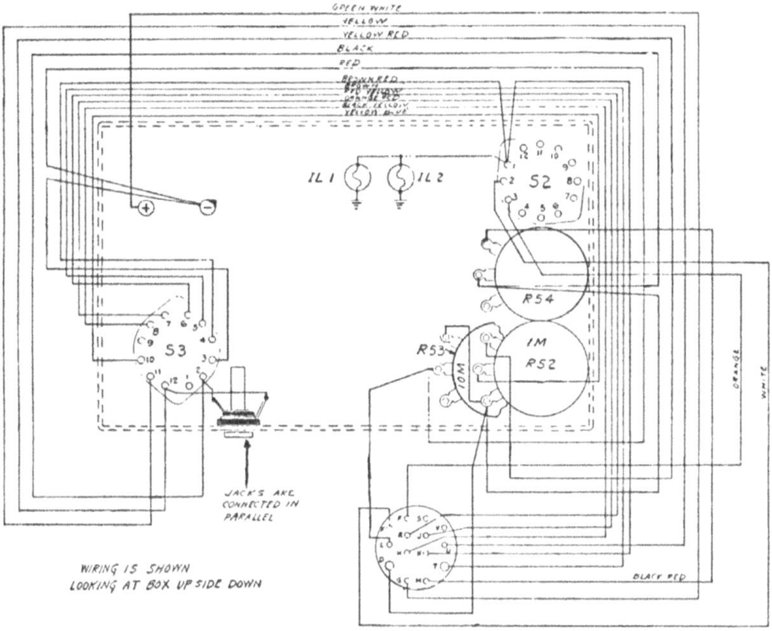

Courtesy Bendix Radio Corp.

FIG. 161—MN-12A CONTROL UNIT WIRING

over letter L to R) of the left-right indicator with a local station tuned in between 550 and 800 kilocycles with the plane's head 10 degrees off the electrical heading of a local station. The compass sensitivity control should be at maximum during this adjustment.

The local receiver dial has markings on it for agreement with the remote dial. The marks correspond to certain frequencies: A-maximum capacity, B-200 KC, C-600 KC, D-400 KC, E-1500 KC, F-minimum capacity.

The complete alignment procedure for the receiver follows:

1. Remove unit from case placing output meter from plug contact No. 4 to ground. Output should be less than 50 milliwatts during all the adjustments. Set the control switch on ANT.

2. Connect generator to V2 (converter) grid and ground and also ground the grid of V3. With a modulated signal at 465 kilo-

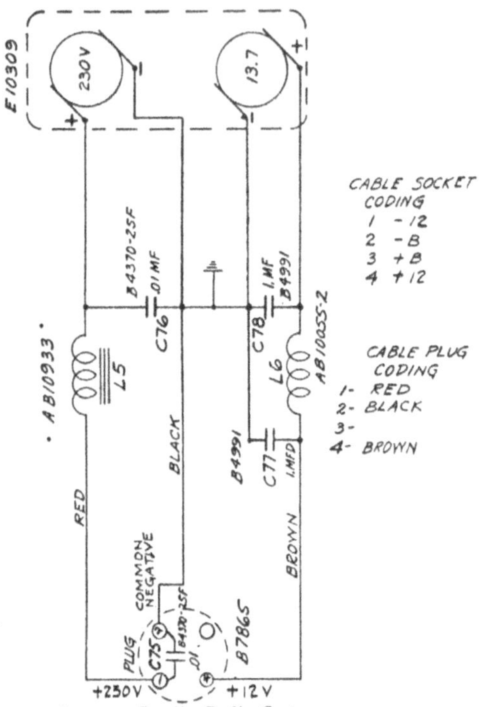

Courtesy Bendix Radio Corp.

FIG. 162—POWER SUPPLY FOR MN-12A COMPASS

cycles C25, C27, C28, C32, C37 and C40 are adjusted for maximum output (Fig. 160).

3. Replace grid cap on V2 and remove ground from V3 grid and attach generator through a 100 mmf condenser to the antenna post of the receiver. With C6 at minimum capacity, the dial set at E and a signal of 1500 kilocycles fed to the receiver the trimmers C9, C18 and C52 are adjusted for maximum output.

4. Retune to C and adjust C60 for maximum output with 600 kilocycles being fed to the receiver (C60 is the upper condenser on the side of the chassis).

5. The high frequency end should again be readjusted carefully as in (3) above.

6. The same process is repeated for the range band starting at 400 kilocycles (D) and adjusting C10, C20 and C53. Condenser C58 (below C60) is adjusted at 200 kilocycles and the process is repeated at the high frequency end once more.

7. With a 465 kilocycle signal fed to the antenna post and receiver tuned to 400 KC the adjustment of L1 is made for minimum output. Its lock is then retightened.

8. Replacing the antenna on the receiver and turning the master selector to COMP, adjust C6 as described above. Tune in a station at the high frequency end of the broadcast band and adjust C61 for maximum right-left needle deflection. Do the same at the high end of the radio range band, using C62 for this adjustment. Figs. 161 and 162 show certain other wiring details.

The compass should then be properly aligned. For the theory of compass operation it is recommended that other sections in the book dealing with this be reviewed.

The following parts are to be found in the receiver accessory equipment with reference to the schematic part numbering, Fig. 159.

R1	63 ohms
R2, R6, R10, R33	250 ohms
R3	.5 megohm
R4	75,000 ohms
R5, R14, R15	2 megohms
R7, R8, R23, R26, R30, R41	35,000 ohms
R9, R11, R32	50,000 ohms
R12, R17, R18, R40	5,000 ohms
R13, R28, R53	10,000 ohms
R16	40 ohms
R19, R20, R22, R42, R43	1 megohm
R21	25,000 ohms
R24, R27, R31, R38, R39	.3 megohm
R25	20,000 ohms
R29, R34, R50, R51, R55	100,000 ohms
R35, R37, R44, R45, R52	1,000 ohms
R36	.2 megohm
R54	15,000 ohms

C1, C2, C3, C7, C8, C49, C51, C65	.1 mf
C4A, C4B, C4C	8 mf
C5A, C5B, C5C	8 mf
C6, C9, C10, C11, C12, C18, C20, C23, C25, C27, C28, C30, C37, C40, C52, C53, C54, C57, C60, C61, C62, C64	50 mmf
C13, C67	.006 mf
C14, C22, C30, C33, C34, C38, C45, C50, C55, C56, C75, C76	.01 mf
C15	250 mmf
C16	15 mmf
C17, C19, C21	25 mmf
C24, C26, C29, C31, C36, C39, C40, C63	100 mmf
C35	150 mmf
C42, C43	500 mmf
C44, C47, C48, C66, C68	.05 mf
C46	.001 mf
C58	200 mmf
C59	400 mmf
C69, C70	10 mmf
C71	5 mmf
C77	1 mmf

V1	Antenna amplifier	6K7
V2	Converter	6L7
V3	Oscillator	6C5
V4	Detector	6R7
V5	IF amplifier	6K7
V6	Audio	6K7
V7	Audio oscillator	6J7
V8	Loop amplifier	6K7
V9	Balanced modulator	6N7
V10	Meter audio	6K7
V11	Output	6K6G
V12	¼ watt, 60 v Neon Lamp	

ARC-6 Lear Direction Finder—

	Band	*Frequency Range*	*Alignment Frequency*
Radio range	1	180– 405 KC	380 KC
Broadcast	2	500–1200 KC	1100 KC
Communication	3	1200–2800 KC	2600 KC
Communication	4	2800–6700 KC	6200 KC
Intermediate		455 KC	455 KC
Audio oscillator		102 cycles	

Power unit: Internal dynamotor
Power consumption: 7 amperes (very approximate)

Power output: one watt undistorted, two watts maximum

Tube complement:	Loop amplifier	—6S7	(V1)
	Balanced modulator	—6N7	(V2)
	RF amplifier	—6K7	(V3)
	Converter-oscillator	—6K8	(V4)
	Audio oscillator	—6F8G	(V5)
	1st IF amplifier	—6S7	(V6)
	2nd IF amp. and AVC	—6B8	(V7)
	2nd det. and 2nd audio	—6F8G	(V8)
	1st audio and B:F:O:	—6P7G	(V9)
	Audio output	—6K6G	(V10)
	Meter output	—6J7	(V11)

	HEIGHT	WIDTH	DEPTH	WEIGHT
Tuner	5¼″	7¼″	9½″	7.3 lbs.
Control unit	4⅛″	4⅛″	2″	1 lb.
IF and AF unit	6⅝″	8¾″	13½″	13 lbs.
Accessories				6.9 lbs.
Added for manual control only				14.2 lbs.
Added for automatic indication only				21.4 lbs.

The radio frequency or tuner unit in the ARC-6 is separate from the intermediate and power unit to which it is connected through a cable (Figs. 163 and 164). The tuner unit also contains the balanced modulator and loop amplifier and the remainder of the compass circuits and its oscillator are in the intermediate and power unit.

The compass is furnished in either the conventional R-L indicator type or with an automatically revolved loop driving an indicator electrically which indicates the direction to the station without 180 degree ambiguity. The unit may also be used as an aural null direction finder. A "distance" meter is provided which shows the relative strength of the signal.

A crystal may be plugged into the tuner panel for crystal control of any one frequency.

Dial calibration may be set by placing the indicator at the white dot on the tuner panel and adjusting the drive shaft so that the tuning condenser is just fully meshed.

The intermediate adjustments are labelled L17 (on the tuner chassis) L18, L19, L21 and L22. A signal at 455 KC is first fed into the 6B8, second intermediate frequency, grid. L21 and L22 are then tuned in the same manner as for the R-3-AB receiver so that both tuning peaks are equally spaced and the same height. A signal generator is then coupled to the first intermediate frequency grid and the remaining intermediate trimmers are adjusted for maximum deflection of the tuning distance meter. Adjustments L21 and L22 must not be touched during this process. The AVC

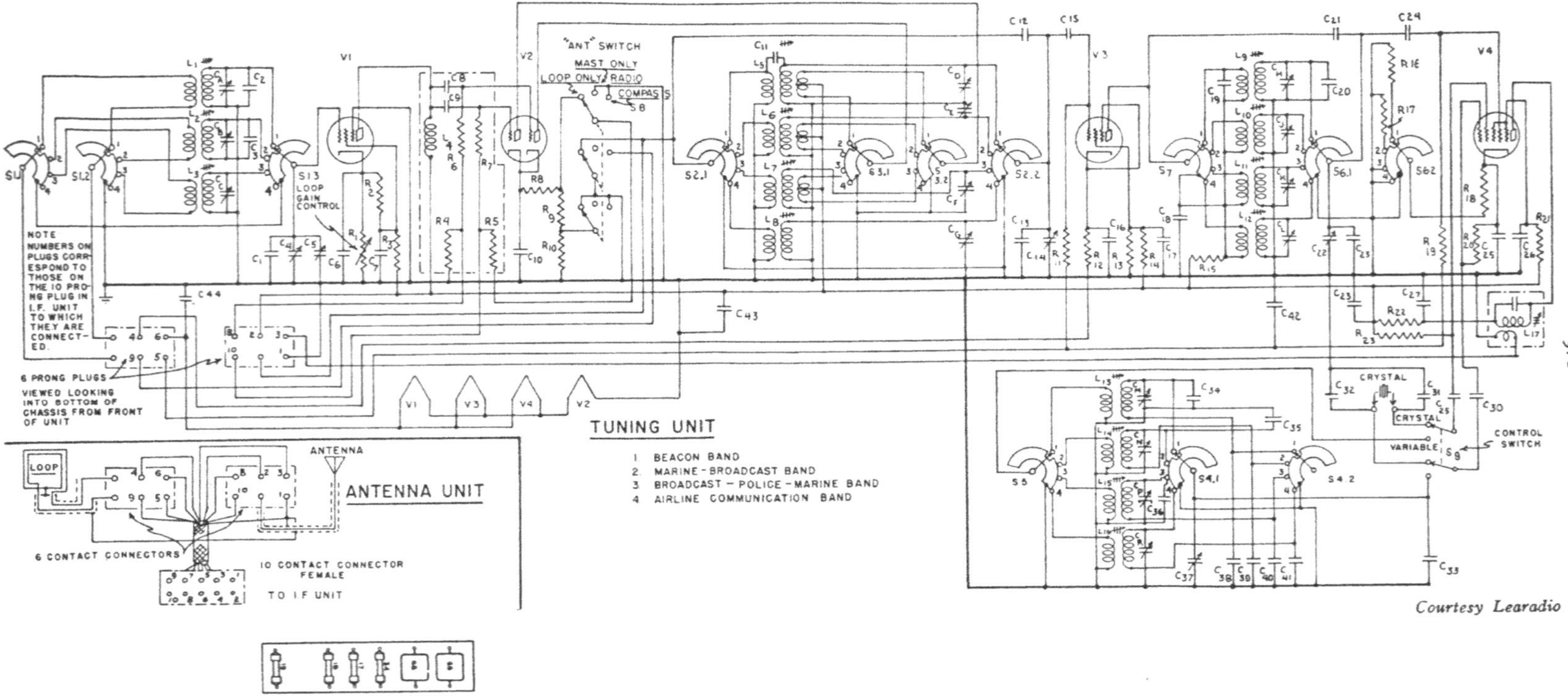

Courtesy Learadio

FIG. 163—LEAR ARC-6 COMPASS TUNING AND ANTENNA UNIT

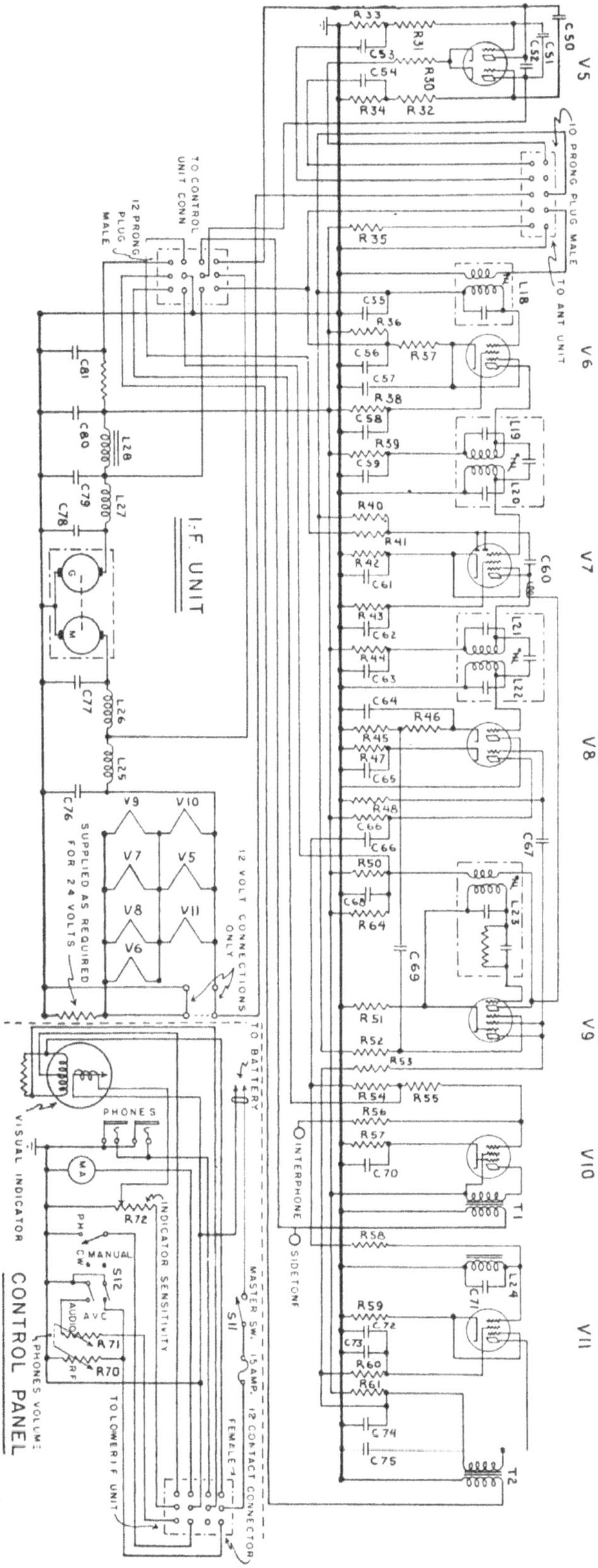

FIG. 164—LEAR ARC-6 COMPASS INTERMEDIATE AND CONTROL UNIT

Courtesy Learadio

switch must be in AVC position during the adjustment, using full receiver sensitivity.

The signal generator should then be attached to the antenna input through a 150 mmf condenser and the receiver should be left in the "antenna" position. Adjustments are then made for maximum receiver audio output as follows:

Band 1 is aligned at 380 kilocycles by adjustment of Cm (osc), Cd (ant) and Ch (RF).

On band 2 at 1100 KC adjust Cn (osc), Ce (ant) and Cj (RF)

On band 3 at 2600 KC adjust Cp (osc), Cf (ant) and Ck (RF)

On band 4 at 6200 KC adjust Cr (osc), Cg (ant) and Cl (RF)

After this the generator should be coupled closely to the loop by means of a few coupling turns attached to the generator and the receiver should be placed in the "loop" position.

On band 1 at 380 kilocycles adjust Ca, on band 2 at 1100 kilocycles adjust Cb and finally on band 3 adjust Cc at 2600 kilocycles. All are adjusted for maximum audio output.

Loop antenna gain adjustment may be made with a screw located under the cap nut at the lower right corner of the compass tuner unit. A strong station should be tuned in, preferably in the number 2 band, and the loop turned for maximum signal. The loop sensitivity adjustment is made so that when switching to either loop or antenna (mast) the tuning meter deflection is exactly the same. These deflections should not show a single bit of difference.

The values of the condensers and resistors in the Lear ARC-6 compass are:

Tuning unit

C1, C33	20 mmf
C2	70 mmf
C3	15 mmf
C6, C7, C16, C17, C18, C25, C26, C27, C28, C42, C43, C44	.1 mf
C8, C9, C15, C24, C29, C30	250 mmf
C10	.01 mf
C11, C13, C20, C12, C21, C23, C32	10 mmf
C19	100 mmf
C31, C34	45 mmf
C35, C36	5 mmf
C38	190 mmf
C39	440 mmf
C40	1250 mmf
C41	2500 mmf

Intermediate frequency unit

C50, C51, C52, C59, C62, C66	.06 mf
C53, C54, C58, C67, C68, C69, C78	.01 mf
C55, C57, C61, C63, C71, C72, C73, C75	.1 mf
C56, C76, C77	.5 mf
C60, C64	250 mmf
C65, C70, C74, C79, C80, C81	8 mf

Tuning unit

R1	25,000 ohms
R2, R4, R5, R10, R13, R14, R21	40,000 ohms
R3	75,000 ohms
R6, R7, R11, R19	1 megohm
R9, R15, R22	5,000 ohms
R12, R18	250 ohms
R16, R17	500 ohms
R20	100,000 ohms
R23	20,000 ohms
R28	2,500 ohms

Intermediate frequency unit

R31, R32, R36, R43, R58, R60	250,000 ohms
R30, R45, R61	10,000 ohms
R33, R34, R49, R53	40,000 ohms
R35, R37	2,500 ohms
R38, R46, R54, R55, R56, R50, R64	100,000 ohms
R39, R42, R44, R63	5,000 ohms
R40, R52	1 megohm
R41	75,000 ohms
R47	1,500 ohms
R48	150,000 ohms
R51	500 ohms
R57	750 ohms
R59	250 ohms
R62	22 ohms

Control unit

R70, R71	100,000 ohms
R72	10,000 ohms
R73	75,000 ohms

ADF-6 Lear Automatic Direction Finder—

Specifications are practically the same as for the Lear ARC-6 except for the following differences.

Power consumption: 8 amperes (approximately) slightly more when loop is moving.

Tube complement:

Loop amplifier	—6S7	(V1)
Balanced modulator	—6N7	(V2)
RF amplifier	—6K7	(V3)
1st Det. Osc.	—6K8	(V4)
Audio oscillator	—6F8G	(V5)
1st IF amplifier	—6K7	(V6)
2nd IF amp. and AVC	—6B8	(V7)
2nd det. and 2nd aud.	—6F8G	(V8)
1st audio and BFO	—6P7G	(V9)
Audio output	—6K6G	(V10)
1st signal amplifier	—6K7	(V11)
2nd signal amplifier	—6R7	(V12)
Thyratrons (two)	—OA4G	(V13–14)
Audio osc. amp. (two)	—6V6G	(V15–16)
Voltage regulator	—VR-150-30	(V17)

Dimensions: See ARC-6

The automatic direction finder uses the essential circuits of the ARC-6 Lear radio compass but in addition two thyratrons control a motor which rotates the loop to the null position automatically. As a sense antenna is used there is no ambiguity.

A remote electrical indicator is used which performs the same function as the autosyn type but operates on a different principle. The loop shaft rotates a slider on a delta connected continuous resistance element. Direct current is fed to the unit and this appears on the three connecting wires between the delta resistor and the three phase field windings on the indicator. The needle of the indicator is attached to a very powerful but extremely small magnet. This magnet takes a position of no torque with respect to the three phase delta field winding and thus indicates the loop position as the loop assumes the correct position automatically from radio signals.

The thyratron tubes effectively replace the right-left indicator of the more conventional compass. Additional amplification is provided as will be seen on Fig. 165. The remainder of the unit, the tuner, is shown schematically in Fig. 166. The thyratrons control the direction of loop rotation to the null. Actual operation of the loop is accomplished by means of the thyratrons through the relays RL1 and RL2.

Manual control of loop rotation may also be secured by the switch in the control unit. This effectively closes the relay circuits.

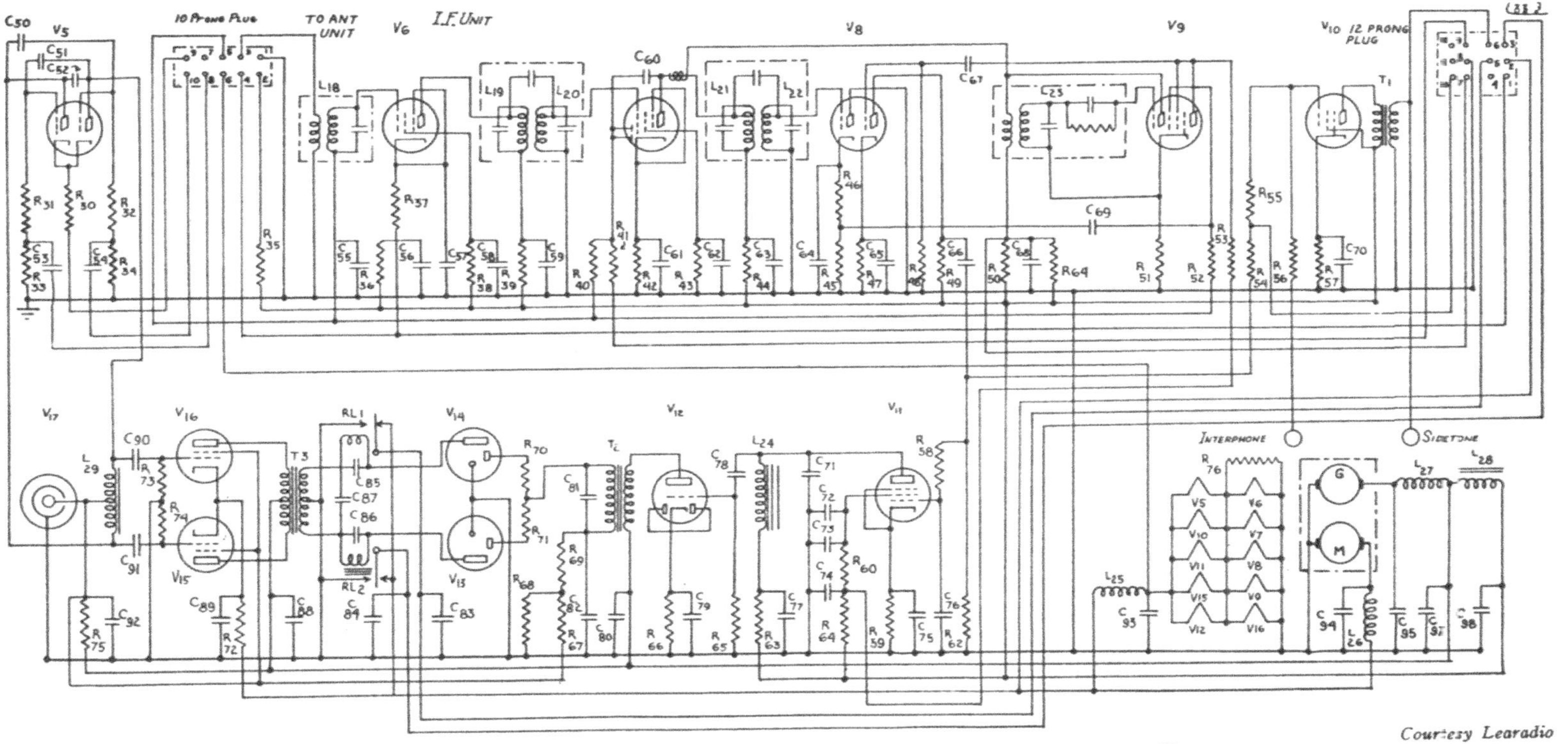

Courtesy Learadio

FIG. 165—LEAR ADF-6 AUTOMATIC DIRECTION FINDER, INTERMEDIATE FREQUENCY UNIT

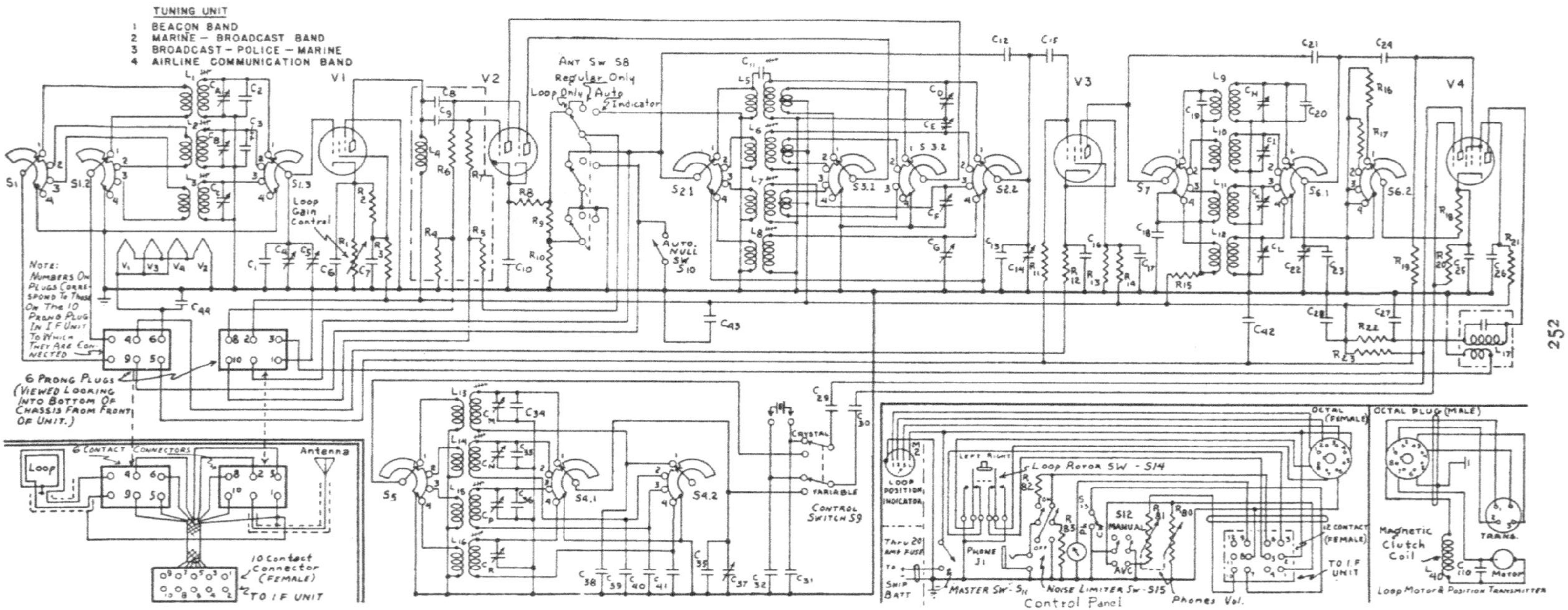

Courtesy Learadio

FIG. 166—LEAR ADF-6 AUTOMATIC DIRECTION FINDER, ANTENNA, RADIO AND CONTROL UNIT

If the automatic feature is in operation the switch will have little control of loop rotation but switch operation will do no damage.

Alignment of the unit is performed in the same manner as for the ARC-6 and reference should be made to that description.

The values of condensers and resistors are the same for the ARC-6 with the following exceptions and additions:

Tuning unit: Same as for ARC-6 both for condensers and resistors.

Intermediate frequency unit: Do not use ARC-6 list but substitute the following:

C50, C51, C52, C58, C68, C59, C62, C66, C81, C90, C91	.06 mf
C53, C54, C67, C69, C76, C78, C87, C95	.01 mf
C55, C61, C63, C57, C71, C72, C73	.1 mf
C56, C82, C83, C84, C93, C94	.5 mf
C60, C64	250 mmf
C65, C70, C74, C75, C77, C79, C80, C88, C89, C92, C97, C98	8 mf
C85, C86	1.5 mf

Same R30 to R57 inclusive as in ARC-6 and not beyond, then

R58, R62	500,000 ohms
R59	750 ohms
R60	200,000 ohms
R61	10,000 ohms
R63	5,000 ohms
R64	100,000 ohms
R65	1 megohm
R66	1,500 ohms
R67, R68, R69	40,000 ohms
R70, R71	20,000 ohms
R72	100 ohms
R73, R74	75,000 ohms
R75	2,000 ohms
R76	50 ohms

Control unit

C110	0.5 mf
R80, R81	100,000 ohms
R82	20,000 ohms
R83	500 ohms

Sperry Mark I Automatic Compass—

	Frequency Range	*Alignment Frequency*
Radio range (Beacon) (X)	195 to 510 KC	450 and 220 KC
Broadcast (A)	540 to 1500 KC	1400 and 600 KC
Intermediate	175 KC	175 KC
Loop modulating frequency	100 cycles	100 cycles

Power unit: Two internal vibrators (One for AC). 800 cycle unit may be substituted
Power consumption: 8 amperes at 12 volts, approximately
Power output: Two 500 milliwatt channels, 500 ohms, through hybrid coil
Tube Complement:

Loop amplifier	—6B8 (V1)
Balanced modulators (two)	—6J7 (V2, V3)
RF amplifier	—6K7 (V4)
Converter-Oscillator	—6A8 (V5)
IF amplifier	—6K7 (V6)
2nd IF-det.-AVC	—6B8 (V7)
Audio output	—6V6 (V8)
Tone amplifier	—6F8G (V9)
Thyratron loop control (2)	—2050 (RCA) (V10, V11)

	HEIGHT	WIDTH	DEPTH	WEIGHT
Receiver	$7^{11}/_{16}''$	$10^{7}/_{16}''$	$19^{5}/_{8}''$	44
Control unit	$5^{3}/_{4}''$	$7^{1}/_{4}''$	$2^{1}/_{2}''$	5
Loop and misc. without tachometer shafts				20 (approx.)

This automatic direction finder or compass indicates the bearing of the station relative to the heading of the airplane automatically. The basic circuit is similar to the other types of radio compasses shown. The method of operation is similar to the ADF-6 previously described.

The bearing indicator is driven by the loop (through a shaft) which in turn is rotated by the alternating current motor M-2 (Fig. 167). The motor is supplied with alternating current secured from one of the vibrator units. The thyratrons short-circuit either one of the output transformers allowing alternating current to flow to the motor due to lowered secondary reactance. The principle is similar to that used for light operation of the 27A marker receiver, except the 27A saturates the reaction primary with direct current and the Sperry thyratrons short-circuit the primary.

Two loops are provided and one is used for the normal loop function. The other loop behaves like a straight antenna but is not subject to rain static. When it is used there will be 180 degree ambiguity but the automatic feature may still be used.

The control unit is equipped with a volume control which is an audio volume control when using AVC but becomes a sensitivity control without AVC in the range band on loop or antenna only.

Switches are provided on the control unit for electrical band change, selection of either loop or antenna operation, and selection of either automatic or manual operation. A switch allows rotation of the loop electrically when not in the automatic position. Thus the following arrangements can be secured—straight receiver (REC) with either loop or antenna, automatic direction finder (AUTO) without 180 degree ambiguity (ANTENNA) or with ambiguity (LOOP).

A corrector cam is built into the control unit for elimination of radio deviations due to the mounting position of the loop and must be cut for each installation.

Alignment is performed in the following manner:

(1) Band A (broadcast) is selected with a 600 ohm resistor load and output meter in the output jack. The generator is coupled to the intermediate frequency amplifier at the converter, V5, and through a condenser .1 mfd without removing grid caps. Adjust cores L-26, L-27, L-24, L-25, L-22, L-23 in turn at 175 kilocycles and go over adjustments at least once.

(2) Place generator on grid of V5, tune to 1400 KC and adjust C-15 (Osc) at 1400 KC. L-19 is then adjusted at 600 KC, and C-15 adjustment repeated.

(3) The generator is coupled to V4 and C-11 and C-13 are adjusted at 1400 KC. L-12 and L-16 are adjusted at 600 KC and the adjustment of C-11 and C-13 is repeated at 1400 KC.

(4) The generator is then connected to the antenna (transformer L-3) through a 100 mmf condenser and C-9 is adjusted through the front panel at 1400 KC. L-8 is then adjusted at 600 KC and the C-9 adjustment repeated at 1400 KC.

(5) With the switches in the "receiver" and "loop" positions, the generator is coupled to the loop by means of an open wire near it. C-17 is adjusted through the front panel at 1400 KC. L-29 is then adjusted at 600 KC and the C-17 adjustment repeated at 1400 KC.

The range (X) band is then selected and the following adjustments are made:

Step (2) is repeated adjusting C-16 at 450 KC. L-21 is then adjusted at 220 KC and the C-16 adjustment repeated.

Step (3) is repeated adjusting C-12 and C-14 at 450 KC. L-14 and L-17 are then adjusted at 220 KC and the C-12 and C-14 adjustments repeated.

Step (4) is repeated adjusting C-10 at 450 KC. L-11 is adjusted at 220 KC and the C-10 adjustment repeated.

Step (5) is repeated adjusting C-18 at 450 KC. L-31 is adjusted at 220 KC and the C-18 adjustment repeated.

L-5 on the A band and L-42 on the X band are adjusted for maximum output in receiver-loop positions. Leave adjustments of R-10 and R-19 strictly alone (adjustable resistors under base).

The low frequency adjustments in each band will require no adjustment as a rule but should be checked occasionally.

When the emergency battery is used, as selected by the switch provided, the compass tubes are turned off to save power.

The Parts for the Sperry compass are:

R1, R14, R20, R24, R29, R106, R107, R109, R110	4700 ohms
R2, R25, R36, R38, R46, R60	470,000 ohms
R3, R4, R31	1 megohm
R5, R6, R74	180,000 ohms
R7	1000 ohms
R8, R11, R12, R17, R21, R27, R43, R59	.1 megohm
R9, R52, R65, R66, R71, R104, R105	10,000 ohms
R13	2700 ohms
R16, R30, R61, R62	330,000 ohms
R18, R37	470 ohms
R22, R33	47,000 ohms
R23	390 ohms
R26	1200 ohms
R28	43 ohms
R32, R63, R64	1.5 megohms
R34, R57	1500 ohms
R35	18,000 ohms
R39, R40, R42, R45	220,000 ohms
R41	2.2 megohms
R44	33,000 ohms
R47, R58, R72, R73	22,000 ohms
R48, R49	68,000 ohms
R53	variable 10,000 ohms
R54	variable 1 megohm
R55	variable 30 ohms
R56	3300 ohms
R68	270,000 ohms
R75	4400 ohms
R101, R102	15,000 ohms
R103	28 ohms
R108	12,000 ohms
R111	2200 ohms
R112	330 ohms
C6, C22, C23, C24, C25, C29, C30, C33, C39, C40, C41, C42, C43, C46, C47, C55, C56, C63, C67, C69, C70, C83, C108, C109, C119, C120, C150, C151	.01 mf
C7	150 mmf
C8	2700 mmf
C19, C71, C84	2300 mmf
C20, C21, C31, C32, C113, C114, C121	Double .1 mf
C26, C27, C28, C48, C49, C50, C51, C52, C53	180 mmf
C34, C54, C99	100 mmf
C35, C36, C37	.1 mf
C38	200 mmf

The Parts for the Sperry compass are:—*Continued*

C44, C45, C96, C97, C103, C104, C112, C122	Double .5 mf
C57, C58, C59, C60, C61, C62, C117	20 mf
C64, C66, C85	3000 mmf
C65, C110	.25 mf
C68, C82, C86	18 mmf
C72	43 mmf
C73	370 mmf
C74	12 mmf
C75	2 mmf
C76, C77, C90	1000 mmf
C78	750 mmf
C79	1350 mmf
C80	.135 mf
C81, C88	50 mmf
C87	10 mmf
C89	5 mmf
C91	220 mmf
C92, C93	Double .05 mf
C94	Double 10 mf
C98	560 mmf
C101, C102, C111, C116	0.5 mf
C105	.05 mf
C106, C107	10 mf
C115	.9 mf
C118	30 mf

AVR 8D and 8E RCA Radio Compass—

	Frequency Range	*Alignment Frequency*
Radio range (X)	200 to 410 KC	380 and 200 KC
Broadcast (A)	550 to 1500 KC	1400 and 600 KC
High Frequency (B)	2200 to 6700 KC	6000 KC
Intermediate	460 KC	460 KC
Audio oscillator	95 cycles	

Power Unit: Internal vibrator
Power consumption: 2.8 amperes at 12 volts
Power output: 700 milliwatts at 300 ohms.
Tube Complement: Loop amplifier (2) 6C6
RF amplifier 6D6
Converter-Osc. 6D7
IF amplifier 6D6
IF and Det. 6B7
Audio 6F7
Output 6D6
Audio Osc. 37
Meter rectifier 84

	HEIGHT	WIDTH	DEPTH	WEIGHT
Receiver 8D or 8E	9″	11 5/8″	16 1/8″	31.3 lbs. (with shocks)
All other equipment (including loop) for operation.				
AVR 8D with NON rotatable loop				25.5 lbs.
AVR 8E with rotatable loop				33.0 lbs.

Radio compass as well as multiband operation is supplied so that the units are most versatile.

There are several points that must be observed in the installation of the equipment which are fully outlined in the RCA instruction book.

The loop cable must not be altered in length as electrical misalignment will occur. The excess cable is usually wound up in a large diameter loop.

The "sense" of the radio compass is dependent upon the location of the loop and the antenna. The antenna must be either a vertical type antenna or a T type with equal sections extending in both directions from the connecting point of the T. The cable connecting the control unit to the left-right indicator must be properly connected. The meter + and − terminals must have the green and yellow, or the red and yellow coded wire attached to them.

The correct connections are shown in Fig. 168 in the small table. Connections as shown are correct provided loop and antenna are both together above or below the airplane. Otherwise with one above and the other below, wire connections are to be reversed.

The pointer of the loop azimuth indicator should point toward the station with the right-left indicator needle in center. If this cannot be accomplished by mounting the loop drive in either the inverted or normal position, an opposite drive worm is supplied to suit the installation.

The powertron (vibrator) must be correctly inserted in its socket with the proper polarity of battery ground (which is usually negative) opposite the ground mark on the chassis. Care should be taken in inserting new units with this in view.

After installation the following adjustments must be made at a point removed several hundred feet from hangars or electric light wires.

The indicator L-R needle should be at center mechanically with no apparatus "on" as adjusted by the screw on the meter's face. With the master switch of the radio equipment, which is on the headphone volume control, turned on and the "radio compass" position selected with the receiver on the A band but not tuned to a station, the button covering the hole on the front of the control unit is removed and the potentiometer adjusted with a screwdriver to electrically center the needle of the L-R indicator.

Placing the indicator sensitivity to extreme counterclockwise position and headphone volume to extreme clockwise position (full on) loop alignment is made. The airplane antenna is connected to a test oscillator after removing the antenna from the receiver and an output meter is connected to the receiver. Signal of as low amount as possible is fed into the antenna so the automatic volume control will not be operating. A separate antenna temporarily placed near the loop may be required. The frequency used must be higher than 1400 kilocycles and free from interference. The fixed mica capacitor of 5 to 6 mmf supplied with the receiver, is connected across the two terminals to which the loop leads are attached and trimmer C-83 aligned after unlocking it, Fig. 169. The condenser is relocked and the mica capacitor removed.

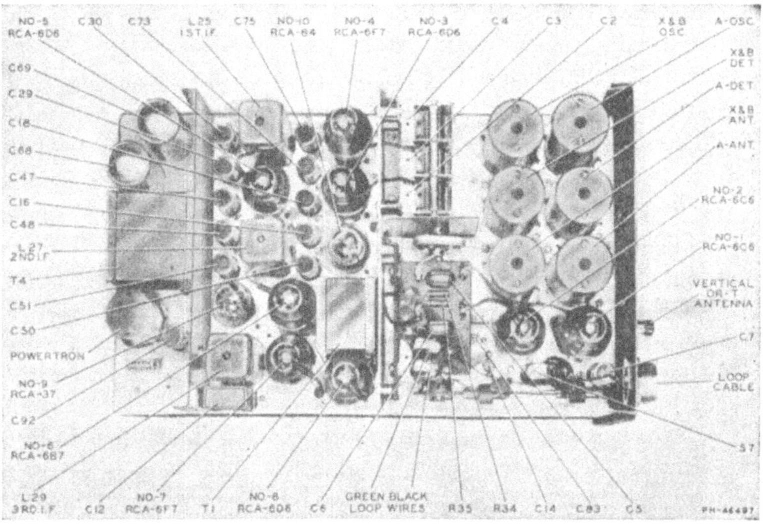

Courtesy RCA Mfg. Co.

FIG. 169—AVR 8 TYPE RADIO COMPASS TRIMMERS

The generator and receiver are retuned on the X band to a frequency higher than 350 kilocycles and lower than 400 kilocycles. The 25 to 30 mmf condenser supplied is attached to the loop terminals and C-14 is adjusted for maximum output. The condenser is relocked and the mica capacitor removed. The loop is now aligned.

After connecting the antenna to the receiver, stations of known direction may be tuned in and the plane pointed to each station. A station at each end of the two bands is carefully checked as follows: The needle should center as the plane is headed to the station with the course indicator sensitivity suitably advanced. As the plane is turned to the right of the heading to the station the indicator should move right. The same should be true at all four frequencies and if on turning the nose to the right from the station the needle moves left in any one case, the loop alignment is not correct and is repeated. If all indications are uniformly in the wrong direction the leads on the L-R indicator are reversed. The newer convention (in a later description of radio compasses) calls for right movement with left turn.

The convention used for needle movement is the same as for the turn indicator of the bank and turn group of airplane instruments. The needle does not point to the station but the station is considered as being fixed at midscale of the meter, thus turning right shows the station to be to the left of the plane's nose (or needle tip). The description of radio compass operation given elsewhere in this book may be reviewed for further details.

The last part of the adjustment is to fix the amount of antenna voltage fed to the receiver on the broadcast band. A weak station is tuned in near the low frequency end of the "A" or broadcast band. As the plane (or loop) is rotated the deflection will be found to reach a maximum where the plane should remain during the following adjustment. The course sensitivity is adjusted to give a sizable reading of about three-quarters full scale. The antenna adjusting screw on the front of the receiver is turned until a point is reached where the deflection reaches the maximum value. A very strong signal, which is between 550 and 700 kilocycles, is tuned in to test the adjustment. The airplane is turned so that there is a slight right or left pointer reading. The receiver is now tuned through the station frequency while watching the needle. It should reach a maximum small deflection as the station frequency is reached and then the deflection should decrease to center scale as the receiver tuning is continued beyond the station frequency. Should the needle move to the opposite side of center upon tuning through the station, the antenna readjusting screw is turned slightly counterclockwise until the undesirable condition disappears. If this does not remedy the situation the mechanical and electrical centering described before should be checked to be sure the needle is at center with no signal being received with the course sensitivity set at minimum.

The equipment gives the choice of selecting the radio compass (RADIO COMPASS), the loop alone (BEACON-WEATHER-LOOP), or the regular vertical antenna (BEACON-WEATHER-

REGULAR). Automatic volume control is used only in the radio compass position. Loop operation can only be secured in the "X" and "A" bands.

A description of the electrical operation is given before describing the receiver alignment procedure.

When using the "antenna" alone with the operation selector in center position, the plate voltage is removed from loop and audio oscillator tubes (Nos. 1, 2 and 9), allowing normal antenna operation.

With the selector in the "loop" position the audio oscillator grid is grounded and the plate voltage to the No. 1 loop tube is disconnected, allowing loop reception through the remaining tube (No. 2), also the antenna is grounded on bands "X" and "A."

With the selector in the "radio compass" position the loop and audio oscillator tubes all function although they are not used on band "B" for any purpose.

Relay S7 (Fig. 168) is operated by selector switch (S8) in the "antenna" and "loop" positions and serves to remove plate voltage from tube No. 1 and grounds the audio grid. The same selector switch removes plate voltage from tubes No. 2 and 9 in the "antenna" position only. This relay grounds the antenna for bands "X" and "A" in the loop position of the operation switch. This explains the control of the tubes in the two paragraphs above.

One of the ganged switches shown below the antenna in the wiring diagram serves to short out the broadcast band antenna balancer for the "X" band. Another part of this same switch shown connected to the antenna eliminates the effect of relay S7 for the high frequency "B" band as far as antenna connections are concerned.

The manual sensitivity control and the headphone volume are ganged together and operated by the knob marked "headphone volume." The selector switch (S8) cuts out the manual volume control in the "radio compass" position if desired, although neither the radio-compass indicator nor loop will work and the antenna alone will be used. The "course indicator sensitivity" should be off except for compass use.

The intermediate amplifier is connected to a signal generator through a 100 mmf condenser to the grid of the 6F7 first detector, grounding the grid to the chassis with a 50,000 ohm resistor for alignment. The receiver is placed on the "A" band at its low frequency end and at a point where no signals are received. The generator is adjusted to 460 kilocycles and trimmers (movable cores) L29, L28,

L27, L26, L25 and L24 aligned in the order named for maximum output (Fig. 169). As small a signal input as possible is used. The adjusting screws are sealed with ambroid, or equivalent cement, after adjustment.

The generator is attached to the antenna post to align the radio frequency circuits through a 100 mmf condenser and the other lead from the generator is connected to the receiver ground. The antenna adjuster, reached from the front of the receiver, is set fully clockwise although its final adjustment will have to be made as described under compass alignment. Condensers C39, C23 and C27

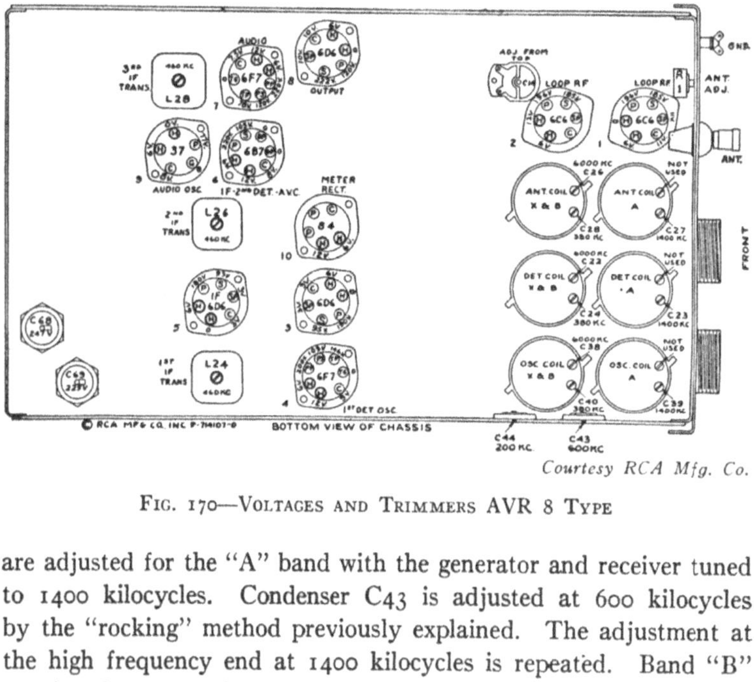

Courtesy RCA Mfg. Co.

FIG. 170—VOLTAGES AND TRIMMERS AVR 8 TYPE

are adjusted for the "A" band with the generator and receiver tuned to 1400 kilocycles. Condenser C43 is adjusted at 600 kilocycles by the "rocking" method previously explained. The adjustment at the high frequency end at 1400 kilocycles is repeated. Band "B" is aligned at 6000 kilocycles with C38, C22 and C26. The "X" band is aligned at 380 kilocycles by the "rocking" method and the 380 kilocycle alignment repeated afterwards. All adjustments are sealed with cement.

Fig. 170 shows the voltages to be found at sockets and Figs. 171, 172 and 173 show further details of the installation.

Most continuity or ohmmeters will burn out the R-L indicator meter of a radio compass. A one-tenth or one milliampere meter may be substituted for the indicator to show that current is flowing in the circuit.

It will be found that some receivers require the following modi-

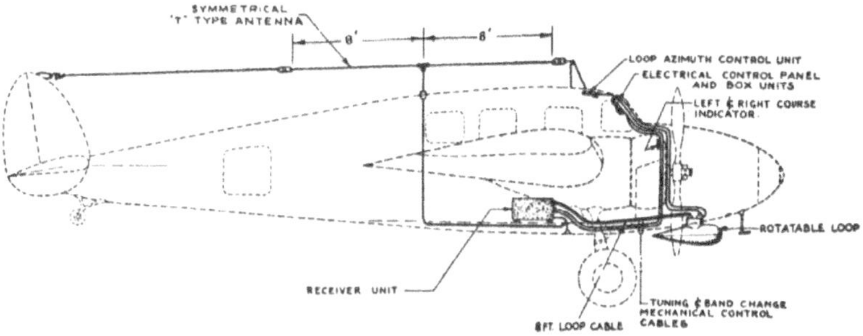

Courtesy RCA Mfg. Co.

FIG. 171—AVR 8 COMPASS—LOOP BELOW

fication where frequency shift with sensitivity control setting has been too great in the "A" band. Resistor R47 (located near L19) is removed and a jumper wire connects the two terminals previously

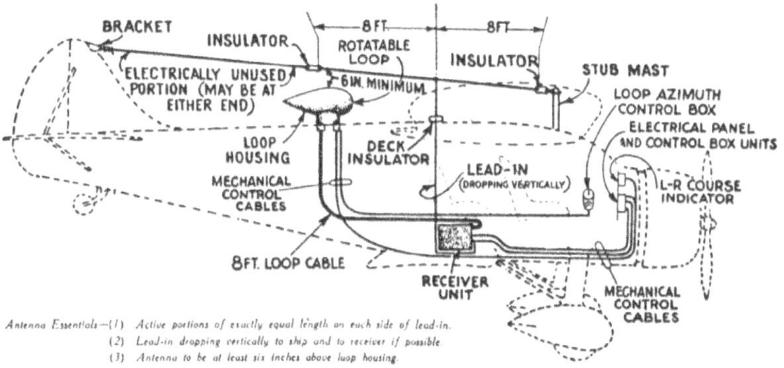

Courtesy RCA Mfg. Co.

FIG. 172—AVR 8 COMPASS—LOOP ABOVE

used. A 1000 ohm quarter watt resistor (R52) is connected across L20 at the base of "A" coil. A 2700 ohm quarter watt resistor (R53) is connected across L18 at the base of the "A" coil.

27A Western Electric Receiver—

	Frequency Range	*Alignment Frequency*
Radio input	75 MC	75 MC
Intermediate	6325 KC	6325 KC
Audio filters	3000, 1300, 400 cycles	Fixed

Power unit: External dynamotor, internal vibrator or internal 800 cycles 115 volt rectifier

Power consumption: Filament voltage 6 or 12 volts with external dynamotor operation. With vibrator 3 amps, 12 volts. At 115 volts 800 cycles, 65 VA

Power output: About 50 milliwatts headphone. About 3 watts AC at 100 cycles for light operation (controlled by about 0.3 watt from the final tube)

Tube complement: Converter 6J7
Osc. and Har. Gen. 6N7
IF amplifiers (2) 6J7
Detector 6Q7
Output 6V6G

	HEIGHT	WIDTH	DEPTH	WEIGHT
Receiver	7 7/16″	5 3/16″	21″	19.3 lbs.

The receiver was the first ultra high frequency receiver in wide use in the United States. It operates on a frequency of 75 megacycles and is a crystal controlled six tube superheterodyne receiver. The receiver is arranged to light one of three lights, depending on the modulation frequency of the transmitter which may be 3000, 1300 or 400 cycles. Fan and cone of silence markers are modulated with 3000 cycles, while approach markers, used for instrument landing, are to utilize the other two modulation frequencies. Three colored lights indicate the modulation frequency. White is standard for 3000 cycles while amber has been selected for 1300 cycles (outer marker) and purple for 400 cycles (inner marker). Many receivers will be equipped only with the 3000 cycle filter for the present.

Another innovation to be noted in this receiver is the internal power supply arrangement, Fig. 174. Either a (9A) 115 volt 800 cycle unit or a (8A) 12 volt synchronous vibrator may be used and a mounting for quick change of units is provided on the receiver chassis. Switching is provided so that the receiver may be thrown to an external power source of six or twelve volts for filaments and 200 volts direct current for plates.

The unit shape is shown in Fig. 175 and is half of the standard width. Two marker receivers may be used in a standard aircraft rack with a small modification of rack catches and backstop pins.

A separate (7A) mounting can be supplied for individual base mounting.

One peculiarity in the electrical circuits is the use of an harmonic generator V6. The crystal drives the first triode of a double triode 6N7 and the output of that part of the tube drives the second triode grid. A neon lamp associated with the circuit tuned to the crystal frequency is employed to keep the excitation constant with variation

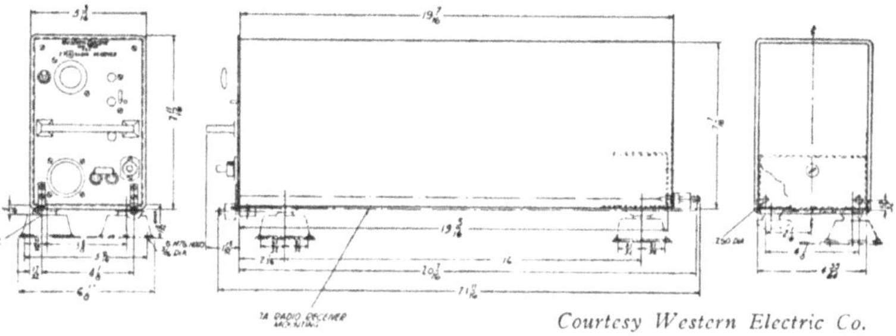

Courtesy Western Electric Co.

FIG. 175—27A OUTLINE DIMENSION DETAILS

in crystal activity. The plate circuit of the first triode and second triode are both equipped with third harmonic tuned circuits so that the crystal frequency is raised nine times total. This gives an equivalent first oscillator frequency of 68.675 megacycles which is 6.325 megacycles removed from 75 megacycles. The intermediate frequency is 6.325 megacycles.

The indicator lights are connected (Fig. 174) in series with a coil on a saturable core reactor. When direct current is supplied to another winding on the core the inductance in series with the lamp decreases and the lamp lights brightly. A feedback circuit is used to bias the first audio amplifier and is incorporated in the second detector. When signal increases, the audio amplifier gain also increases, causing the light to reach full brilliance with relatively small signal increase. No relays are required to operate the lights. This contributes greatly to receiver reliability.

A fuse is used in a front panel holder and the position in which the fuse is inserted selects the type of operation desired. For twelve volt operation the fuse is a ten ampere, twenty-five volt fuse and for 800 cycle operation it is a three ampere, 250 volt fuse. Fig. 176 shows a typical installation diagram.

Antenna and preselector tuned circuits can be tuned through

holes provided under a slip cover on the front panel. Oscillator and intermediate adjustments are made as in any normal receiver. Particular attention must be paid to tuning of the oscillator circuits to assure that the third harmonics are being found in each case. Maxi-

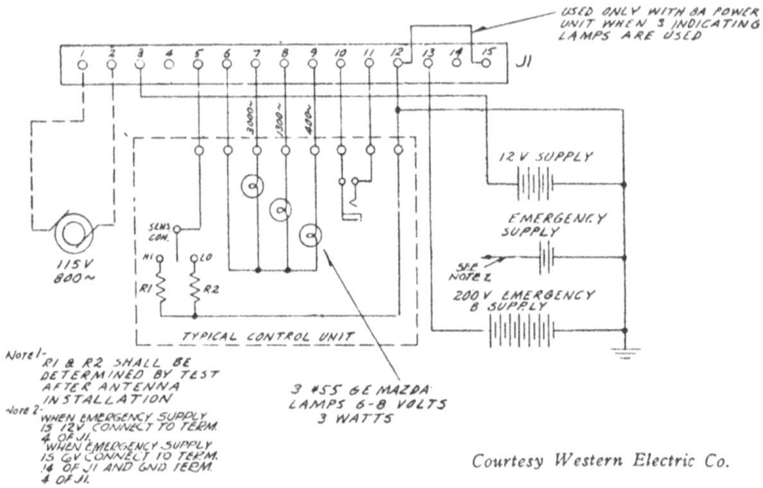

Courtesy Western Electric Co.

FIG. 176—EXTERNAL WIRING 27A RECEIVER

mum drop will exist across the plate dropping resistor (R3) of the first detector when the oscillator circuits are properly tuned. This can be measured across the terminal board.

Voltages measured to ground are as follows (within plus or minus about ten per cent):

POINT	VOLTAGE	POINT	VOLTAGE
Plate V1	220	Plate V4	175
Screen V1	110	Cathode V4	3
Cathode V1	6	Plate V5	220
Plate V2	220	Screen V5	230
Screen V2	105	Cathode V5	13
Cathode V2	2	Each plate V6	220
Plate V3	220	Ter. 7 (8A unit)	13
Screen V3	105	Ter. 7 (9A unit)	57
Cathode V3	2	Ter. 4 (8A unit)	13

In addition the 800 cycle alternating voltage between terminals No. 2 and 3 is six volts and between terminals No. 1 and 8 on the 9A power unit is 116 volts.

The following values of parts are found in the 27A Western Electric receiver:

PART	DESCRIPTION
C1	5 mmf
C2, C3, C4, C5, C6, C7, C10, C13, C16, C21, C26, C28	.006 mf
C8, C25	50 mf
C9, C11	.01 mf
C12, C22, C23	Two 8 mf, one 5 mf
C17	100 mmf
C18	10 mmf
C14, C15, C20	15 mmf
C24	1.75 mf (400 cycle)
C27	12 mf (1300 cycle)
C29	12 mf
C101	2 mf
C102, C107	.5 and .1 each
C103, C104, C106, C108	.006 mf
C105	.0075 mf
R1	1,500 ohms
R2, R17, R29	100,000 ohms
R3, R7, R11, R12, R28, R30	1,000 ohms
R4, R8, R35	10,000 ohms
R5, R9	300 ohms
R6, R10	75,000 ohms
R13, R15, R16	.25 megohm
R14, R27	2,000 ohms
R18	.5 megohm
R19	600 ohms
R20	100 ohms
R21, R22	120,000 ohms
R23	2 megohms
R24	1 megohm
R31	25,000 ohms
R32	50,000 ohms
R33	50,000 ohms (1300 cycle)
R34	50,000 (400 cycle)
R101	5,000 ohms
FL5, FL6, FL7	Sat. reactors
V1, V2, V3	6J7 tube
V4	6Q7 tube
V5	6V6G tube
V6	6N7 tube
V201	5W4 tube
X1, X2, X3	Rectox units

QUESTIONS ON CHAPTER IX

1. The choke L15 in the Western Electric 9B receiver is for:

   ( ) a. Automatic volume control. ( ) b. Reduction of higher audio frequencies. ( ) c. For removal of radio frequencies beyond the detector.

2. The inductor L14 in the 9B receiver high voltage lead is:

   ( ) a. To provide better audio gain. ( ) b. To filter out dynamotor noise from the power unit. ( ) c. To provide coupling to the headphones.

3. The coils L6 to L11 in the 9B are replaced in modern receivers with resistors in order to save:

   ( ) a. Cost and weight. ( ) b. Amplification otherwise lost. ( ) c. Voltage drop in the chokes.

4. The Western Electric 9B and 9D receivers differ in that:

   ( ) a. The 9D has a continuously tuned antenna circuit. ( ) b. The 9D is a superheterodyne. ( ) c. The 9D employs a ballast lamp.

5. The 6B7 in the Western Electric 17A is:

   ( ) a. Both a detector and a converter. ( ) b. Both a converter and oscillator. ( ) c. Both an intermediate and audio amplifier.

6. The element R15 in the 17A receiver is:

   ( ) a. A plate voltage dropping resistor. ( ) b. An automatically variable resistor to reduce noise. ( ) c. Not a resistor but a quartz crystal.

7. The vibrator in the AVR10 receiver is a:

   ( ) a. Synchronous vibrator, rectifying its own high voltage. ( ) b. Non-synchronous type using a tube as a rectifier. ( ) c. Static and noise reducing device.

8. The AVR10 volume control not only changes the first 6F7 cathode bias but:

   ( ) a. It also reduces the signal from the antenna. ( ) b. The bias of the second 6F7 as well. ( ) c. It also controls the audio gain.

9. Filters are supplied in the AVR10 vibrator unit, not only to filter the high voltage circuit, but also:

   ( ) a. The low voltage, to furnish filtered DC for the rectifier. ( ) b. Protection to the receiver filaments. ( ) c. To prevent vibrator noise from disturbing the receiver.

10. The resistor in series with the headphones in the RCA AVR7 receiver:

( ) a. Prevents overloading. ( ) b. Reduces static crashes by limiting the peak audio output. ( ) c. Acts as a dummy output load for all testing.

11. The vibrator in the AVR7 is of the:

( ) a. Synchronous type. ( ) b. Non-synchronous type. ( ) c. Half wave type.

12. Intermediate transformers in the AVR7G are tuned with:

( ) a. Movable iron core. ( ) b. A standard condenser trimmer. ( ) c. Variable coupling between primary and secondary.

13. The CW oscillator is employed in the AVR7G to:

( ) a. Beat incoming signals to the intermediate frequency. ( ) b. To operate a tuning meter. ( ) c. Beat incoming continuous wave signals such as code signals in the intermediate amplifier.

14. The automatic volume control in the AVR7G controls:

( ) a. The IF, RF and converter amplifiers. ( ) b. The RF and converter amplifiers only. ( ) c. Only the intermediate amplifier.

15. The volume control in the AVR7G controls:

( ) a. Cathode voltages of the RF, IF and converter tubes. ( ) b. Only the audio volume, the AVC handles the rest. ( ) c. Negative grid voltages of the RF, IF and converter.

16. The Western Electric 20A receiver used the two diodes of the tube for:

( ) a. Automatic volume control and detection separately. ( ) b. Both AVC and detector as they are joined together. ( ) c. Limiting the static peaks.

17. The 20A uses the ballast lamps E2 and E3 to:

( ) a. Hold the filament voltages constant. ( ) b. Hold the filament current constant. ( ) c. Light the dial scale for these are dial lamps.

18. The tube E1 in the 20A is used for:

( ) a. Diode rectification. ( ) b. Reducing high radio voltage inputs. ( ) c. Completing the grid circuit.

19. The antenna circuit of the 12D Western Electric receiver is:

( ) a. A series tuned circuit. ( ) b. A standard parallel tuned circuit. ( ) c. Not tuned.

20. Microphone signals may be fed to terminal 6 so that the 20A receiver final tube will act as an interphone amplifier:

( ) a. This is so. ( ) b. The choke L9 would make this impossible. ( ) c. The sidetone circuit is provided for additional headphones.

21. The Western Electric 14B receiver uses a:

( ) a. It is not equipped for such purpose. ( ) b. Manually operated airport traffic control selector. ( ) c. Magnetically operated airport traffic control frequency selector.

22. The relay in the output of the ES198 receiver is to:

( ) a. Provide interphone service. ( ) b. Cut off the receiver output during transmission. ( ) c. Limit audio output on static bursts.

23. The Kurman relay near the ES198 antenna circuit is to:

( ) a. Transfer from antenna to loop. ( ) b. Cut off the receiver during transmissions. ( ) c. Operate the 278 kilocycle airport control switches.

24. The W4A receiver employs automatic volume control in the range receiver for:

( ) a. Future purposes as it is shorted out. ( ) b. Keeping the range signals at a comfortable level. ( ) c. Accentuating the signal level change.

25. The day-night switch in the W4A is:

( ) a. Manually controlled at all times. ( ) b. Operated by a separate switch from the selector switch. ( ) c. Operated automatically by a cam on the crystal selector.

26. The Bendix RA2B receiver employs a frequency change motor that:

( ) a. Travels in the shortest distance to a new frequency. ( ) b. That turns only in one direction until its selector open sector comes opposite an open contact and a gear clutch is opened by its solenoid, the motor coasting to a stop. ( ) c. Moves the selector arm to a new frequency when the pilot hearing the correct signal, lets go his selector push-button.

27. If one filament opens up in the RA2B receiver another filament is bound to cool off immediately as:

( ) a. Filaments are series wired in pairs. ( ) b. The other filament will receive twelve volts and burn out. ( ) c. This is not true.

28. The Bendix RA-4C receiver utilizes the 6L7G as a:

( ) a. Converter only. ( ) b. Converter and oscillator. ( ) c. CW beat oscillator.

29. The RA-4C utilizes two output amplifiers for:

( ) a. Additional audio gain. ( ) b. Reduction of plate power consumption. ( ) c. Providing separate amplifiers for pilot and co-pilot.

30. When a plug is fully inserted in the audio jack in the RA-4C it will:

( ) a. Cut off receiver signals to both pilot and co-pilot. ( ) b. Cut off receiver signals to either pilot or co-pilot. ( ) c. Not cut off signals to either man.

31. Only one output jack is provided on the Bendix RA-6A receiver:

( ) a. So that only one of the output channels can be tested at the receiver. ( ) b. As there is only one output tube. ( ) c. Because both output tubes are in parallel.

32. The Bendix MN-12A radio compass employs:

( ) a. Only one intermediate amplifier tube which has four tuned circuits between it and the preceding tube. ( ) b. The neon tube to reduce large input signals and not to control oscillator stability. ( ) c. Two loops at right angles to one another and no vertical antenna in normal compass use.

33. The band switch motor in the MN-12A is reversable:

( ) a. So as to reduce switching time to an absolute minimum. ( ) b. So as not to twist the pigtail connections off the switches. ( ) c. It is not reversable.

34. As a plane equipped with a right-left indicator needle such as on the RCA 8E compass passes a radio station well to one side the needle will:

( ) a. Cross center scale as the station is passed. ( ) b. Reach maximum deflection at the time of passing. ( ) c. Swing across the scale and back indicating the station has been passed.

35. The course indicator of the RCA 8E compass can be reversed in direction of needle movement in a turn by:

( ) a. Reversing the two meter leads. ( ) b. Reversing either of the two meter pairs. ( ) c. Reversing connections to the battery.

36. The Western Electric 27A receiver employs three filters in the output circuit to:

( ) a. Provide for reception of three modulation frequencies. ( ) b. Reduce static interference. ( ) c. Provide spares in case of failure of one filter.

37. The 800 cycle power unit for the 27A receiver:

( ) a. Converts DC to AC for operation of the indicator lights. ( ) b. Converts DC to AC for plate supply. ( ) c. Converts 800 cycles to DC for plate supply alone.

Chapter X

RADIO WAVES AND STATIC

The paths taken by radio waves explain some of the peculiarities of radio transmission. For aircraft purposes, the radio waves are divided into three frequency groups.

1. Low frequencies (200 to 400 kilocycles)
2. High frequencies 2500 to 20,000 kilocycles)
3. Ultra high frequencies (50 to 500 megacycles)

The frequencies between the low and the high frequencies are chiefly used for commercial broadcasting. Their only use to aircraft is in some direction finding done upon these broadcasting stations.

The frequencies between the high and the ultra high frequencies have no great application in most aircraft service.

The frequencies above 500 megacycles are not sufficiently developed to be used at present as apparatus to generate and receive these frequencies is in the experimental stage.

The Heaviside Layer—The Heaviside layer is so named for the English physicist, Sir Oliver Heaviside. It consists of a layer of ionized gas above the earth at a distance of roughly fifty miles. There are actually several layers, some of which are found lower and a few higher than this figure. For the present purpose, the chief point of interest may be explained by assuming that there is one layer.

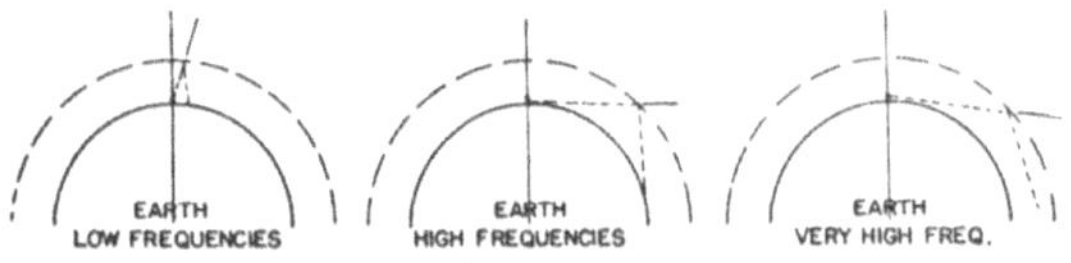

FIG. 177

Fig. 177 shows a radio wave emanating at high frequency from a station on the surface of the earth. It will be seen that when the wave direction to the Heaviside layer approaches the zenith overhead, the wave is no longer reflected but passes through the layer to outer space and is lost. Reflected waves strike the earth as shown.

Waves that are nearly horizontal to the earth's surface at the point of transmission may not be reflected directly back to the earth.

If waves of all frequencies were reflected equally and if the layer remained at a constant height it would be very easy to predict the operation at any radio frequency. However, the height of the layer varies appreciably with the time of day and the season. When the sun sets, the ionization begins to disappear at the bottom of the layer first. Every time there is a major sunspot the bottom of the layer is pushed far down toward the earth and badly broken up.

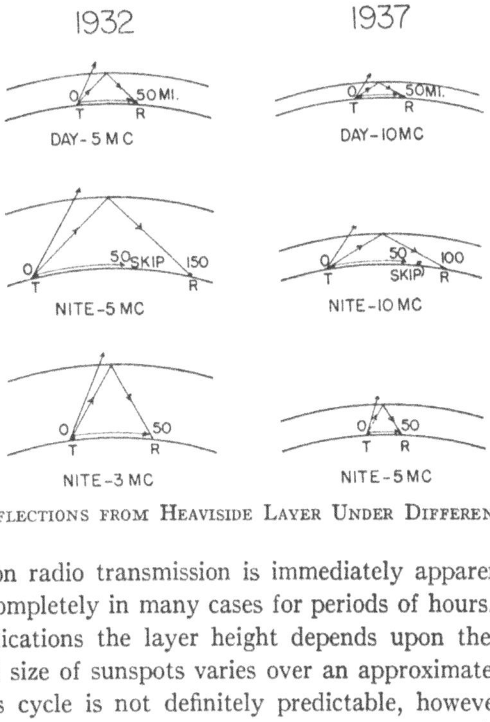

FIG. 178—REFLECTIONS FROM HEAVISIDE LAYER UNDER DIFFERENT CONDITIONS

The effect on radio transmission is immediately apparent in loss of reception, completely in many cases for periods of hours. To add to these complications the layer height depends upon the year. The number and size of sunspots varies over an approximate eleven year cycle. This cycle is not definitely predictable, however a sunspot maximum was apparently passed late in 1937. At that time the average layer was as low as it will be until some time (approximately) in 1947. In 1942 or 1943 the layer ought to reach its maximum average height.

Examination of Fig. 178 will show the behavior of two frequencies in 1937 and 1932. Each frequency is the mean of the aircraft daytime or nighttime frequencies. Each United States airline has one or more groups of these day and night frequencies.

It will be seen that in 1932 the direct ray at 5000 kilocycles, which roughly follows the earth's surface for fifty miles or so, blends into the reflected skywave during most of the daylight hours. In the early evening, when the layer is rising, a gap appears giving poor signals between about fifty and one hundred fifty miles from the station. Stations and planes separated by less than fifty, or by more than one hundred fifty miles will have good reception. In air transport service it is very greatly desired to have the region completely covered from the ground station out to about 250 miles. Thus, when the late afternoon or early evening condition occurs, a change is made to the night frequency which will not show this gap in its transmission path. This is the reason for using a pair of frequencies, one daytime and one night, over the twenty-four hour day.

Under the conditions in 1937, the day frequency could be much higher and would have started to show gaps, or skip, at night. A night frequency could have been employed which would be virtually of the same frequency as the day frequency in 1932. It is obvious why in 1937 the day frequency was used for long periods of months without resorting to night frequency at all.

It is obviously impossible to change frequencies with the sunspot cycle and with the season: first, because of the expense of crystal change involved and second, because of the necessary number of frequency assignments. The best solution would be the addition of another frequency of approximately eight to ten megacycles for use under certain Heaviside layer conditions.

The high frequency spectrum is unique in its behavior in that there will always be an optimum frequency at any time of day which will give uniform coverage from the ground station to the airplane. This frequency will also give signals of maximum strength. Thus airplanes equipped with these frequencies are afforded the best circuit possible with the least required power. As powerful aircraft transmitters are heavy and far more expensive than small ones, the choice of frequency is very important.

The low frequencies, on the other hand, are not high enough so that there will ever be a gap in the transmission path. Sometimes there is a tendency toward skip so that signals may be weakened, but not absent, in certain regions. In this case the signal will merely become increasingly weaker as the distance from the station increases. At night, when the layer becomes higher, the signals at a distance increase substantially in strength. Thus the range of reception is about twice as great at night as in the daytime. The effect of reflections is not as pronounced in the daytime but at night may cause some periodic fading of signal.

Fig. 179 shows three possible paths for a signal to travel from a transmitter to a receiver. Path 1 is the so-called ground wave or a wave that seems to follow the surface of the earth for some distance. At high frequencies, this wave is attenuated rapidly with increasing distance. The path labeled "2" is the ordinary reflected wave which carries most of the energy picked up by the receiver. The path "3" wave goes all the way around the earth and sometimes produces an echo in the received signal as the time for its travel is appreciable.

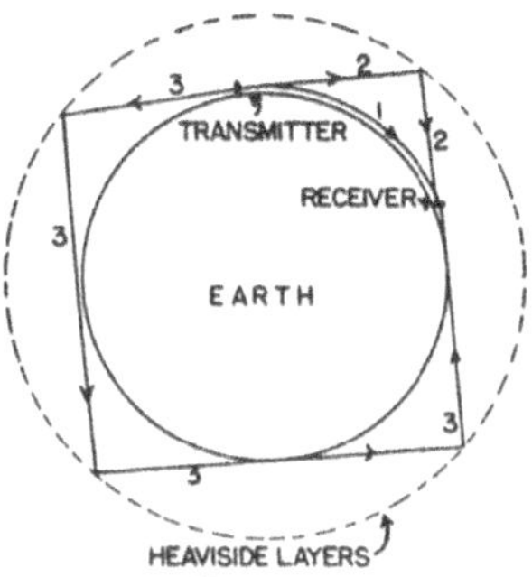

FIG. 179

Ultra High Frequencies—The ultra high frequencies show quite a different behavior. Frequencies of approximately fifty megacycles and higher are never reflected back to the earth. Further these frequencies show no tendency to bend appreciably (be refracted) around the surface of the earth. Mountains and other obstructions will cut off transmission when within the "line of sight" from the ground transmitter to the airplane. This means that the airplane must gradually rise as it leaves the station in order to obtain good transmission and reception. In level country, with the ground transmitter antenna at a height of about fifty feet, the plane must be at 1000 feet above ground at about fifty miles from the transmitter; 3000 feet high at a distance of about seventy-five miles; 10,000 feet high at 140 miles; 15,000 feet high at 160 miles and about 25,000 feet high at 230 miles.

As a matter of fact the waves will bend slightly and two curves are shown in Fig. 180, which give the required height of the airplane at various distances over flat terrain. The line of sight curve is the actual free line of sight while the curve including refractions shows the maximum ranges over which fair transmission may occur. The amount of refraction will vary considerably with the humidity of the air and other variables. Attempts to gain greater transmission distances, well below the line of sight, will require something like a million times increase in power to cover fifteen per cent further distance, for example.

This limitation of the ultra high frequencies is a great advantage as well as a disadvantage. The advantage is the lack of long range interference and complete lack of static. The chief disadvantage is that frequencies are hardly suitable for transmission from one ground station to another or to planes at low level at great distances, unless geographical conditions show that a line of sight can be maintained.

Fading—Fading of signals is produced in several ways. The most general reason for fading is the combination of two sky-reflected waves or a sky and direct, or ground, wave. If the length of the two signal paths differ by one-half wavelength and the two

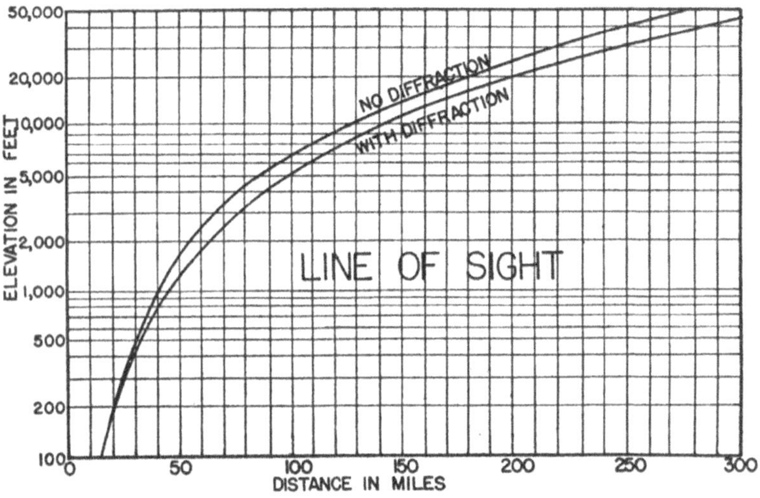

Fig. 180—Transmission Distances on Ultra High Frequencies

signals are nearly equal in magnitude, they will virtually cancel one another and no signal will be received. As the Heaviside layer is not smooth but rolls like the sea, this type of fading may occur between two stations which are fixed in location. Fading is usually not very severe on the low frequencies because of little reflection but may be exceptionally severe at high frequencies. Fading is only found at ultra high frequencies due to the transmitted wave being reflected from the intervening ground or from the side of a building near the transmission path, etc., and will be encountered only if the distance between the transmitter and receiver is changing.

Occasionally, when a very severe sunspot occurs the whole high frequency spectrum will fade out completely for periods of a few minutes to several hours. It will sound as if the receiver has been suddenly turned off and static which originates at even a short distance is not heard. The farther north the station location (in the northern hemisphere) the worse and the longer will be these complete fadeouts. These fadeouts have recently been named "Dellingers," for the scientist in the United States Bureau of Standards who is investigating these phenomena.

Fadeouts have little effect upon reception in the low frequency band and do not have effect at the ultra high frequencies. Apparently they are due to complete breaking up of the Heaviside layer, i.e: intense ionization. They occur in greatest severity in the peak of the sunspot cycle. Between 1930 and 1936 they were not of sufficient length or consequence to be noticed, however in late 1936, 1937 (especially), and in 1938 they caused severe interruptions in high frequency communication.

Interference—Two transmitters located at a distance from one another may not interfere with each other at either locality but a condition of severe interference may result at points between the stations, particularly half-way between, where both stations can be heard equally well. Thus the day frequency channels used by aircraft are segregated into regions of use as well as by airline company. Transcontinental airlines use two or three different day frequencies, one in the eastern part of the country, one in the central part and one in the western part. Each frequency is used by all aircraft and ground stations in the particular region. Thus the circuit is similar to a party telephone line and unless transmissions are carefully scheduled and supervised there will be considerable interference. The same is true for the night frequencies used. For this reason, the ultra high frequencies are ideal for plane to ground radio contacts as each station covers a given area and beyond that area the station does not cause interference. One ultra high frequency, for all times and regions, only need be employed. The necessity for changing frequency is done away with and the cost of crystal control apparatus is reduced.

At the low frequencies it will be found that as transmission conditions improve at night, interference between stations becomes fairly severe. Range stations are operated on a rather limited group of frequencies, which assignments must be repeated over the country. While the distance of reception may be sufficient in the daylight hours it is too great at night. The tendency is therefore to reduce the power of stations but their daylight range may be impaired by so doing.

Static—The most general form of static is due to electrical discharges in the atmosphere accompanying thunderstorms. The level of general static discharge in the atmosphere is very severe in equatorial latitudes. Centers of such disturbance may be located quite remotely from the actual receiving station and static will disturb reception thousands of miles from its source.

The low frequency bands are by far the most affected by static. In equatorial latitudes these disturbances may render low frequency communication almost impossible over distances of a few miles.

At high frequencies, the disturbance is much less pronounced except when relatively nearby thunderstorms (with lightning) occur. The higher the frequency the less will be the disturbance. The ultra high frequencies are almost entirely free from any of this disturbance. Airplanes in a thunderstorm of great severity will not be bothered by static and even nearby lightning discharges will cause but a slight click in the headphones. For this reason the ultra high frequencies are very desirable.

Man-made Static—Man-made static comes from a number of sources. There are several sources on airplanes which can usually be traced to intermittent contacts in metallic parts of the plane and sparking produced on commutators and relay contacts, such as those used in the generator control boxes. There are three ways of overcoming such interference: shielding, bonding and filtering.

Shielding—Shielding consists of enclosing wiring and apparatus in a metallic shell. The thicker the shell the better will be the shielding. In aircraft practice, the wall thickness must be held to a minimum to conserve weight. Most machines such as generators, magnetoes and motors, have a metallic frame which encloses the machine except for points where leads must be brought out or inspection holes for servicing, changing brushes, etc., are provided which break the shielding. It is at these points that great care must be taken (Fig. 181). A metal coverplate which is making poor metallic contact to the machine may allow serious leakage of disturbing radio interference fields. Such covers should be secured with good contact to the machine and the contact surfaces should be kept clean.

There is a collar on most magnetos (Fig. 182) through which the wire leads run to the distributor. This collar is of metal and fits over the metallic frame of the machine. Leakage at this point is usually severe and every effort is made to keep the fit of the collar to the machine tight and clean. Nevertheless, oxidation will cause the contact to become poor with time and noise level from this source will gradually increase.

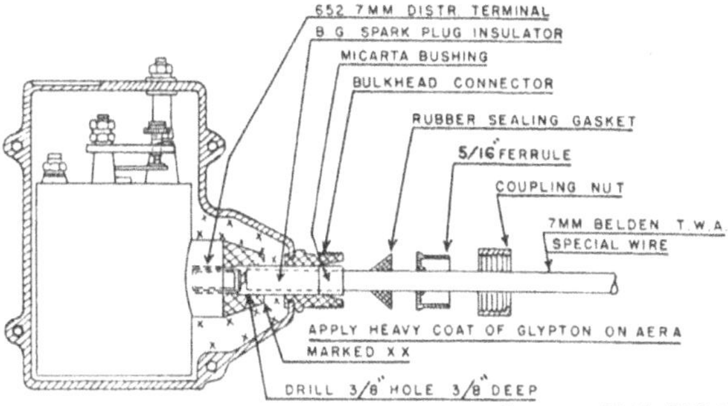

Courtesy UAL-TWA

FIG. 181—BOOSTER COIL SHIELDING

Wires are generally enclosed in either a non-flexible aluminum conduit or a flexible metallic hose. There is relatively little danger of leakage along the hose as long as the material is unbroken. Noise may be radiated at the point of connection of the shielding hose to

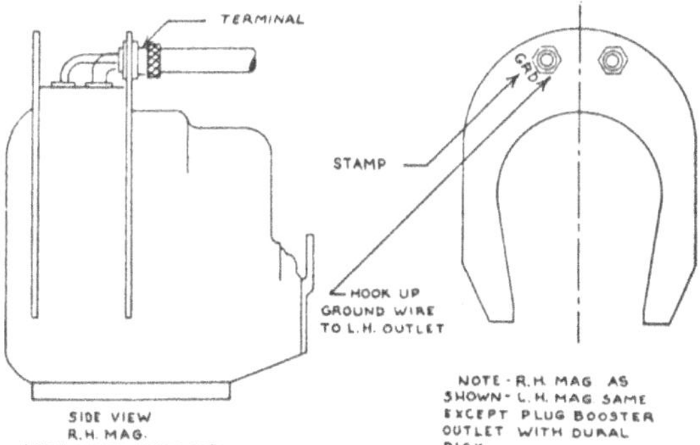

Courtesy UAL-TWA

FIG. 182—MAGNETO COVER SHIELDING

the machine or at the point of connection to the other metal conduits. The hose must be soldered into a metallic ferrule at each end and a nut placed on each ferrule to secure the hose to the threaded bushing of another conduit or machine cover. When aluminum braid and fittings are used, there is only one process of connection of the end rings to the hose and this is by means of crimping. As the connection is flexed it becomes poor and the shielding no longer serves its

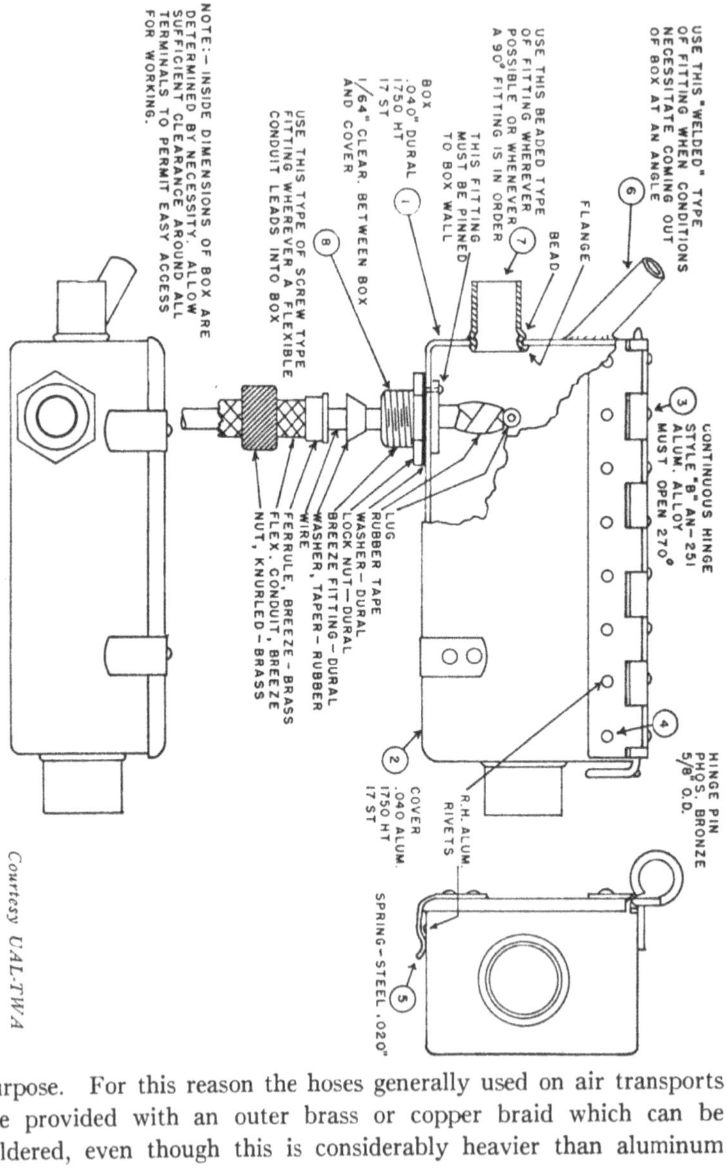

FIG. 183—TERMINAL BOX PRACTICES

purpose. For this reason the hoses generally used on air transports are provided with an outer brass or copper braid which can be soldered, even though this is considerably heavier than aluminum braid (Fig. 183).

Some idea of the magnitude of leakage that is present with a long crack may be gained from the fact that a crack one-sixteenth inch wide, about an inch long will allow almost as much leakage as a round hole about three quarters of an inch in diameter. Thus small holes or very short cracks need not cause concern but long cracks may cause trouble even though very narrow.

In all-metal airplanes the shielding and bonding are not as important when inside the fuselage as they are in a fabric covered plane, as the metal skin cover does not allow radiation of noise to the external antennas.

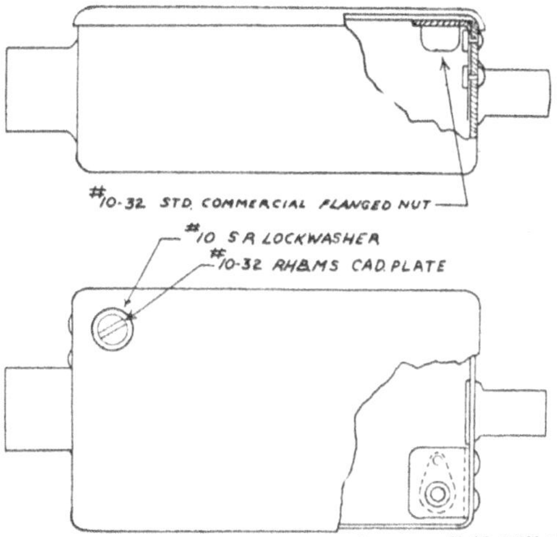

Courtesy UAL-TWA

FIG. 184—WATERTIGHT TERMINAL BOX

One of the chief sources of leakage is through the covers of terminal boxes in the electrical system. The covers should be well lipped so no open crack appears around that end of the box (Figs. 183 and 184).

In summary, shielding must be nearly watertight, particularly where long slots occur. Pin hole or small hole leakage is not particularly important except where water protection is necessary (Fig. 184). If a long slot is almost inevitable, as in the case of terminal box covers, a good overlap of metallic surfaces, even though not in contact, is beneficial.

Bonding—Bonding is the attachment of shielding to the metallic structure of the airplane. A conduit should be bonded at intervals of not over two feet at the most (Fig. 185). Such bonding must be clean so that the resistance of the bond is practically negligible. Long bonding leads are very much poorer than short ones as they exhibit far more reactance and resistance. The reason that one ground per conduit is not satisfactory is that radio waves may occur on the conduit as they do on an antenna. If interference is the radiation source it will cause noise in receiving apparatus.

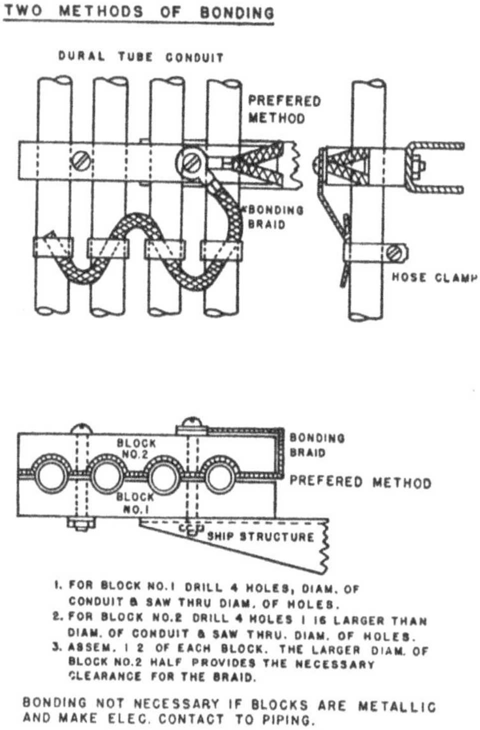

Courtesy UAL-TWA

FIG. 185—BONDING OF PIPING

The electrical connection at the end of the flexible hoses is very important and, as covered above, the shield should be soldered to end connections and the nuts securing the shield to the conduit should be kept bright, clean and tight.

Certain parts of the airplane itself may cause noise due to two non-bonded elements making intermittent contact. Thus a non-bonded engine cowling will cause interference if it rubs on its metallic supports. Such parts should be connected with a heavy short braid across the gap at points of mechanical contact. In some cases it will be found that cowling is well insulated from the supports with leather or rubber bumpers. In this case bonding is necessary for the cowling may pick up an electric charge as the plane flies through charged rain and the like. The charge will leak to the plane causing structure noise.

Most bearings, even though filled with oil, make good contact and need not cause concern. The main engine structure should be well bonded to the plane with heavy braid at its attachments. A bad

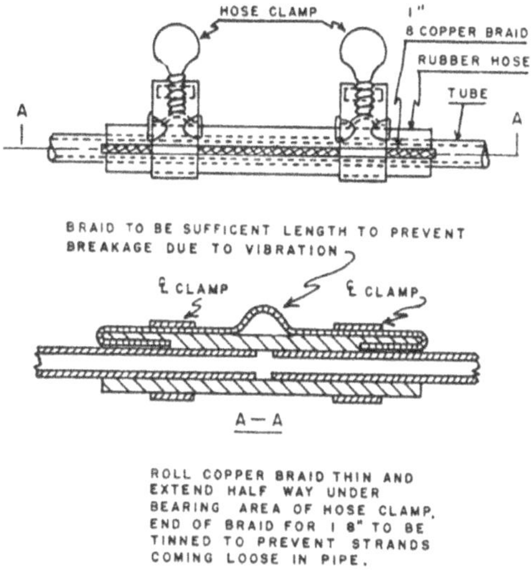

Courtesy UAL-TWA

FIG. 186—BONDING HOSE CONNECTIONS

ball bearing in a dynamotor, however, can cause interference. Metallic tubes are connected with hoses of rubber at some points as in Fig. 186 and should be well bonded.

Control cables used to move the airplane control surfaces must often be bonded, as in Fig. 187, and so must hinges on ailerons and movable tail control surfaces. On large transports of metallic construction it will usually not be as necessary to bond the control cables themselves. Cable bonding should be avoided as the bonding must be very carefully arranged so that it cannot cause jamming of control wires under any condition.

In summary, bonding consists in frequent "ground" connections (to the metallic plane structure) between all metallic parts and conduits unless sufficient riveting assures good contact, as in the wing and fuselage construction.

Filtering—Filtering is necessary when a source of interference, such as sparking brushes, is encountered. The heavier the current the greater will be the interference. Shielding and bonding will not be sufficient unless a great weight of each is added to reduce such interference. The alternative is to place a filter in the leads of the machine causing disturbance. Thus the interference is stopped very near to its source and causes no difficulty.

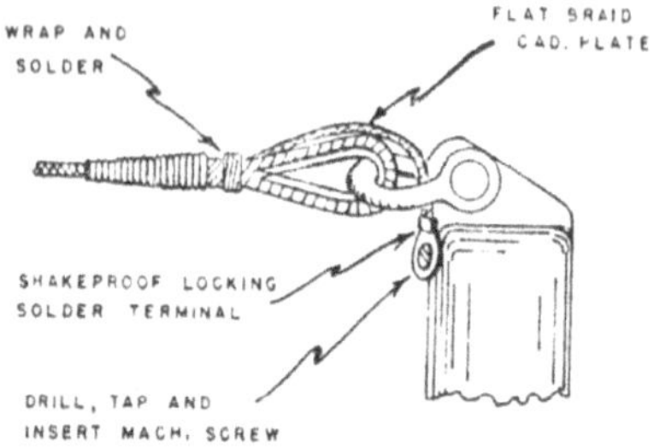

FIG. 187

While starters carry very heavy current and set up a good deal of interference, this occurs at a time when radio reception is not desired and last only for a few seconds. This does not mean that these leads may be unshielded for they are connected to other electrical apparatus either directly or through capacity between wires at switches, in terminal boxes and in conduits. It does mean that a filter to reduce noise from this source is not required.

Another type of filtering that must be employed is the condenser-resistor series combination across sparking points of relays. The method is equally important to reduce this sparking from the standpoint of contactor life alone.

At ultra high frequencies, it will be found advantageous to place a small filter, such as a simple inductor, in magneto ground return leads, for example. These leads carry no current unless the magneto is grounded but the high frequency interference may be carried by the wire and passes down the lead by virtue of the capacity of the lead to its shield.

In summary, filtering is usually required where sparking occurs.

Other Interfering Sources—Interference that exists on the ground at the receiving station is very important because the power of the aircraft transmitter must be sufficient to deliver a signal which will override this electrical noise. The usual expedient is to locate the receivers at a point remote from the airport where such noises are at a minimum. This requires controlling the remote receiver over and delivering its signal through a telephone line. While this rental expense is high it is far cheaper than carrying a larger transmitter on the airplane.

Were it not for static, there would be little difficulty in maintaining communication with very low power. An example of the possible increased power requirement is shown in the following example. At a certain point a signal was received which had a field strength of five microvolts per meter. The noise at that location from all man-made machinery was 2.5 microvolts per meter. Under certain conditions the man-made noise reached a peak of ten microvolts per meter and even under these conditions the location was considered good. To maintain the same signal to noise ratio as before, the distant transmitter would have had to be increased in power from its original fifty watts to a new value, delivering four times the voltage or sixteen times the power, or 800 watts. In summer static condition, the general static level might be twenty-five microvolts per meter. Thus the power to override this would require a transmitter (under the above conditions) of 5000 watts.

Naturally it makes little difference at what point the transmitter is located as far as noise is concerned and as long as a good antenna can be erected. Thus transmitters are found on or very near to the airports. The only restriction is in antenna height so as not to cause any obstruction to flying.

Precipitation Static—Rain static is the most common form of "precipitation" static. Rain, hail, snow or ice particles may cause the same type of static. Dust clouds cause a similar kind of interference.

Apparently there are two main causes of interference from rain. The first is from minutely charged particles of rain, snow, etc., hitting the antenna and discharging through the aircraft receivers to cause interference. The second, seemingly of equal importance, is the electrical noise created by discharge of electricity in the form of a corona from the airplane structure at sharp points.

Particles of rain, snow, sleet or dust are usually charged to a certain voltage. As the plane passes through these charged particles it will collect these charges as they strike the metal fuselage. The actual charge will be the sum of all the small charges picked up so that potentials of several hundred thousand volts may exist on the structure. St. Elmo's fire or corona will exist as the discharge of this high potential takes place into space. The higher the speed of the plane the more particles will be picked up each second and the longer and brighter will be the corona. Speed is therefore an important contributing factor to the amount of rain static. Occasionally there will be a rain static disturbance on the ground when a high wind blows rain across antennas. This is also true in dust conditions, particularly in dry desert country, especially where antennas are erected very close to the ground.

Weather conditions have quite an effect upon rain static as the static usually occurs near a cold front mass of air and where there is some turbulence. The condition is particularly pronounced just under clouds that are forming. Rain static areas may be relatively small or may extend over thousands of square miles. Rain static will vary in intensity with altitude but may be sufficiently great to disturb reception at any reasonable altitude.

There have been two solutions for the problem of rain static interference. The first is to shield the antenna with metal so that the electromagnetic radio waves may enter but electrostatic charges on the metal shield will not affect the antenna but pass to ground over the plane structure.

The second method is to prevent the airplane from discharging in a corona into space and causing interference. This has been done by attaching a ten foot length of resistance cord and wire to the tail cone of the plane. The resistor has about a half megohm resistance or more. Beyond this resistor, and attached to it, is a length of about ten feet of very fine wires, a few thousandths of an inch in diameter, braided together. These fine pieces of wire have sharp points and will allow discharge to take place much more readily than from any other part of the plane. Thus the corona discharge will take place at a point well removed from the airplane antennas and will cause little interference. The purpose of the resistor is to prevent the discharge interference from returning to the plane. The direct current which starts the discharge flows evenly through the resistor and in sufficient quantity to assure discharge.

Shielded Loops—The electrostatically shielded antenna referred to previously is required on all transports in the United States. It consists of an antenna made in the form of a coil of wire (loop). This loop is wound inside a tube bent in the form of a ring (Fig. 188). If the ring were continuous, the wire inside would be shielded completely and no signal would be received. For this reason the ring has a sector cut out of it which provides a gap of about one hundredth of the circumference. (The usual loop diameter is about ten inches.) The shielding is no longer perfect and magnetic radio waves may easily enter it. Rain particles that hit the metallic surface cannot affect the wires and the current created by the raindrop charging the airplane merely flows to the metallic skin over the outer surface of the shield. The reduction of rain static by this means is equivalent to reducing it by a factor of at least 10,000 times in many cases. Clear reception is often provided with the loop when reception is impossible on other antennas.

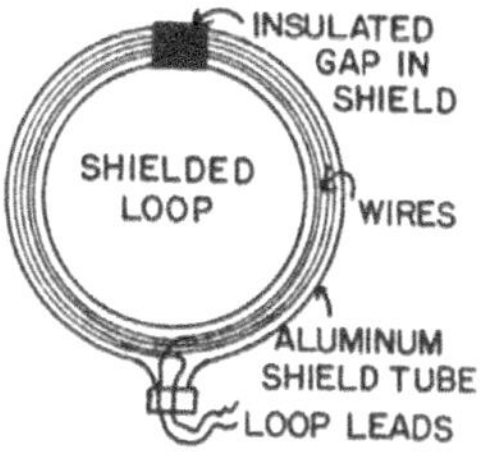

FIG. 188—ELECTROSTATICALLY SHIELDED LOOP ANTENNA

Sand Static—Sand static differs somewhat from snow, sleet and rain static in that it never causes corona on the plane and is apparently much more difficult to reduce with the electrostatic shield just described. One method that has been found effective in several cases is to enclose the loop in a large streamlined housing. This housing is covered with parallel strips of metallic covering which are insulated from each other except at one point where all are joined to the plane structure. The magnetic radio field penetrates between the strips while the charged particles striking the surface drain off to the plane structure through the strips. Further research is being carried on for the best methods of reducing rain static in all its types. The loop is being used extensively. Even so there are conditions, particularly in warm rains, where reception is prevented even when using a loop. Presumably the addition of a discharge wire will allow reception under all conditions if the loop is used too. Many aircraft are equipped with both devices.

QUESTIONS ON CHAPTER X

1. Aircraft use night and day frequencies because:

   ( ) a. The Heaviside layer does not reflect the night frequency as much at night. ( ) b. The lower night frequencies carry further at night. ( ) c. Uniform coverage without skip is desired at all times.

2. Ultra high frequencies have an advantage because:

   ( ) a. Their range is fairly well fixed regardless of time of day. ( ) b. They cannot be used between two ground stations. ( ) c. Obstructions along their path have small effect.

3. Fading at high frequencies and over short intervals is caused by:

( ) a. Signals being reflected back to the transmitter. ( ) b. Static interfering with reception. ( ) c. Out of phase combination of direct and sky waves.

4. Fadeouts for long periods during times that reception otherwise would be good are caused by:

( ) a. Rolling of the base of the Heaviside layer. ( ) b. Sunspot disintegration of all Heaviside layers. ( ) c. Severe reflections from the Heaviside layer.

5. The use of a single high frequency in a certain region has the advantage of:

( ) a. Eliminating interference between stations in that region. ( ) b. Preventing interruption of ground station communication by planes. ( ) c. Placing planes and ground stations on a constantly watched frequency.

6. Ultra high frequencies are better than lower frequencies because:

( ) a. Ordinary static causes negligible interference. ( ) b. They have a much greater range. ( ) c. They are reflected from the Heaviside layer better.

7. In shielding a wire the most serious type of noise leakage is through:

( ) a. Poor "ground" connections. ( ) b. Small perforations through the shield. ( ) c. Extremely thin and fairly long cracks.

8. The most important part of bonding is to:

( ) a. Prevent leakage through long thin cracks. ( ) b. Fastening parts adequately together electrically. ( ) c. Prevent chafing of metallic surfaces against non-metallic surfaces.

9. Filtering is primarily concerned with:

( ) a. The reduction of noises due to metallic parts of the plane rubbing together. ( ) b. Reducing leakage through long narrow cracks. ( ) c. Prevention of noise travelling along wire outside of the unit producing interference.

10. Man-made electrical noise may disturb ground receivers miles from the interference source with the radiated interference alone but will disturb aircraft reception less seriously because:

( ) a. The plane will leave the disturbance area rapidly. ( ) b. Such disturbance will not effect the plane at all. ( ) c. Such radiated interference will only travel along the ground.

11. Rain static may be best reduced with:

( ) a. A well-shielded antistatic loop. ( ) b. A static discharge wire at the tail of the plane. ( ) c. Both of the above together.

12. Increasing the transmitter power one hundred times will deliver one hundred times as much voltage at the receiving point.

( ) a. This is true. ( ) b. No. The increase would be ten times in voltage. ( ) c. No. The increase would be 10,000 times in voltage.

CHAPTER XI

DIRECTION FINDING

Either directional transmission or reception is used for the following:

1. Radio ranges
2. Airplane direction finders and radio compasses
3. Ground direction finders

In order that the operation of these devices be understood it is essential that antenna directionality be explained.

Vertical Antennas—A single vertical antenna is said to have a non-directional pattern. This means that signals sent from it may be heard equally well in any direction. The antenna pattern may be shown in two ways as illustrated in Fig. 189 (A and B). At the

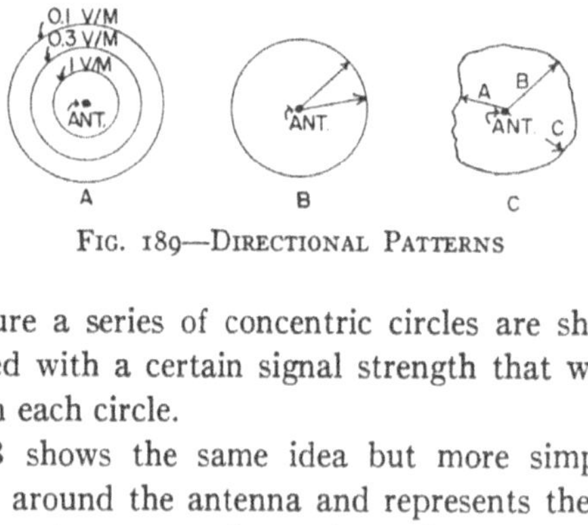

FIG. 189—DIRECTIONAL PATTERNS

left of the figure a series of concentric circles are shown and each circle is labelled with a certain signal strength that will be received at any point on each circle.

Fig. 189B shows the same idea but more simply. A single circle is drawn around the antenna and represents the directionality of the antenna. Any arrow drawn from the antenna to any point on the circle will have the same length in this case. These arrows are called vectors. Their strength is proportional to signal strength in the direction of the vector.

Suppose that for some reason transmission to the west of the station was somewhat poorer than in any other direction. The remainder of the antenna pattern is often not a perfect circle. The pattern in Fig. 189C shows this condition. A number of vectors have been drawn and the length of the vector "A" to the west is much shorter than the vector in any other direction. Reception to the west is weaker by the ratio of the vector lengths.

Nothing has been said about distance as it is assumed that the measuring points are always equidistant from the transmitting point. Signals will decrease in strength with greater distance in any direction from the station.

Such a directional pattern as that just shown in Fig. 189C may not hold strictly for all distances as the pattern may change slightly with distance due to intervening objects. For most purposes the general shape will be of principal interest.

Radiation—It is desirable to think of the action of a radiated wave in space by means of a hydraulic analogy. A wave on the water represents little motion of water continuously in the direction of wave travel but it does represent a relatively great movement of water vertically, up and down. The movement of water along the surface will be neglected as it is not essential to wave motion.

Suppose that a lead ball were suspended on a string at the water's surface. If this ball is moved up and down it will create waves which will move outward from the ball. It is necessary to move along at the wave speed to stay on the crest of a wave. The water in the wave is not moving away from the ball but the wave crest is doing so.

Consider that the wave crest is the most positive voltage point on the wave and the trough is the most negative point on the wave. A point half-way up the wave slope is equivalent to zero voltage. Radio waves move at a rate of 186,000 miles per second.

A receiving antenna is analogous to the action of a float which may move only up and down. Energy may be obtained from the receiving float if it is attached to a mechanical load. Naturally as the distance from the transmitting point increases, the wave energy will be spread over a greater circumference. Thus the greater the distance from the transmitting point, the less energy may be received at any given point. The total energy around the circumference is about the same, however, except for some losses in transmission.

Radio waves are due to alternating current and voltage in the transmitting antenna structure. When waves from two antennas are radiated, the instantaneous wave voltages are important. Voltage at any one instant of time from one source may be compared with another voltage at that instant from another source. Further, if both waves are of the same frequency, they will always bear the same relation of instantaneous voltages at any time. If the two frequencies are different they will bear a constantly changing relationship of voltages with time.

A wave traveling through space will cause a variation of the voltage at a receiving antenna from positive to negative and back to positive again as one full wave travels past. A radiated voltage may be positive, negative or zero voltage at a given instant (or at any intermediate voltage). Thus an antenna might radiate maximum positive voltage at one instant. Ever after that it is possible to treat this voltage as having a positive identity as it does not change in polarity but moves. Thus a radiating antenna, using this convention, will radiate first a positive voltage, a quarter cycle later it will radiate a zero voltage, a quarter cycle still later it will radiate a negative voltage, etc. The alternating current cycle is then plus, zero, negative, zero and plus again, but with a smooth change.

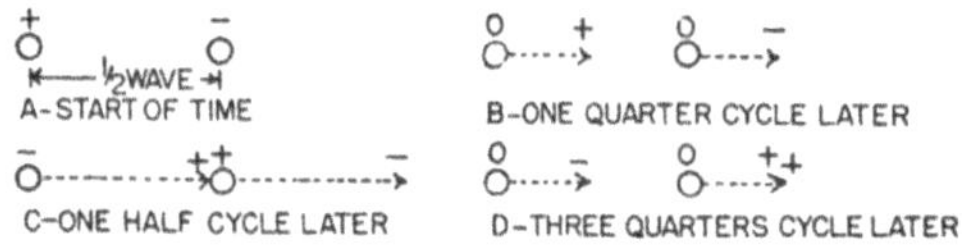

FIG. 190—WAVE COMBINATION

Patterns from Two Antennas—With this simplified idea, it is possible to visualize readily fairly complicated directional antenna structures. Suppose that two antennas are spaced exactly one-half wavelength apart, Fig. 190A. Suppose that at a given instant radio transmitters are started at both points with a positive peak at one antenna and a negative peak at the other, Fig. 190B. The positive voltage then travels from its antenna to the other. As this positive voltage reaches the second antenna the voltage at the second antenna is positive, as it has changed one-half cycle during the time for travel, Fig. 190C. Thus the positive voltage joins with a second positive voltage from the second antenna and both travel on together, Fig. 190D (think of the hydraulic analogy). Assuming that they are equal, the net result on any antenna that they strike will be twice that of either alone. Thus, in effect, the two antennas radiate along the line drawn through both.

However, on a line drawn at right angles to the line connecting the two antennas, the combined radiation is nothing as the positive voltage from one cancels the negative from the other, Fig. 191. As before, they both radiate opposite voltages at the same instant. The distances from either antenna to this line at right angles are identical. Thus the two voltages will cancel. If any two antennas at less than one-half wave separation are supplied with opposite voltages (180 degrees out of phase) at the same time, they will radiate no power at right angles to the direction of the line connecting the two antennas and drawn from the center point outward. The power radiated in the direction of the line connecting the two antennas will always be a maximum.

FIG. 191
SIGNAL CANCELLATION

Loop Antenna—A loop antenna produces a directional pattern, as shown in Fig. 192. The reason for this is explained in the light of what has been said. If the loop is erected with its plane vertical, current flows in one direction around the loop and the current must be flowing up one vertical wire, across the top and down the other, i.e: flowing in opposite directions in the vertical wires. The radiated voltages will thus be opposite in polarity. On a line at

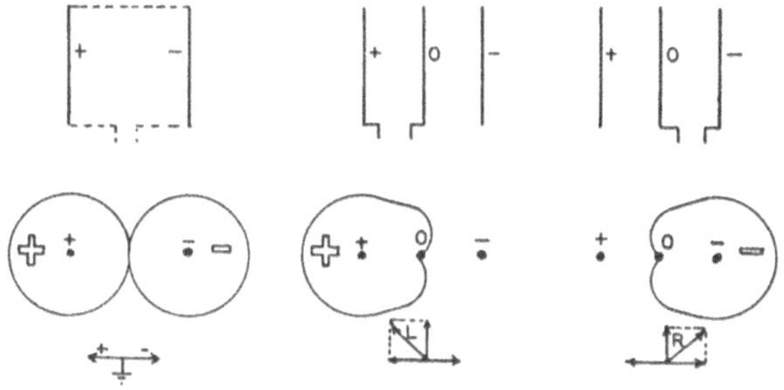

FIG. 192—COMBINATION OF LOOP AND VERTICAL ANTENNA TO SECURE RIGHT AND LEFT CARDIOIDS

right angles to the loop face, the voltages radiated from the two wires will cancel and there will be no voltage received along this line.

Receiving and Transmitting Antennas—Patterns may be shown for any antenna and apply equally to transmitting or receiving. Actually, it does not make any difference whether transmitter and receiver are interchanged as equal results would be secured as far as directionality is concerned. An infinite number of patterns may be obtained by changes in spacing and phase difference between two radiating antennas.

There is one other conception of directionality involving projected length. An antenna radiates (or receives) a signal over its entire length. The antenna is not efficient off its ends due to small projected radiating area. Thus a vertical antenna radiates upward poorly.

Vertical Paired Antennas—This type of antenna has advantage over the loop as it does not radiate from the horizontal wires which are present in the loop. Paired vertical antennas are referred to as TL or Adcock antennas and do not radiate energy wastefully (upward) to the Heaviside layer, Fig. 193. Further, as these TL antennas radiate little signal upward to points above the station they show a cone-shaped volume of reduced signal strength above the station. At low levels the cone is small in diameter while at high

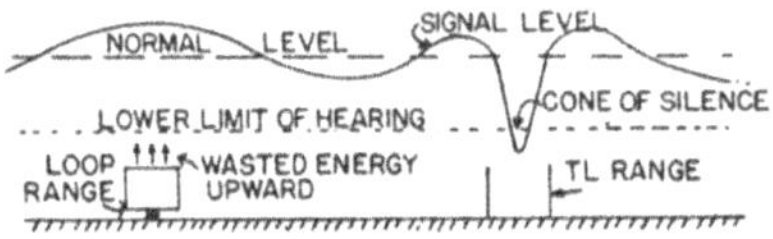

FIG. 193—CONE OF SILENCE

levels the cone is large in diameter. The cone has roughly about the same diameter as the elevation in feet above the station. This zone of reduced signal strength is called the "cone of silence." If a pilot flies over the station he will have been reducing his volume control for a comfortable signal because the signal strength has been steadily increasing as he nears the station. The signal will become inaudible, with this reduced receiver sensitivity, as the plane passes over the antennas if the sensitivity control setting is not disturbed.

Two Paired Antennas, Radio Ranges—When two sets of antennas are erected in the form of a square, as shown in Fig. 194, it is possible to produce two directional figure of eight patterns by supplying signals to each of the two diagonal pairs of towers. A radio range operates with this arrangement and the diagonal tower pairs radiate signal alternately. A relay accomplishes this transfer and operates so rapidly that, in positions where the signals from each pair of towers are the same, it is impossible to tell by ear that switching is taking place.

Assuming that the towers are located at the four cardinal points of the compass, from the center of the antenna arrangement, the North and South antennas will operate together, producing strongest signals North and South and weakest signals East and West. Immediately after a signal has been sent from these towers, the signal is sent East and West by the corresponding towers, giving weakest signal North and South.

Fig. 194 shows the two patterns of the antennas involved. The North and South towers are keyed so that they send a long interval of tone modulated signal, pause for a second, and then send a short

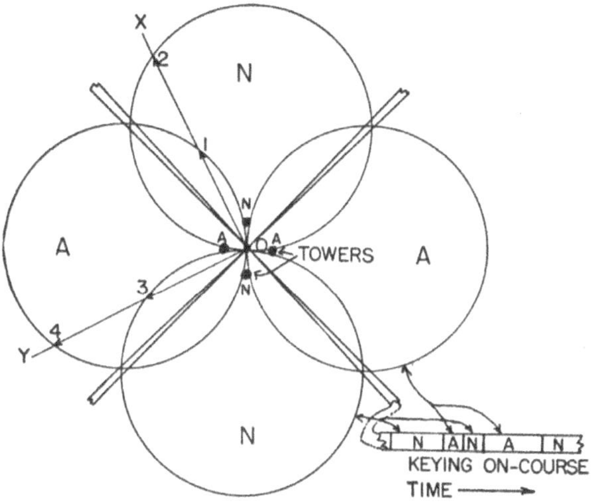

FIG. 194—RANGE KEYING

interval of signal. This is the code letter "N." Whenever these towers are not sending a signal, the transmitter is connected to the East and West towers. As a result the East and West towers send a short and then a long signal. This is the code letter "A." The interlocking of these two signals is shown in Fig. 195 which shows the signal received aboard an airplane at point "X" on Fig. 194.

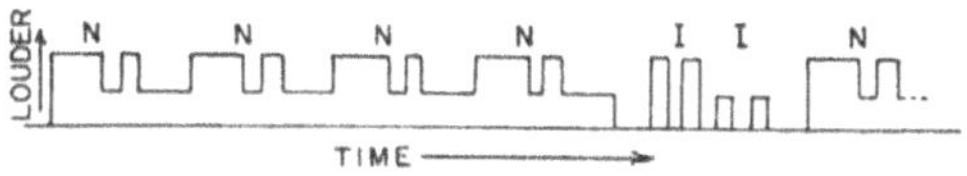

FIG. 195—SIGNAL AT POINT X ON FIG. 194

There are four lines shown on Fig. 194 which are the course legs or "on-course" paths. The signals from either set of antennas are of the same volume level on these lines.

The ratio of the "A" signal to the "N" signal can be found by comparing the vector lengths taken in any one direction from the station. The line O-X on Fig. 194 shows that the "N" signal is about fifty per cent louder than the "A" signal by the ratio of vectors O-1 to O-2. In the same manner the signal on line O-Y shows that the "A" signal is about twice as loud as the "N" signal by the ratio of vector O-3 to O-4.

Fig. 196 shows the signal zones that are passed through on a trip around a radio range station on a circle at any distance from the station. Starting from a point north of the station, around the circle the signal will be a pure "N" signal as the "A" signal strength is too low to be audible. Continuing, the "A" will soon be heard weakly in the background during pauses in the "N" signal. As the north-east on-course is approached, the "A" signal will become louder and

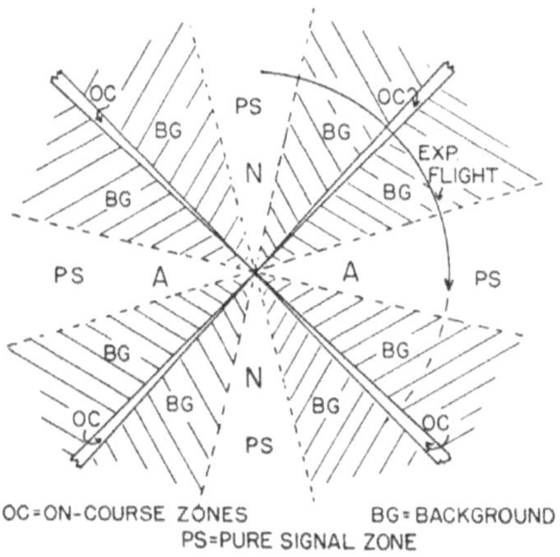

FIG. 196—RANGE ZONES

louder until it is equal in strength to the "N" signal. When equal, the on-course zone has been reached. The on-course zone will have some width as it is not possible to distinguish exceedingly small volume differences by ear. This is the reason that the on-course will seem to be about two degrees wide in arc. The linear width of the on-course zone widens at increasing distance from the station.

Upon continuing through the on-course, the "N" will diminish in strength until the background "N" signal drops below audibility.

Progressively the increasing background zones, on-courses, decreasing background zones and pure signal sectors will be intersected.

The two background zones occupy about two thirds of each such sector between two on-courses (see Fig. 196), and a pure signal zone occupies the remaining third of the quadrant.

Twilight Zone—The term "twilight zone" is loosely used to refer to a nebulous zone immediately bordering the on-course. Both "A" and "N" signals are supposed to be nearly the same in volume and still there is to be actually some audible difference with a distinguishable off-course signal. This zone is not subject to definition as its extent depends tremendously on individual interpretation of just how much difference in the two signals may be tolerated before the signal turns into a background zone signal. For this reason the term "twilight zone" is considered useless and the on-course, pure signal, and background zones only will be considered. The background zone edge depends on the volume control setting as well as the acuity of the ear. If reasonable volume is used the background zone will cover the approximate areas shown. It can be defined as the area from the point when an opposite quadrant signal is just heard to the point where an on-course is heard.

Identification Signals—Each range station is identified at about thirty second intervals so that the pilots may know the station to which they are listening. The identification signals are composed of two code letters. The "A" and "N" transmission is completely interrupted for the entire period of station identification. After a pause the two identification signals are transmitted from the paired towers which were producing the "N." This is followed by another pause and the "A" towers send the same two letters. After a short pause the interlocked "A" and "N" signals continue.

The signal is always sent from the "N" towers first, so by comparing the strength of the first and second identifications the comparative "N" to "A" ratio may be determined. The two identification signals will be of equal loudness when on-course. If they were interlocked like the "A" and "N" signals it would be impossible to identify the station when on-course. For this reason the signals are sent entirely separately with a pause between them.

Headphone Volume—The ear has remarkable qualities which makes it possible to distinguish the "A" and "N" signals through a large amount of static and other noises. The ratio of these two signals must be determined when flying a radio range and the ear has certain peculiarities in this respect which must be recognized. At extremely low volume levels which are just above the threshold of audibility, the ear cannot easily distinguish differences in the volume of the "A" and "N" signals. In fact it can only do about one tenth as well as it can at higher, but comfortable volume levels.

When flying on the edge of the on-course, it is very desirable to detect small differences between the "A" and "N" signals so that the volume level should be set high enough to give a loud but not unpleasant signal.

Some receivers will have their detectors operating on the square law at very low volume levels while at higher volume levels the detector will be more nearly linear. Under these conditions it may be that some point of volume output will be found where the combination of the characteristics of the ear and the detector permit greatest distinction between the "A" and "N" volume ratios. This can be determined quite easily by listening to a signal which is slightly off the on-course zone and by adjusting the volume to give the best audible ratio between the signals. This approximate volume level should be used thereafter for accurate flying of radio ranges.

The volume level used in orientation when a background signal is being used should be high but not uncomfortable so that the background zone will extend over as great an area as possible, regardless of the type of detector.

Overloading—Every radio receiver has an overload point at which the signal no longer gets louder as the sensitivity control is advanced. In fact the signal will begin to drop as the sensitivity control is still further advanced. When this is pronounced, the weaker signal entering the receiver may actually sound louder than the stronger radio signal producing the overload.

Thus if a strong "N" overloads the receiver substantially, a weaker "A" signal might not actually overload the receiver. Thus the weaker signal would sound louder. When this happens a reversed signal is heard, giving a wrong indication.

In testing any receiver before using it, the receiver sensitivity (volume) control should be slowly advanced until the headphone volume becomes very loud and then begins to decrease as the sensitivity is further increased. The receiver should never be used at headphone volumes even approaching this amount of output as the chance for signal reversal becomes greater. In fact, the receiver volume control should purposely be retarded after first tuning in a station to make very sure that the signal does decrease with sensitivity retardation. It is possible for a station to be tuned in with an overload present with disturbing results if the condition is not corrected.

Signal Strength—The signal from a radio range will vary in strength as the airplane circles the station at a constant radius. Naturally the ratio of the "A" to the "N" signal will change, as was shown previously, but if only the louder signal is used for strength comparison there will actually be a stronger signal in the central part of each quadrant than on-course.

It is necessary in the solution of some types of orientation problems to fly on a given course and to determine whether the signal is increasing (building up) or decreasing (fading out). For such purposes the plane is usually flown parallel to the bisector of the quadrant in which the plane is flying.

An example of a flight on the line 1-2 is shown in Fig. 197 on a course on which the volume is increasing. Lines 3-4 and 5-6 show the same thing except that the increase on each for a given flying time is not the same.

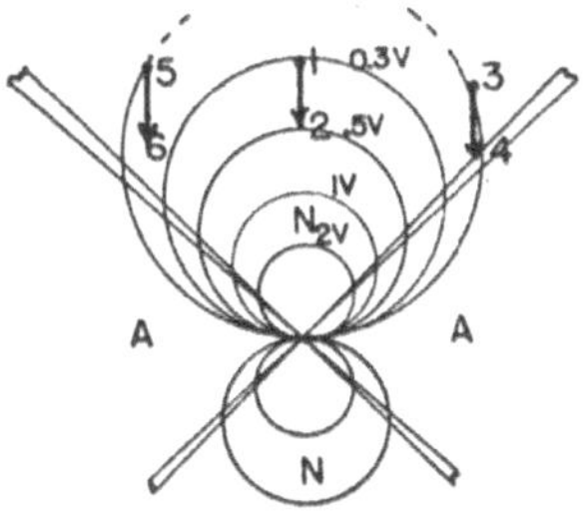

FIG. 197 — VARYING DEGREES OF SIGNAL INCREASE OVER EQUAL DISTANCES

Squeezing — Unfortunately all radio ranges are not so adjusted that the four course legs are at ninety degrees to one another. There are two ways of adjusting the courses so they do not bear the ninety degree relationship.

The first and easiest method is to "squeeze" the courses, as shown in Fig. 198. The way in which a squeeze is accomplished is merely to transmit less power from the "A" towers than from the "N" towers in this case.

A new and special condition is found which shows a decrease in signal strength when the plane is flown parallel to the "N" bisector and in the general direction of the station. Either the line 7-8 or 9-10 shows this condition. When flying line 11-12 or 13-14 the increase occurs as would be expected. Careful study of Fig. 198 shows why this condition exists. If there is any wind blowing, the line 15-16 shows a further possible exaggeration of signal change when flying generally toward the station.

There is a method of getting around this difficulty. It will be noted that under the conditions where this "false fade" or "false build-up" may occur, the background signal can be heard. Therefore under conditions where the background signal is audible, it is by far the best procedure to listen for the ratio of "A" to "N" signals and then determine whether the ratio is becoming larger or smaller. If the ratio is approaching unity it will definitely be known that the plane is approaching an on-course leg.

Bending—The second method of changing the relation of the course legs is known as bending. In bending a leg, the voltages in

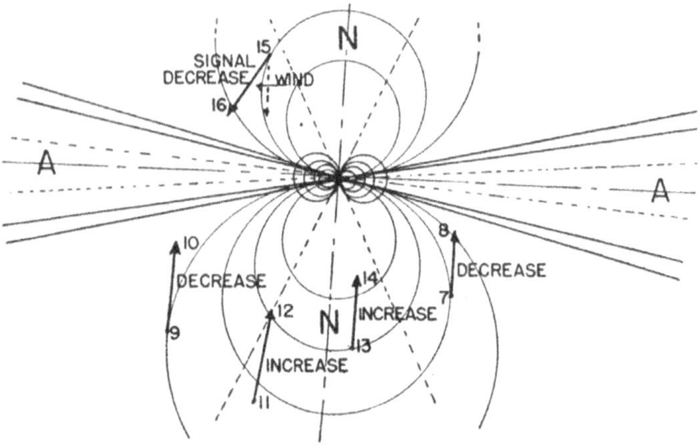

FIG. 198—COMPARATIVE RANGE FIELD STRENGTHS

the diagonal towers are adjused so that they are not exactly 180 degrees out of phase. Thus the pattern is distorted and the effect shown in Fig. 199 results. A bend can always be recognized by the fact that the reciprocal courses which would normally extend through the station are not in alignment. A squeeze and a bend are shown in Fig. 199, as both may be made together when desired.

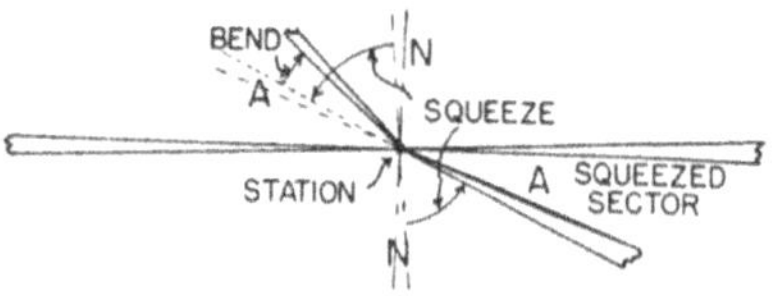

FIG. 199—BENDS AND SQUEEZES

Bracketing—When an airplane intersects a radio range course it is not usually possible to turn the plane to fly down the known course leg as there may be an unknown cross wind blowing. The wind forces the pilot to alter the heading of the plane so that the magnetic course flown and the magnetic bearing of the on-course will not exactly agree. The pilot, after crossing the on-course, may be required to change the heading of the plane substantially to intersect the leg once more. The heading of the plane is then altered to leave the leg again and by reducing the amount of turn each time that either an off-course signal or an on-course signal is heard, the heading required to fly the leg will be "bracketed," as in Fig. 200.

If the heading of the leg is not known the plane should make a standard turn of as much as 180 degrees immediately after crossing the leg. If the leg is not intercepted before the turn is completed the turn is stopped at 180 degrees and the plane flown until the leg is again intercepted. Immediately upon intercepting the leg, the plane is turned in the opposite direction from the first turn and on a heading about sixty degrees less than the first turn until the off-course signal

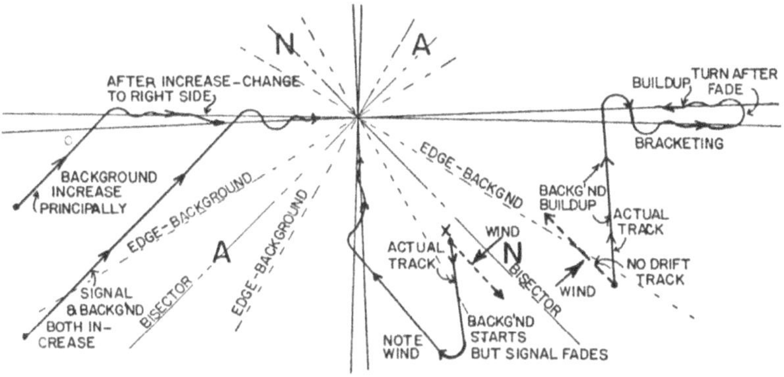

FIG. 200—BRACKETING AND TRUE FADE ORIENTATION SOLUTIONS

is heard again. Each successive turn is reduced until the leg is bracketed. Fig. 200 shows several such brackets.

Orientation—As there are two "A" quadrants and two "N" quadrants on each range, it is impossible to determine by these signals in which of the two quadrants the plane is located. There are two things that make orientation most difficult. First, a preconceived idea that the location is known before starting the problem and second, being very sure of the fade or build-up before actual verification. The solution to the difficulties is to listen to the signal very carefully and hear what is happening, rather than to hear what is desired to be heard. A build-up or fade of signal may require a rather long period to establish definitely.

Another similar difficulty exists upon intersecting a leg. The most difficult part is not to assume the bearing of the leg but to bracket the leg as if it were not known. A full 180 degree turn when necessary for the first bracketing will almost always save time in establishing the heading necessary to fly the on-course.

The "true fade" system of orientation is undoubtedly the best and simplest system in use today. It will give correct orientation solutions with squeezed courses, in heavy winds and even on ranges whose legs are unknown.

The initial heading, after being "lost," is one parallel to the bisector of the quadrant whose signal is being heard (or North in an "N" and West in an "A" quadrant of an unknown range). Either one of two things will occur: (1) the general signal level will increase or decrease, or (2) the background signal will increase or decrease. A direction of flight is then established parallel to the bisector as before but in the direction giving the increase of either general signal level or background signal level. If the background signal can be heard its change should be used with little or no attention being paid to the general signal level change with one exception. If the general signal level fades rather rapidly and the background signal does not change appreciably, the direction of flight must be changed 180 degrees. This will then result in an increase of both background and general signal level (see Fig. 200, plane at X).

Upon intersecting a course leg it is bracketed in the normal manner and the direction of the station on that leg is established by listening to the fade or increase of general signal level.

Several correct solutions of orientation problems with this method are shown in Fig. 200.

Multiples—The reflection of waves from the Heaviside layer causes fading or increasing of signals, as has been previously explained in the preceding chapter. For this same reason the presence of mountain ranges, rivers and the like will cause the patterns of the transmitter to be uneven and somewhat scalloped. This will produce multiple on-course signals which will sound exactly like the true on-course except that they will usually be bounded by identical quadrant letters. In some cases the multiple will be like the true course or even with reversed signals. For this reason it is necessary to fly through the on-course upon first intercepting a range leg to be sure that the true leg has been found.

If a multiple has the correct letter quadrant signals on both sides of it, there is no method by which it can be ascertained that this is a multiple. The easiest solution to the difficulty is to fly down the multiple and at the termination of it a quadrant letter will be heard. By the usual means of bracketing, the required turn must take the plane to the true course eventually. Most multiples lead generally toward the station and occur not more than about twenty degrees to either side of an on-course at the very most.

Bends—The presence of a bend in a leg is due to the same topographical irregularities which cause multiples. By flying the bracketing method no great difficulty will be experienced.

False Cones—Where a radio range signal crosses an irregularity in terrain, such as a deep canyon, there may exist a false cone of silence. As the radio waves pass over such an irregularity some of the waves will be reflected back from the far wall and upward. The reflected waves serve to cancel the direct waves when over the canyon and cause the signal to disappear momentarily. There is little excuse for confusing these with a true cone of silence as they are not accompanied by the substantial surge of signal just before the cone and by the substantial surge following the cone, with a rapid decline in signal level beyond that. If a true cone of silence is accidentally missed by half a mile or so at an altitude of several thousand feet above the station, it may give an indication very much like the false cone as the surges will not be found. In this last case, a course leg will be crossed at about the point of minimum signal with accompanying quadrant signal change.

False cones may change in character at different times of day. Additional Heaviside reflection may obscure the false cone at night.

Simultaneous Ranges—With the loop or vertical radiator (TL) radio ranges, it is necessary to cease transmission of directional "A" and "N" signals during the voice broadcasts. Under conditions where the plane needs the continuous operation of the range, it is necessary to discontinue the voice broadcasts until the plane has finished using the range. The new type of range, now widely installed, is known as a simultaneous range and it may broadcast voice and send directional course signals at the same time.

Simultaneous stations are equipped with five towers. The four corner towers send out a signal which is actually only one sideband of a normal transmission, while the center tower radiates the carrier continuously. Thus the combination of the directionally transmitted sidebands and the non-directional carrier gives the same two figure of eight patterns required for range operation.

The directional sideband results in modulation in the receiver of about thirty per cent when on-course. The voice signals are modulated on the center antenna carrier in the normal manner, but at about fifty per cent modulation. Both voice and range signals are mixed together in the receiver. There is no keying on the central antenna tower so that the voice is not interrupted, regardless of the position in the quadrant in which the plane is located.

It is necessary to separate the range directional signals from the voice weather broadcast. The tone used on all radio ranges is 1020 cycles and the voice covers a range from 100 to about 4000 cycles, except some voice is eliminated near 1020 cycles. Two filters are used in aircraft which were described and shown in Fig. 51. A small part of the voice near 1020 cycles is lost when listening with the filter which excludes the range signals. When listening to the range, the voice signals are practically excluded by the sharp range band pass filter.

Visual Radio Ranges—There are no low frequency visual ranges operating at the present time but several have been tried. Visual ranges are being used, however, for ultra high frequency instrument landing equipment. Instead of keying the two loops as in the conventional range, each loop or pair of TL towers is modulated with a low frequency. Frequencies of seventy and ninety cycles have been used for such purposes.

As the ear cannot compare the intensity of two such signals, two filters are arranged in the receiver which will pass the seventy and ninety cycle frequencies separately. Each of the modulation frequencies is then fed to a single meter whose needle normally rests at the center scale. When the seventy cycle frequency, which may represent the "A" signal, is stronger than that corresponding to the "N" signal, the meter will deflect to one side. When the ratio is reversed it will deflect to the other side. Thus a single meter indicates the quadrant signal and its amount, relieving the pilot of the necessity for use of headphones. Although the ear is a remarkable filter in selecting the correct signal from a general static background noise, electrical filters for such visual indicators have been developed which are remarkably free from static.

Loop Direction Finders—Direction finding is done with a loop antenna on aircraft. The loop is composed of a number of turns, usually thirty in high impedance loops and about ten to twelve turns in low impedance loops. The greater the number of turns (and diameter) the greater will be the voltage induced in the loop in almost direct proportion. A transformer is used with the low impedance loops that steps the voltage up to almost the same value as with a high impedance loop. The usual diameter of loops is from six to eighteen inches.

If a loop is mounted on an airplane it is possible to turn the loop so that the direction of the range, or other radio station, is at right angles to the plane of the loop. At this point the signal will practically disappear and this is called a "null." There are two nulls, one opposite either face of the loop.

The null is used for direction finding. The loop is usually rotatable as this is more convenient than turning the airplane. Naturally it is impossible to tell which of the two nulls gives the correct direction to the station as they have a 180 degree relationship.

A simple diagram (Fig. 201) shows the solution of the 180 de-

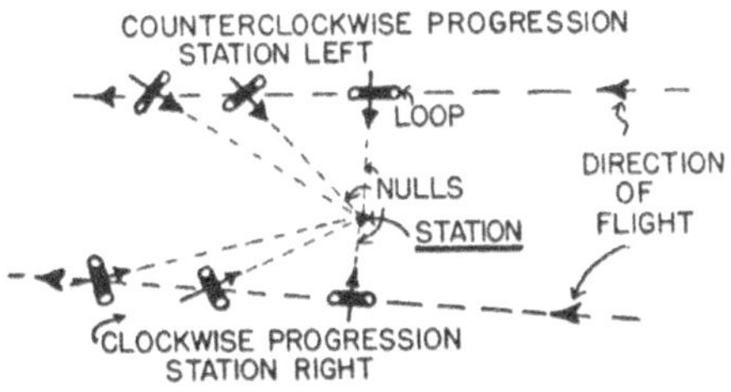

FIG. 201—SOLUTION OF 180 DEGREE AMBIGUITY WITH NULL ON ONE STATION

gree ambiguity on a single station. If the null progresses clockwise the station is to the right.

Direction finding maps are issued by the government as an aid to plotting bearings.

Radio Compasses—There are a number of direction finders which indicate visually instead of aurally. They operate by combining two figure of eight patterns with 180 degree ambiguity present, or by combination of a figure of eight and a circular pattern.

Fig. 192 shows the patterns of a figure eight and a circular pattern superimposed. When the signals are combined electrically, the circular pattern combines in phase with one of the figure of eight patterns and out of phase with the other. The combination is shown and the heart-shaped pattern is called a "cardioid."

Right-Left Indicator Types—The usual radio compass has a keying oscillator which alternately produces two cardioids, as shown in Fig. 192. A meter is also arranged in the output of the receiver in combination with the same keying oscillator which shows the amount of signal received from each alternate cardioid (Figs. 202

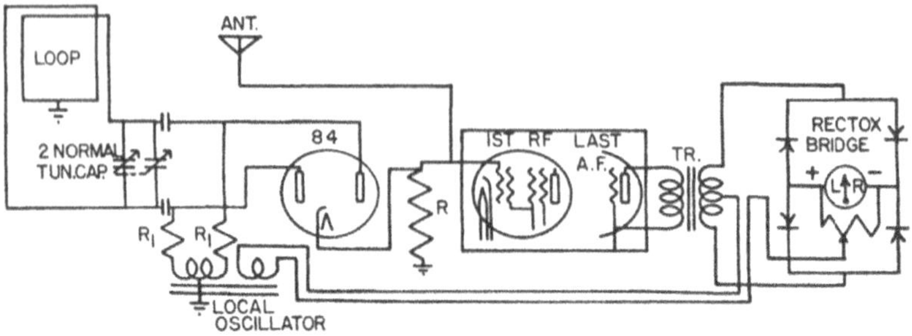

FIG. 202—RADIO COMPASS CIRCUIT (OLD TYPE)

and 203). Thus the needle will show the station to be either on the right or on the left side of the airplane. If the airplane makes a right turn the needle will move to the right if the station is ahead. If the station is behind a right turn will give the opposite needle indication.

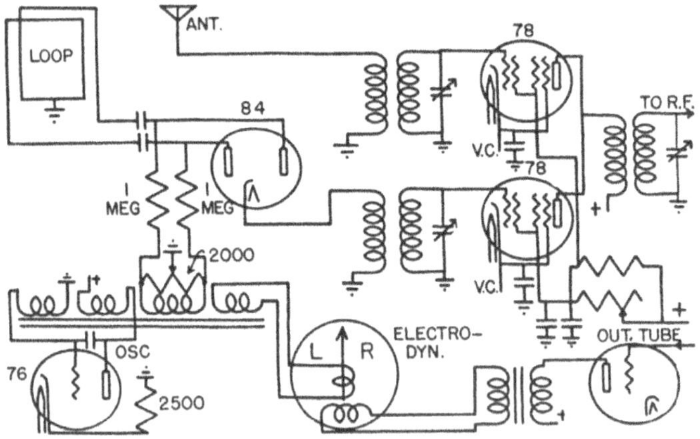

FIG. 203—RADIO COMPASS WITH ALTERNATING CURRENT INDICATOR (ELECTRO-DYNAMOMETER)

This is shown in Fig. 204. Thus by quickly determining the needle motion in a small turn, the direction of needle movement will show whether the station is behind the plane or in front of it. Thus the 180 degree ambiguity is removed.

The convention used for needle movement is based on the older convention of having the needle move right for a right turn from a heading to the station. The new convention is the exact reverse of this so that the needle "points" to the station (right turn from station —left movement). Any R-L indicator can be changed by reversing the meter leads. The new convention makes needle movement agree with that of the 360 degree automatic radio compass. Pilots should determine which convention is used on each installation.

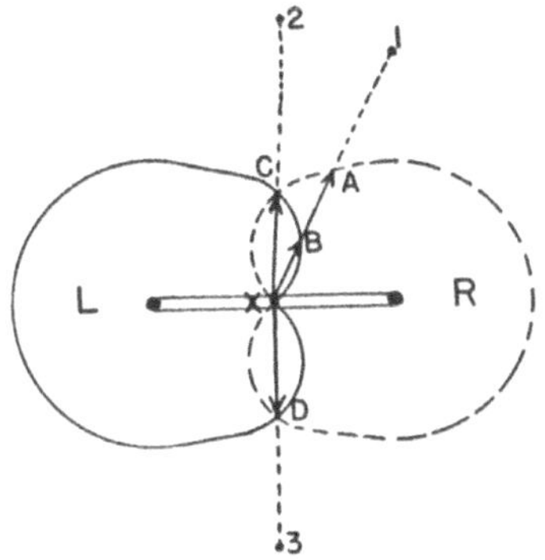

FIG. 204—METHOD OF SECURING NEEDLE DEFLECTION IN COMPASSES

Sperry Mark I Type—The usual radio compass indicates on a right-left meter which shows the direction with respect to the heading of the plane, or if the loop is rotatable, with respect to the loop face. The same principle has been modified somewhat in a compass (Sperry type) in which the output of the receiver, in combination with the keying oscillator, drives a motor which turns the loop so that its face is at right angles to the station at all times. The loop is attached through a tachometer shaft to the cockpit bearing indicator. The indicator will therefore indicate the direction of the station correctly at all times with no necessity of turning the plane to solve the ambiguity, as there is none. As the receiver is controlled by an automatic volume control, it is only necessary to tune in the station and identify it aurally before using the visual automatic indicator. The Lear ACR-6 compass operates on this same general principle.

Rain Static Effect—Direction finders, using a vertical sense antenna, are subject to rain static interference. The Sperry Mark I type is equipped with a second loop mounted inside of and permanently at right angles to the first loop. In rain static the operation is the same as before with the exception that 180 degree ambiguity is present due to the figure eight pattern of the second loop rather than the circular vertical antenna pattern used except in rain static.

Simon Radio Guide—This direction finder utilizes two loops which are mounted at right angles to one another. It operates in exactly the same way as a radio range but is on the airplane and is used for receiving instead. This type not only has 180 degree ambiguity but also has two 90 degree ambiguities as well. A sense antenna with a circular pattern is used for ambiguity determination with a switching arrangement so that when a button is pushed the needles, one attached electrically to each loop through two receivers will show by their movement which of the four bearings is being used.

The receiver is a dual receiver with one half attached to each loop. The sensitivity of the dual receivers must occasionally be adjusted when approaching or leaving a station and adjustment is provided which allows setting the two receivers to exactly the same sensitivity. Indication of direction is given with two crossed needles at the point of intersection of these needles. Thus if the indication is to the right of vertical the needles will cross to the right of the instrument's center line. The instrument is also calibrated so that the station's direction, in degrees off the plane's heading, may be read up to about fifty degrees on either side. In some installations the complete dual loop is rotatable.

European Revolving Loop—Another type of radio compass (IT&T-RC5) employs a loop which is rotated at about 400 revolutions per minute. As the loop rotates, the signal increases and decreases as each null is passed. An indicator shows the heading to the station by combining the fluctuating receiver output with a frequency generated on the loop rotating shaft.

The indicator is a motor in which the receiver output is connected to the rotor and the alternating voltage from the loop shaft is fed to the stator. The rotor assumes fixed positions of no torque and continuously indicates the direction of the station but with 180 degree ambiguity. This type of direction finder is used extensively in Europe. Extremely high-powered, low frequency broadcasting stations allow operation over long distances.

Radio ranges are practically unused, except for instrument landing, in Europe. All of their radio navigation is done by direction finding on the ground or in the plane.

Errors in Direction Finders—A number of errors are common to most direction finders. There are others found in addition to these with the radio compass of the visual type.

Deviation—Any direction finder mounted on an airplane will have errors or deviations due to defraction or reflection of radio waves by the structure of the plane. Mounting position of the direction finder will have an influence on the amount of deviation. As a rule the direction finder loop will be mounted well forward on the airplane so that the length of loop rotating control shaft will be short, thus eliminating backlash and making rotation smooth. In addition, it is very advantageous to keep the leads from the loop to the receiver as short as possible to reduce losses in transmission lines connecting the loop to the receiver.

There is a common belief that bearings taken on stations in directions to the rear of the plane will be subject to more error than those that are taken over the nose of the plane. This is wrong as any calibration will show.

In some instances loops have been mounted in a nose shell made of insulating material. In this case the loop is so close to the wall behind it that large errors occur which may be as great as fifty degrees in some positions of the loop. Such errors will cramp the calibration scale so noticeably that the loop calibration is useless. An error of a few mechanical degrees of resetting will thus produce enormous electrical errors. Bearings are therefore unreliable over a rather wide sector of the scale. Such loops may still be used for homing by setting the loop on a null over the nose of the plane. Bearings to the rear may be used also.

In the normal installation of the loop of a direction finder, either below the plane or on top if it, the maximum deviation will be about eight to fifteen degrees and will occur at forty-five degrees either side of the nose or the tail. Thus at a mechanical rotation of forty-five degrees either side of the nose, the radio bearing will be from fifty-three to sixty degrees, depending on the amount of deviation. Bearings at 135 mechanical degrees from the nose will also give radio bearings of from 120 to 127 degrees from the nose, depending on the maximum error. The deviation curve will normally be very symmetrical, particularly if the loop is mounted on the center line of the fuselage, either above or below it. The error at zero, 90, 180 and 270 degrees will be very small.

It is difficult to calibrate a loop on the bottom of the plane (or on the top for that matter) with the plane on the ground due to radio errors involved. For this reason it is best to calibrate the plane in actual flight. About the easiest method of making this calibration is to fly the plane to a point about fifty miles from a range station and select a road, for example, which heads directly to the range station. The plane is then flown parallel to and over the road. The directional gyro is set to zero degrees. The null bearing is marked on the temporary bearing scale of the rotating device.

The plane may then be flown across the road at angles which are increasing about fifteen degrees per step and the full calibration may be secured by taking a null when crossing the road each time. An occasional check should be made to be sure that the gyro has not precessed. Bearings should be taken at every fifteen degree point, both for bearings ahead and astern. This calibration usually requires about two hours.

There are two ways of incorporating the deviation in the bearing indicator. The first and easiest method is to calibrate the scale with

Courtesy TWA

FIG. 205—DIRECTION FINDER SCALE INCLUDING DEVIATION IN ITS CALIBRATION

the proper amount of squeezing and widening in accordance with calibrations so that the pointer will read the correct radio bearings (Fig. 205). The second method is to utilize a correction cam in the device which will either advance or retard the needle with relation to the drive shaft. Even scale calibrations will then give the correct radio bearing.

An advantage in this last method is that an azimuth scale can be placed around the fixed scale which shows the relative headings. This azimuth scale may be rotated so that the magnetic heading being flown appears opposite the zero of the fixed scale. In this manner the pilot may rotate the azimuth scale so that bearings may be read directly in magnetic direction to (or from) the range station being used.

If a cam correction is not used, the bearings on the indicator must be added or subtracted from the magnetic heading. In order to facilitate addition or subtraction, some bearing indicator scales will be found with red calibrations on the left half of the scale. The bearings will run from zero to 180 degrees counterclockwise. The bearings on the right side of the scale are on a green background and run from zero at the nose to 180 degrees clockwise (Fig. 205). Red bearings are subtracted from the compass heading while green bearings are added to the compass heading. This simplifies calculations for the pilot in comparison to scales which run from zero to 360 degrees clockwise for in this case calculation of bearings would involve numerous additions and subtractions of 360 degrees.

Deviation cams or adjusted calibrations will hold very closely for individual planes of the same exact make with identical loop installations.

One final method of taking care of deviation is to place vertical copper bars near the loop and electrically correct for the errors. The vertical bars may also be in the form of a U below the plane skin and with the ends bolted to the skin. One U is placed on either side of the loop and parallel to the center line of the plane. By regulating the size of the U the calibration may be made to agree very closely with the mechanical rotation. This method has been used to some extent in Europe.

Loop Mounting Position—As a rule there is a choice of either mounting the loop above or below the plane. There are some advantages to mounting it below the plane, as will be seen.

(1) Loop below and used for homing: When the loop is mounted below the plane and is used for homing the null will give way to a strong signal when the plane is very close to the station and no null will be found until the plane has passed beyond the station. The reason for this is apparent when it is realized that the loop has a null for stations at right angles to its face, both horizontally and vertically. As the loop passes over a station the signals come from below the loop and in this position the loop shows maximum pickup of signals. As soon as the plane passes over the station and somewhat beyond, the signals come to the loop essentially at right angles to the plane of the loop again and the null may be found.

(2) Loop above the plane and used for homing: In this case

the null is maintained right across the station. The radio waves coming from the radio station bend over the nose of the fuselage structure causing signals to arrive at right angles to the loop face at all times.

(3) Loop below the plane used on maximum: When the loop is used with its plane in the direction of the station, in flying in rain static conditions for example, a cone of silence will be heard over the TL type of station. A cone of reduced signal exists directly over the station and the operation of the loop is very much like that of any antenna. If the plane passes over a loop range station however, there will be no cone of silence as neither plane nor ground loop possesses a cone of silence characteristic.

(4) Loop above the plane on maximum: In this condition the loop will not show a definite cone of silence over any station. As the plane approaches the station the signal will not rise in value in the normal manner due to the signal coming more and more from below the plane and being refracted over the nose to the loop. The normal surge before and after the cone are greatly reduced, therefore. The cone of silence is not a place of no signal but merely of reduced signal. Thus as the plane passes over the station the signal level will change somewhat but not enough for positive aural identification of a cone.

For the reasons given above it will be seen that there is some advantage to be gained by mounting the loop beneath the plane. The cone of silence is normal with TL stations. When homing the null disappears over the station, giving away to a strong signal, showing that the plane is passing over the station. These two advantages are quite worth while from a pilot's standpoint.

One other effect will be noticed with a loop when homing and most particularly near the range station where the leg narrows.

As the plane is turned slightly so that some signal is heard, the signal may be an "A" on a slight right turn and an "N" on a slight left turn, for example, even though the plane may be on-course. This effect is due to the fact that the loop is more nearly on a null for signals emanating from one set of towers of the range in one position and more nearly on a null for the direction of signal arrival from the other set of towers in the other position. The effect is noticed only very near the null and the remaining sector of the loop will give the correct signal. This is not very important as the quadrant signal is usually of little importance when homing.

Changes of Signal Strength—The use of a loop during the solution of an orientation problem when using "true fade" or other orientation systems is risky in that changes of signal level are not to be relied upon. If the plane is flown on a course on which an increase of signal might normally be heard, it is quite possible that the plane may be changing the angular relation of its head to the station. This will cause the signal to increase or decrease due to the effect of the loop's directionality alone. False results may be easily secured. The best plan is to use the loop orientation system, described elsewhere in this chapter.

Mountain Effect—Mountain effect is defined as the disturbance of mountains upon the arrival direction of the radio signals due to reflection from the faces of a mountainside or ridge. This, it will be recollected, caused multiple courses of the radio range. The combination of the direct and reflected signals arriving at the plane will cause nulls to be wide and erratic and will also cause violent movements of radio compass needle indicators.

It will often be found that when a plane passes over a high ridge that the aural null will widen greatly or disappear for a few seconds. The radio compass needle will also show very erratic indications under these circumstances.

Night Effect—Reflections from the Heaviside layer, particularly at sunrise and sunset and during the night as well, will cause nulls to widen greatly and will cause erratic indication of radio compasses. This is very much the same in its method of formation and in its effect as mountain effect. Nulls will often widen to 100 degrees or more while in the same position in daylight the null might be but a few degrees in width and very steady.

The Heaviside reflection in the broadcast band is much more severe than at the lower radio range frequencies. For this reason, the higher frequencies in the broadcast band will be unusable for the usual direction finder or radio compass much of the time and particularly at night.

Interference Between Stations—When two stations interfere with one another, it is very difficult to take a bearing on either one. This may be done aurally very much more easily than with the visual radio compass. If the tone modulation of the ranges differ, a good bearing may be taken aurally but unfortunately the tone modulation on all ranges is now the same so that considerable difficulty exists. The visual compass is almost unusable under these conditions as it may indicate a bearing which is in a direction having nothing to do with the direction to either of the stations which are interfering with one another.

Static Interference—Static interference causes much the same difficulty as interference between stations when using a visual compass. Under heavy static conditions the visual indicator will tend to indicate the direction of static arrival periodically, rather than the direction of the station. Static will also cause difficulty with the aural null direction finder as well, as it makes it difficult to distinguish the null.

Radio Compass Error—In addition to the foregoing, the following errors will occur with radio compasses which, it will be remembered, also utilize the loop in some form.

1. Interference between stations may cause radio compasses to indicate what appears to be a good bearing whereas the bearing will be the electrical resultant of the two signals. Naturally the bearing will be incorrect. When using a radio compass it is well to listen at intervals for such interference.

2. Failure of part of the antenna system or internal electrical trouble in the receivers will cause erroneous indication. The indicator will usually be abnormally sensitive or sluggish or the indicator will constantly indicate a bearing even when the plane is turned.

3. Rain static may cause any radio compass which utilizes an open wire antenna for removal of ambiguity, to cease functioning.

4. Interference from static or man-made sources will cause the usual radio compass to indicate a direction to a station somewhere between the direction of signal and static arrival.

5. Radio compasses will show an oscillating movement of the indicator with "A"-"N" keying when taking bearings on range stations with the plane located well within a quadrant. This is due to the change of signal level from the station and the resultant disturbance of the compass circuits. Further, if the background signal is weak, as when near the center of a quadrant, the compass will indicate the direction of static arrivals during the intervals between the quadrant letters. Simultaneous ranges will not give this effect, however, as the carrier is on continuously.

Radio Compass, General—In summary, the radio compass is a very useful device from the pilot's standpoint and saves the fatigue of listening to signals and static for long periods. Certain errors are present in compasses which must be known to the pilot for intelligent use of the equipment. On the other hand the aural null direction finder is extremely simple and therefore reliable. It is also far less expensive. Most radio compasses today employ a switch by which the compass may be used as an aural null direction finder under conditions where the compass is faulty or is operating under adverse conditions, such as with rain static or severe interference.

Ground Direction Finding—Ground direction finding is most difficult, particularly on the high frequency band. To date, no completely successful high frequency direction finders have been built that compare with those at the lower frequencies. Plain loops are practically useless for high frequency purposes.

The only kind of direction finder which has been found at all reliable is the type depending upon a very carefully and symmetrically placed set of two vertical antennas. This direction finder is arranged to rotate to a position where the phase of the signal voltages from the two antennas is alike, as indicated on an oscillograph. An aural null may be observed with the two antennas feeding the same receiver with the antennas connected so as to cancel signal, but this requires that the antennas be very carefully tuned and test nulls should be taken at intervals on a known transmitter located nearby. In either case there is 180 degree ambiguity unless a third sense antenna is brought into play, which is located symmetrically between the other two. When the sense antenna is attached to the receiver the whole antenna structure is rotated ninety degrees and it is determined whether the signal increases or decreases when the sense antenna is added. When this is known the direction to the airplane may be determined.

If four fixed antennas are used they may be located like the towers of a radio range and at the center, where the two sets of feed wires cross from the diagonal towers, there is a device called a goniometer. It consists of the two antenna coils, one for each of the two diagonal towers, wound on a square form at right angles to one another and with a minimum of coupling between them. The receiver is attached to a rotatable armature in the center of the coil assembly and located so that it is absolutely symmetrical to windings on the stator or antenna coil structure. As the rotor is turned, a point of minimus signal will be found. This minimum or null depends on the relation of the rotor to the two stator coils and gives the direction of the plane from the station with 180 degree ambiguity. A sense antenna may be used to remove this ambiguity but as a rule this type of null direction finder is used in connection with another direction finder and a fix determined by triangulation made on the airplane. Serious errors in direction will occur in such an arrangement at unpredictable times. The four antenna system should be rotatable so that signal arrival is always bisecting the angle between towers rather than using a goniometer.

These methods are rather slow. To take bearings requires several seconds during which time the plane's transmitter must radiate continuously.

Automatic direction finders are being developed which will automatically show the direction of the plane. The operation of these direction finders is such that the goniometer is replaced by an electrical keying system and sense is added in the method of electrically keying the antennas. This, coupled with cathode ray tube indication, does show whether bearings are steady or not which is a great advantage. The Western Electric direction finder of this type shows great promise.

The effect of the Heaviside layer upon signals is primarily responsible for the difficulty in high frequency direction finding. Under good radio conditions, and especially during the day, the direction finder may work well. At night the direction finder may be erratic at times, especially so at sunset and at sunrise. Usually the direction finder will give the worst results between thirty miles and two hundred miles from the station. At shorter distances and at greater distances it is easier to obtain reliable bearings.

In all events the structure of the antenna is of the greatest importance. Antennas must be vertical, spaced accurately and free from nearby reflecting objects.

QUESTIONS ON CHAPTER XI

1. A signal strength vector shows the:

   ( ) a. Signal contour and all points on the vector are equal. ( ) b. Direction but not the magnitude of the signal. ( ) c. Relative strength of signal in a given direction from the station.

2. Two antennas are a half wavelength apart and radiate a positive maximum voltage at the same instant at the same frequency. The pattern is:

   ( ) a. A figure of eight with the maximums at right angles to the line connecting the antennas. ( ) b. A cardioid. ( ) c. A figure of eight with the maximum in the line of direction through the two antennas.

3. Vertically paired antennas (TL) are not the same as loops because:
( ) a. Loops do not radiate a figure of eight pattern. ( ) b. Loops radiate both upward and outward. ( ) c. The vertical sections of loops do not behave like the TL in radio range uses.

4. A "background zone" is one where:
( ) a. Both "A" and "N" signals are heard but not at the same strength. ( ) b. A pure quadrant signal, such as an "N" only is heard. ( ) c. The signal is rapidly fading out.

5. The "twilight" zone refers to:
( ) a. A zone bordering the on-course of thirty degrees in arc. ( ) b. A zone of a few degrees in width at the outside edge of the background zone. ( ) c. A nebulous zone of indeterminate borders.

6. "Overloading" will sometimes:
( ) a. Turn a background "N" into a background "A." ( ) b. Aid in flying an accurate on-course. ( ) c. Give quicker true indication of a fade as the plane flies from the station.

7. A "squeezed" range differs from a "bent" range in that:
( ) a. No courses bear 180 degree relationships on a squeezed range. ( ) b. One or two pairs of through courses do not bear a 180 degree relationship on a bent range. ( ) c. Both squeezing and bending are really the same.

8. "Bracketing" refers to a method of:
( ) a. Reducing corrections in heading alternately right and left until a course leg has been determined. ( ) b. Added corrections in heading in one direction until an on-course is intercepted. ( ) c. Determining the width of the on-course zone.

9. If a two course range were constructed with its legs east and west and the "N" signal was north of the courses, a plane in an "N" zone could fly south until the course was intercepted. The direction to the station after bracketing the leg could be determined:
( ) a. By noting whether the "A" was on the right or left. ( ) b. By waiting until the cone marker was heard. ( ) c. By listening for a signal increase or decrease.

10. Multiples are:
( ) a. False on-courses bordering on the true on-course. ( ) b. False on-courses usually found near the quadrant center. ( ) c. Duplication of station identification signals.

11. A false "cone" is one where:

( ) a. A cone is heard which is exactly like a real cone. ( ) b. Signals fade out and in but are not accompanied by surges before and after the fade. ( ) c. The on-course gives way to a multiple.

12. Simultaneous ranges are so named because:

( ) a. Voice and range signals may be transmitted together. ( ) b. Range signals and identification signals are transmitted together. ( ) c. Four courses may be transmitted simultaneously.

13. Loop direction finders use one shape of pattern while "radio compasses" use another. These patterns are respectively:

( ) a. A figure of eight and a non-directional one. ( ) b. A cardioid and a figure of eight. ( ) c. A figure of eight and a cardioid.

14. In the Sperry type of compass:

( ) a. The loop rotates automatically with its face toward the station at all times. ( ) b. The loop stands still but the indicator rotates automatically to point to the station. ( ) c. The right-left needle is used to solve ambiguity.

15. Deviation in a direction finder is not important because:

( ) a. The error amounts to only part of a degree as a rule. ( ) b. This is not true, it may amount to fifteen degrees or so. ( ) c. The calibration scale may be merely rotated about fifteen degrees permanently when the constant error usually existing has been found.

16. Most accurate bearing can be taken with a loop over fairly great distances when the loop is mounted:

( ) a. Either on top of or beneath the plane. ( ) b. On top of the plane only. ( ) c. Below the plane only.

17. Better cone indications will be heard when flying a range with the loop:

( ) a. Either on top or beneath the plane. ( ) b. On top of the plane. ( ) c. Beneath the plane.

18. Mountain and night effect are caused respectively by:

( ) a. Terrain and Heaviside layer. ( ) b. Mountains and darkness. ( ) c. Heaviside reflections in both cases.

19. Ground direction finding is done on the high frequency bands because:

( ) a. It is more accurate than at lower frequencies. ( ) b. These frequencies are better adapted for usual communication. ( ) c. Direction finding apparatus is easier to build.

20. Ground direction finding is better over water than over land because: ( ) a. Transmission is more uniform over sea water and there are no large irregularities of surface. ( ) b. Heaviside conditions are less severe. ( ) c. Direction finders need only cover a smaller arc, when such apparatus is located along the coastline.

CHAPTER XII

ULTRA HIGH FREQUENCY EQUIPMENT

It is interesting to note that some of the earliest work ever done in physics in showing that radiation was possible was done on ultra high frequencies. The ultra high frequencies were generated by simple spark gaps with great inefficiency. In fact it is surprising that any results were obtained.

Before the advent of vacuum tubes of a commercial nature some twenty years ago, the usable radio spectrum extended from about ten kilocycles up to about 500 kilocycles. These frequencies were generated by mechanical alternators up to about 200 kilocycles and with arcs or sparks up to 500 kilocycles. Receivers employed galena crystal detectors and no amplification was employed. Tubes immediately opened the region from 500 kilocycles up to frequencies of some 2000 kilocycles and frequencies above this were considered rather useless.

Every succeeding year the commercially used frequency spectrum was increased and it was found that the high frequencies were of great importance. In 1929 two airlines proposed and installed apparatus for communication on what are now the day and night airline frequencies. Very few people actually considered that the frequencies were usable and great opposition was encountered. Between that period and today the high frequency spectrum has become crowded and there are not sufficient frequency assignments to take care of the thousands of uses to which radio can be put.

The first serious consideration given to the ultra high frequencies dates back to about 1933 and since then many commercial installations have been made. Some experimental installations have been used at frequencies as high as 3000 megacycles. The majority of vacuum tubes commercially available today are usually limited in their highest frequency, with reasonable efficiency, to about thirty to seventy megacycles maximum. Some tubes of recent origin will operate efficiently at 300 megacycles. It was only in 1938 that the Federal Communications Commission made permanent assignments from thirty to 300 megacycles. Assignments made above 150 megacycles are probably of a more temporary nature. Whereas the main limiting factor in the past has been national and international interference, it is believed that no serious interference will exist on frequencies above seventy megacycles outside of the local area. Thus an ultra high frequency may be used over and over again without great danger of interference. Even if equipment is mobile, as in the case of aircraft, a single frequency may serve a whole transcontinental air transport line.

There are commercially used ultra high frequency instrument landing systems in Europe which have been operating since 1937. It is now possible to predict the use of ultra high frequencies in the United States for use by airlines and itinerant flyers. The discussion will be divided into the following groupings:

1. Markers
2. Airport traffic control
3. Company communication
4. Radio ranges
5. Weather teletype
6. Instrument landing
7. Above-ground altimeters

Markers—At the present time there are a number of low frequency non-directional marker stations operated on the same frequency as the nearest radio range, one leg of which passes over or near the marker station. When the plane approaches these marker stations a signal will be heard faintly within about five to twenty miles of the station. The range signal is still far stronger than the marker. No difficulty is encountered in following the range as the marker emits only identification signals at widely spaced intervals (twenty to thirty seconds). As the plane passes over the marker the signal will be somewhat stronger than the radio range and then will die out at a point about five to twenty miles beyond the marker. Thus, by timing the interval over which the marker was heard and dividing this time by two, it can be estimated at what time the marker was passed. This process is not very satisfactory but in it is the basis for all markers, i.e: some mark along the range leg for computation of speed and for estimating time to the range station. With the advent of direction finders on planes in the United States it is now possible to secure the same information at most points by taking a bearing on a station to one side of the airway.

It is possible, with ultra high frequencies, to build directive antennas that are very efficient and occupy a small space. The directionality may be so great as to allow a station to project a signal over an arc of but a few degrees or so with an unusable signal in all other directions. If such an array of antennas is located on the ground, so that it directs a signal upward, it is possible to project a thick fan of signal across the radio range leg. The fan may be several miles wide and several hundred feet thick, as measured in the direction of flight along a range leg.

It is also possible to project a cone of signal upward from a point and this cone can be regulated in shape to have the contour of a cylinder at altitudes from 1000 to 10,000 feet with the top and bottom tapering to a point. The top of this cigar-shaped cone is limited by power from the transmitter while the bottom is limited due to the physical size of the antenna structure which is about twelve feet square. The signal is detectable by the aircraft receiver for about 1500 feet through the cone in any direction of flight when between the two altitudes mentioned.

Types of Markers—There are three general types of markers:

1. Airways markers (fan-type, FM)
2. Range station markers (cone-type, Z)
3. Instrument landing or approach markers (fan-type)

Airways markers are located along the airway over which the radio range leg passes. There may be one or more on each leg, usually there are two. One is about thirty miles and the other about ten miles from the station. These fan markers are installed on many airways.

The coding of these markers is uniform no matter whether one or more are installed. If the north leg or the leg in the northeast quadrant of the radio range has a fan marker it will be identified by one dash, followed by a pause and then one dash and so on. The next leg clockwise is identified by groups of two dashes separated by pauses. Thus the southwest leg will have three dashes and the northwest leg will have four dashes. These dashes may be seen visually on the marker light on the instrument board or heard aurally, as the pilot selects.

The usual cone of silence over range stations is a negative rather than a positive indication. For some time it had been felt that a positive indication of passing the station would be helpful and Z type or cone of silence type markers are being installed at all radio range sites.

Instrument landing systems will employ one or two approach markers. At the present time it is believed that two markers will be necessary to aid in timing the approach to the field. One marker, which is referred to as the "inner marker," will be installed at or very near to the approach edge of the field. The other marker which is called an "outer marker" will be installed along the approach leg and within three to five miles of the field. The outer marker may be installed at the point of last high obstruction, such as on a ridge.

Fig. 206 shows the arrangement of markers which might be installed at a typical airport.

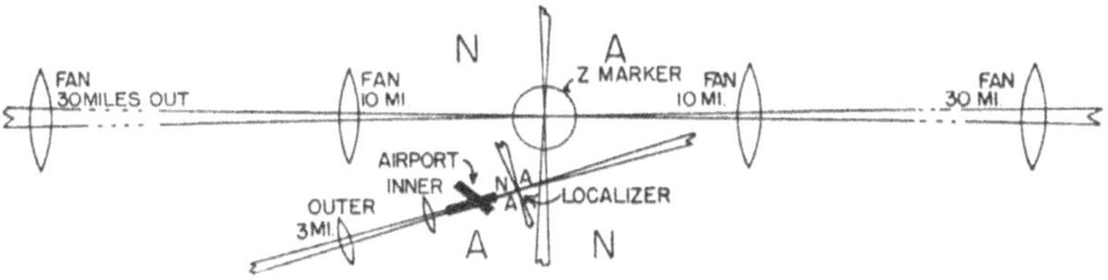

FIG. 206—LOCATION OF MARKERS ON AIRWAY AND INSTRUMENT LANDING SYSTEM AT AIRPORT

As has been described in the chapter on aircraft receivers, the markers will indicate on a series of lights on the instrument board of the plane. The white light will be used for the FM and the Z type markers. The amber and purple lights will be reserved for the outer and inner markers.

Airport Traffic Control—At the present time air transports call the airport traffic control tower on the company's alternate high frequency. By this is meant the frequency not being used for company transmission at the particular time of day or night. Thus the night frequency will be used in the daytime and the day frequency at night. The airport answers the plane on a frequency in the radio range band of 278 kilocycles. Transmission from the airport control tower is very limited in heavy static conditions and on good winter nights a great deal of long distance interference is present.

A regulation has been issued by the government that few more assignments will be made on this frequency and that the traffic control towers will be operated on ultra high frequencies as soon as it is possible. By 1941 it is expected that some of these installations will be under way and that aircraft will be equipped with an ultra high frequency transmitter and receiver for airport traffic control purposes. These will operate at about 130 megacycles. There will be two or three frequencies for this control work and one will be assigned to the principal airport used by commercial planes and the others will be assigned to all smaller airports in the same vicinity from which there might be interference if all were operated on the same frequency. In addition to these there will be other frequencies used by government aircraft at their own airports.

Thus 1940 to 1942 will find marker receivers and airport traffic control transmitters and receivers on airline transports. Further, the transmitter may be equipped to operate in the present day and night bands and serve as a second or auxiliary transmitter. Larger transports will probably carry separate transmitters for the high and ultra high frequency bands.

Company Communication—The advantages have already been shown of ultra high frequency transmission and reception between aircraft and ground stations, and of course, aircraft and aircraft. Due to the amount of work and money that will be expended by the airlines in equipping their planes with the two receivers and transmitter, as outlined above, it will probably not be possible to add the company two-way ultra high frequency communication before 1942. When this time comes, the transmitter normally employed for high frequency communication will probably be discarded. Communication, however, will be maintained on those frequencies. When required, transmitters normally used for airport traffic control will be capable of shift to the present day and night frequencies. A new transmitter will be installed for company communication at any one of four frequencies between 140 and 143 megacycles also capable of operating between two and twenty megacycles. A companion receiver for this same service will also be installed and will take the place of the company high frequency receiver. Reception on the high frequencies (two to twenty megacycles) will be taken care of by this new ultra high frequency receiver by a frequency shift.

Transmitters used for this purpose will be of about twenty watts output. This power will give more reliable, static free transmission than a much more powerful transmitter at high frequencies.

Company communication will probably move to higher frequencies as communication on these high frequencies will be devoted entirely to long distance (or emergency) transmissions. Frequencies from six to twenty megacycles may be used as the skip zone from thirty to 100 miles or so from the station will be adequately served by the ultra high frequencies. Large transports will probably have a 100 watt, separate high frequency transmitter.

Radio Ranges—An experimental program is being carried on by the government on ultra high frequency radio ranges. These ranges will relieve the interference now present on the low frequency band and the frequency congestion. A series of twenty to thirty frequencies can be used repeatedly over the country. Night effect will be absent as no reflections will occur from Heaviside layer. Bends and multiples will be reduced somewhat and especially bends near the station which cause difficulty at present.

When this program is instituted a receiver will be added to replace one low frequency radio range receiver and may be equipped for direction finding at the ultra high frequencies. This program will be under way in 1940 in the sixty-three megacycle range.

Weather Radio Teletype—It may be that by 1945 it will be considered profitable to equip transports with a tape or page radio teletype. A frequency between sixty and sixty-six megacycles will probably be used for this service. This will greatly supplement the broadcasts of weather information and furnish printed records which may be retained for reference.

Instrument Landing—There is little doubt as to the elements of an instrument landing system that will be in use soon. The system will be used for instrument approaches before it is used for complete landings.

A number of instrument landing systems have been installed in the United States at Newark, N. J., Chicago, Ill., Cheyenne, Wyo., Oakland, Calif., Cleveland, Ohio, and Los Angeles, Calif. These have been installed by the airlines for the express purpose of providing a proving ground. A vast amount of experimental work has led up to the present development which is all based on the original concepts as fully demonstrated at Newark, N. J. by the Bureau of Standards in 1933 and 1934. At that early time the glide path was being operated at the ultra high frequencies. Today's system utilizes a full complement of ultra high frequency instrument landing equipment. It is also very similar to the system adapted from the original work at Newark by European countries. In Europe there are many dozens of instrument landing systems in daily operation.

There are three elements necessary for air transport instrument landing systems. These elements are: First, two markers which have been described; second, a runway localizer which is nothing more than a radio range with one leg directed down the center of the runway; third, a glide path which may be provided by the same transmitter and antenna that is giving the runway localizer signal, or probably later, by a separate transmitter.

Runway Localizer—The runway localizer is visual and provided by the modulation of two directional signals which overlap to give an on-course signal. The localizers being used in the United States are very much narrower than either the radio range at low frequencies or the ultra high frequency localizer used in Europe. In fact the visual localizer indication is sufficiently exact for the pilot to determine that his position is within about plus or minus twenty feet on either side of the exact center line of the runway. Thus if a landing is to be made on a narrow runway seventy-five feet wide, it can be ascertained before contact is made whether both wheels will be on the runway proper or not.

This sensitive indication also plays a very important part in the initial approach to the edge of the field. If a plane were flying by visual instrument along the localizer leg and on-course width was a substantial number of degrees (four or five), it would be impossible to obtain an exact heading as the instrument would not indicate an off-course amount that was readable until the plane had moved far to one side or the other of the on-course zone. Thus the flight would involve a series of approximations in correcting the heading. If a bracketing system were used, it would be possible to obtain the heading with great enough accuracy but on one side and not in the center of the on-course zone. There are two important reasons why bracketing does not apply. The first is that there is no time for it nor is there time for the less accurate method of attempting to find the heading by flying an on-course. The second reason is that great accuracy is required and the side of the on-course cannot be established electrically as accurately as the center of the on-course.

If, on the other hand, the width of the on-course zone is a few tenths of a degree, as it is in the present installations, the needle indicator will be in almost constant motion unless the plane is held close to the correct heading. Thus needle movement indicates at all times to the pilot the correction to be made. The indicator needle moves to the right of center as the plane moves slightly to the right of the on-course. It will be obvious immediately that a correction to the left is necessary. Further, the rate of needle movement of the indicator indicates the amount of correction needed. Thus the sharp course accomplishes two definite purposes. It not only shows that a correction is necessary but it also indicates the approximate magnitude of the correction being made.

It was thought for some time that a broader course would be satisfactory if a radio compass were used to obtain the heading to the station. The movement of the radio compass needle would indicate the amount of correction required and the movement of the course indicator would show at what point to make the required correction.

There are two difficulties with this method. The first is due to the fact that there may be a slight cross wind, which even though not of serious consequence on the field, might have a higher velocity above the field at an altitude of several hundred to a thousand feet. Thus the heading supplied by the radio compass would be misleading for, if it were followed, the plane would drift to one side of the runway localizer course. Sufficient time would not be allowed to correct the heading of the plane by slight rotation of the compass loop, particularly as the correction might vary with altitude.

The second drawback would be the difficulty in following two visual instruments whose indications were alike but whose information is not alike. Thus can be seen the necessity for a sharp on-course.

Glide Path—The glide path is quite another problem. As the plane approaches the airport, it is an excellent idea to have a marker which shows the minimum safe elevation and the elevation which should be maintained at every point along the approach. Such a device is a glide path.

In Fig. 207, the shape of the signal pattern radiated from the transmitter at the far end of the approach runway is shown. There is a nest of such patterns for different signal intensities and receiver sensitivities. Suppose that the contour of field strength shown is 20,000 microvolts per meter. Suppose also that the receiver sensitivity on the airplane has been so adjusted that a needle, in the same instrument case as the runway localizer vertical needle, is exactly horizontal. When less signal is received the needle will drop downward toward the bottom of the case. When more signal than the amount specified is received, the needle pointer will rise toward the top of the case.

As the plane approaches the nose of the pattern, the needle will rise toward horizontal position. The plane is then forced to either ascend or descend to stay on this line of constant signal. If the plane descends it will follow the line shown. As this is continued, the wheels will finally touch the ground on the runway. The vertical needle is assumed to be held vertically in center as the localizer course is flown.

The shape of the contours of weaker field strength are well separated at a distance from the field, as shown in Fig. 207. Nearer the ground the contours become closer together. The nearer the

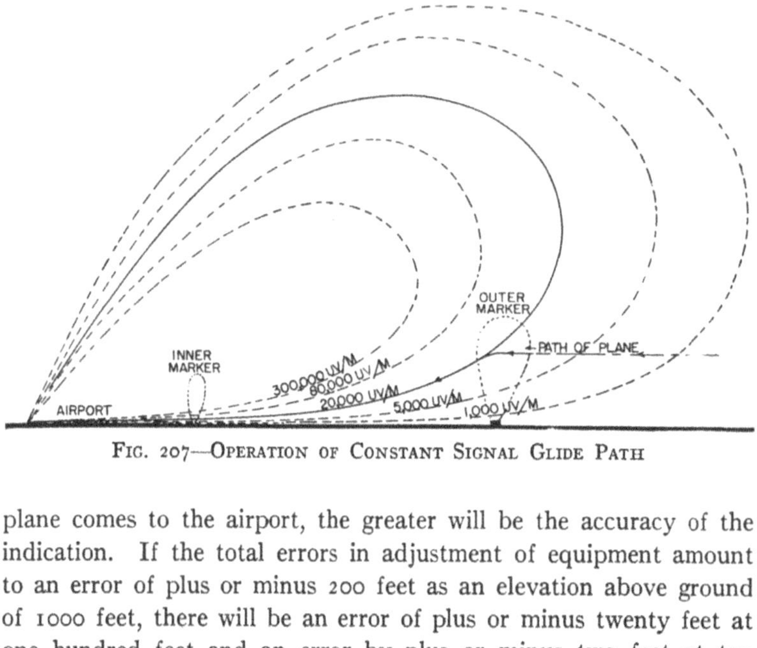

FIG. 207—OPERATION OF CONSTANT SIGNAL GLIDE PATH

plane comes to the airport, the greater will be the accuracy of the indication. If the total errors in adjustment of equipment amount to an error of plus or minus 200 feet as an elevation above ground of 1000 feet, there will be an error of plus or minus twenty feet at one hundred feet and an error by plus or minus two feet at ten feet. Thus a variation in the sensitivity of the receiver or of the output of the transmitter would result in a very small altitude error near the ground where accuracy is most important.

The two most likely sorts of error are in the transmitter falling off in power or in the receiver falling off in sensitivity. Either of these errors will cause the airplane to come in too high rather than too low. It is very improbable that the sensitivity of the receiver or the power of the transmitter would increase appreciably from any electrical cause. Safety is thus afforded.

A check over the outer marker will readily show any error in the glide path setting. For example, the error will be assumed to be plus or minus 300 feet at this altitude. Thus if the altitude as registered on the dual altimeters in the plane is between 1200 and 1800 feet, it will be known that it is safe to continue with the glide path. The importance of a marker to indicate this check point can be seen.

The glide path is somewhat curved as can be seen from the figures but it must be remembered that the horizontal scale of the drawing is very much reduced in proportion to the vertical scale. Thus when the plane enters upon the glide path at 1500 feet above the ground the rate of descent is fairly constant, until a point is reached over the approach end of the runway where the path be-

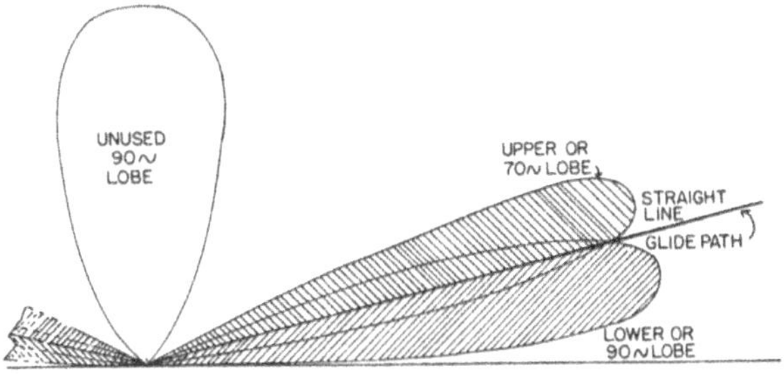

FIG. 208—STRAIGHT LINE GLIDEPATH USING TWO ANTENNA STRUCTURES—ONE EMITTING TWO LOBES

comes flatter, causing the rate of descent to be reduced and cushioning the landing.

The straight line glide path has been investigated and there are several means of producing it. One method is not by virtue of signal strength but by projection of two patterns at a flat angle. The signal from one pattern in Fig. 208 might be modulated at seventy cycles and the other signal might be modulated at ninety cycles. The glide path would be indicated by a horizontal needle actuated in the same manner as the vertical localizer needle. The straight line glide path seems most desirable. It has the advantage of better clearance at the edge of the airport (but not at a distance) and the advantage of a constant rate of descent.

The main difference in opinion in the United States at present is whether the glide path signal should be separately transmitted on a frequency near to the localizer frequency or whether both should be transmitted from the same antenna and rely upon the combined strength of the two localizer modulations. In Europe and in the present installations in the United States a single transmitter antenna array is used. This is simpler and less expensive but the separate glide path has the advantage of allowing the location of its transmitter to the side of the runway on or near the field. The point of contact of the plane can therefore be pushed nearer the approach end of the field and the plane may roll further and pass to the side of the glide path transmitter location.

There are a half dozen other systems for instrument landing, some utilizing altimeters and markers with a radio compass as the chief means of obtaining heading.

Above Ground Altimeters—The principle of operation of the above ground altimeter is very interesting as it utilizes a measurement principle not heretofore described.

The transmitter of the above ground altimeter operates on a very high frequency, about 500 megacycles. Instead of transmitting on a given frequency, it changes frequency at a very rapid rate. This is accomplished with a condenser in the oscillator circuit which rotates at a speed sufficient to change the frequency over a complete cycle of shift in about one hundredth of a second. The frequency is changed about twenty megacycles upward and then returned to the original frequency for every rotation of the condenser. Thus the rate of change amounts to an equivalent continuous rate of 4000 megacycles change per second.

The transmitter is attached to a very small half wave antenna (only about a foot long) under the plane and out under one wing about ten feet from the center line of the fuselage. A receiver is attached to a similar antenna located in a similar position under the other wing. The antennas are pointed at one another so that minimum pickup is secured for signals from transmitter to receiver direct. Frequencies of the order of 500 megacycles are rather well reflected from the ground and still better from water. A signal from the transmitter is radiated downward and is reflected back to the airplane again. In the meantime the transmitter frequency has changed slightly by the mechanism explained above and a beat note will result in the receiver, due to the frequency difference in the direct signal frequency from the transmitter and the reflected signal.

Radio waves travel at about 186,000 miles per second or roughly 1000 million feet per second. The time required for a signal to reach the ground and return from an airplane at 500 feet above ground will be one millionth of a second, remembering that the signal must travel twice 500 or a total of 1000 feet. The transmitter changes frequency at a rate of 4000 megacycles per second or 4000 cycles in one millionth of a second. Thus a beat note of 4000 cycles results. The beat will be eight cycles per foot of altitude.

The frequency indication is shown on a meter which, instead of being calibrated in cycles per second, is calibrated in feet above ground. A double scale may be used which might indicate altitudes up to 1000 feet on one scale and from zero to 5000 feet on the other scale. The change is merely in the full scale frequency indication, giving better visual accuracy in reading at lower altitudes above ground.

As the signal transmission toward the ground is directed over a rather large area, the altitude above ground indicated will be the average of the elevation covered by a cone of about thirty degrees. When the plane passes over a building at very low altitudes there will be a fairly accurate indication but at higher altitudes the difference will not be pronounced as the building is but a small part of the area averaged in the indication.

Equipment Complements—There will undoubtedly be some changes made in equipment during the transition to ultra high frequencies. The equipment, shown in Fig. 209, probably will be found on large transport aircraft.

Thus the airliner of 1945 will be equipped with seven to ten receivers and three transmitters (counting the above ground altimeter) in contrast to the airliner of today with four receivers (including the marker) and one transmitter. The important part which radio plays in the operation of an airline can be gathered from this program which is now taking place. The exact dates are reasonably close approximations only and are indicative of the trend in equipment. Smaller aircraft will not carry as complete equipment as the larger four engine transports.

QUESTIONS ON CHAPTER XII

1. Non-directional low frequency markers are:

( ) a. Fan markers. ( ) b. Z markers. ( ) c. Operated on radio range frequencies (200-400 KC).

2. Ultra high frequency markers having a circular and elliptical patterns respectively are:

( ) a. Z and FM markers. ( ) b. Fan and FM markers. ( ) c. Fan and cone of silence markers.

3. The "glide path" is:

( ) a. A line in space. ( ) b. A path provided to indicate the deviation to right or left. ( ) c. A curved surface of equal signal strength.

4. A "runway localizer" is:

( ) a. A radio range leg extending down the runway. ( ) b. A marker localizing the end of the runway. ( ) c. The combination of glide path and marker.

5. A "minimum safe altitude marker" is another name for a:

( ) a. Approach marker to the runway. ( ) b. Runway localizer. ( ) c. Glide path.

6. Above ground altimeters utilize a:

( ) a. Transmitter whose frequency varies. ( ) b. Transmitter with varying amplitude of modulation. ( ) c. Transmitter and receiver that operate alternately.

7. Communication with the ultra high frequencies will carry further than at low frequencies when:

( ) a. Below the line of sight. ( ) b. The Heaviside layer is high. ( ) c. Thunderstorm static is encountered.

8. Airport traffic control towers originally used:

( ) a. High frequency receivers and low frequency transmitters. ( ) b. Low frequency transmitters and receivers. ( ) c. Low frequency receivers and high frequency transmitters.

9. The two most desirable reasons for the use of ultra high frequencies for radio ranges are:

( ) a. A reduction of number of ranges and elimination of static. ( ) b. Lighter weight of airplane equipment and elimination of night effect. ( ) c. Producing straighter courses and elimination of interference between ranges.

10. High frequencies of the same order as now used for company communication will probably be used for the following purpose when ultra high frequencies are adopted:

( ) a. Long range and stand-by transmission. ( ) b. Contacting planes at very high altitudes. ( ) c. Radio ranges.

Chapter XIII

INSPECTION AND MAINTENANCE

The two subjects of inspection and maintenance are closely allied. If flying personnel are acquainted with the most probable sources of trouble for the particular equipment in use, it will aid in intelligent and quick diagnosis of trouble. It is considered an excellent plan to make up a chart of troubles and their remedies to be carried by flying personnel for quick reference. In this manner well over half the trouble that occurs aloft may be either permanently or temporarily repaired.

Servicing—A definite program is established on most airlines for inspection and servicing of radio equipment. Tube changes are made at definite times and apparatus is cleaned and readjusted. The intervals of servicing depend on the equipment used and it will always be found that there is a maximum and a minimum time at which service operations should occur. An example will show why this is true.

Tube life in aircraft receivers will vary somewhat with the location of the equipment as vibration plays a part in tube operation. A recent analysis of three years of reports on about thirty-five transport aircraft showed that the greatest frequency of tube failures occurred at about 100 flying hours. The number of failures per hour dropped off rapidly beyond this time. At about 300 hours the curve of failures with time gradually increased. Another determination was made as to the condition of tubes removed at various times and it was found that certain of the types were becoming electrically inferior to new tubes at about 800 hours.

Thus there must be some best time of change.

If the tubes were changed every 100 hours it is obvious that the number of failures in flight would be greatly increased as the maximum in the failure cycle would be reached each 100 hours. If the tubes were left in service for 1000 hours it would be found that the operation of the receiver would be slightly below standard even though the number of failures for the 1000 hour change cycle would be lower than at any smaller number of hours. Thus a period was established of about 1100 flying hours between tube changes and most satisfactory results were obtained.

Occasionally receivers should be allowed to exceed the change time so that a check on the correctness of the timing can be made. Other receivers should have their tubes changed at a shorter period for the same purpose.

Another factor of some importance is establishment of the tube change at a time when some other important maintenance procedure is being undertaken. The tube changes in the case mentioned are made at the second change of the airplane's engines (about 1100 hours). Further, as engines are changed occasionally before their full time, it is well to specify the change at every second left engine change, for example. Thus a few receivers will have their tubes changed before the full time of two normal engine changes and these receivers serve to supply the tube change data for the tube check at the shorter number of hours.

The Civil Air Regulations impose certain restrictions on the operation of the plane with defective radio equipment and it is extremely important that the plane can be serviced at any station through which the plane passes. For this reason spare tubes for each type of equipment used, or other spare equipment, must be maintained at each station on the airline.

The time of transmitter tube change will not necessarily occur at the time of receiver tube change. Usually transmitter tubes have longer life than receiver tubes. It may be found necessary to change the brand of a tube to a different type for the best service as the particular vibration conditions under which the tube operates may cause premature failures of one brand. Another brand may not suffer premature destruction because of slight differences in mechanical construction, filament tension, etc.

Tubes fail for two main reasons. The first is element short-circuiting. The elements in the tube, such as two adjacent grids, will touch or vibrate against one another. The second cause is opening of welds of the tube elements which will cause noise. Loose parts vibrate together and where a connection to the element is made through a weld, extreme noise or absolute failure will result. When servicing a receiver that has been reported noisy, it is important to tap all tubes with a small rubber hammer if operation of the receiver seems other than normal on the bench. If this does not disclose the trouble, the receiver should be allowed to operate for several hours before looking for the trouble again. Some "opens" occur under high temperatures only and sufficient time must be allowed for the tube to reach its maximum temperature.

In many cases it may seem easiest to change the tubes without locating the trouble. This should not be done until all other methods fail, for the trouble which is so elusive may not be in the tube at all. Good airline servicing will show a reason for each type of maloperation. If changes are made at random and the receiver operates after many such changes, it will be impossible to know the real cause. It is only by carefully establishing reasons for failure and analyzing them that future service work can be intelligently planned.

The complete unit of radio equipment should be thoroughly cleaned and inspected at some predetermined interval. All shield cans should be removed and every covered unit should be opened so that all parts may be seen, repaired and cleaned. This should not be done too often as the major inspection of a unit may in itself be the cause of a failure in flight. If procedures are thorough and the unit is carefully inspected before reassembly by a second man not working on the equipment, the chance of failures due to servicing will be small. Most aircraft radio equipment is especially designed for ease of shield removal and other inspection.

Records—A minimum number of forms for record purpose should be used but these forms should be very complete. It is well to service all equipment for major overhaul at one location. At this place a record should be kept by a card file of the history of each piece of equipment with its troubles and periods of major service. The card will also show the plane in which the equipment is located, for equipment is usually exchanged between planes at times of overhaul and for replacement of defective equipment. Spare equipment with its location is also shown as major units will be stocked at frequent intervals along the route.

Records should show the operation of only the major pieces of removable equipment such as transmitters, receivers, control units, power units, etc. Some airlines carry this record with the unit, particularly where major service is performed at several places.

The other necessary record is an inspection record, an example of which is shown in Fig. 210. The record reveals that the plane number and date are given at the top of the sheet. The service man's initial or clock number is shown at the bottom of the sheet. The man who checked the installation when the plane arrived at a terminal and the initial of the service man who checked out the equip-

TRANSCONTINENTAL & WESTERN AIR, INC.
SYSTEM RADIO MAINTENANCE
RADIO INSPECTION REPORT
W4A EQUIPMENT

PLANE NO.________ STATION________

DATE { IN________ OUT________ PILOT { IN________ OUT________ FLIGHT { IN________ OUT________

PART A

	IN	OUT		IN	OUT		IN	OUT
Power Unit No.____			Ants & Spares____			T. M. Oprn.____		
17A Rec. No.____			Jumper & Bea. Ant.____			Jack Box Oprn.____		
Cont. Unit No.____			Emer. Bat. Seal____			Spch. Mon. 4 freqs.____		
Transmitter No.____			2 Headphones____			All rec. outputs____		
W4A Rec. No.____			2 Microphones____			Dum. Ant. Cur. Day____		
W5DC2 T. U. No.____			Oprn. of all Relays____			Dum. Ant. Cur. Nite____		

PART B

	IN	OUT		IN	OUT		IN	OUT
Trans. Ckd. by____			W4A Rec. Ckd. by____			Cont. Unit Ckd. by____		
Frequencies____	ED | WD	EN | WN	SW Oprn. (Day-Nite)____			Freq. Selector____		
Osc. Grid Cur.____			LW Oprn.____			Vol. Controls____		
Doub. Grid. Cur.____			T. M. Commutator____			Output Meter____		
Amp. Grid Cur.____			5335 Relay Assy.____			Autolyn Calib.____		
Plate Voltage____			Plugs & Cables____			Bea. Tuning Switch____		
Total Current____			Emergency Dial____			On-Off Switch____		
Plugs & Cables____			17A Rec. Ckd. by____			____		
Rack & Shocks____			LW & SW Oprn.____			Dial Lights____		
5335 Relay Assy.____			Calibration____			Power Unit Ckd. by____		
Trans. Seal Brkn. by____			Emer. Bat. Voltage____			Plgs. Cbls. & Mtg.____		
____			Dial Light____			____		

PART C

	IN	OUT		IN	OUT		IN	OUT
Installation Ckd. by____			W4A Relay____			T. U. Recep. Insp.____		
Insp. of all fuses____			Fwd. Jnctn. Box Wrng.____			All Units Saftied____		
____			Rear Jnctn. Box Wrng.____			License No.____		
____			Arresters (3)____			____		

PART D

PILOT'S REPORT________

FOUND________

CORRECTION________

IN check by________(signed) at________A. M. P. M. OUT check by________(signed) at________A. M. P. M.

FINAL APRON CHECK BY________(signed) at________A. M. P. M.

Original WHITE copy—Retain. BLUE copy—Dispatch Daily Maint. Dept., Kansas City, Mo.
PINK copy—Dispatch Daily to Communications Dept., Kansas City, Mo.

Courtesy TWA

FIG. 210—INSPECTION REPORT

ment about an hour before departure time appear. These two inspections are necessary, for when the plane arrives it must be ascertained which pieces of equipment must be changed or repaired before departure. The second "out" check is necessary to insure that the work has been done and that the equipment is in operating condition.

As a plane may spend a day or more at one location, the second inspection is necessary to assure operation even though no service work has been done. The equipment will be given better service as the man responsible for checking "in" the equipment is held personally accountable for the fact that service was needed so that no delay will occur before departure time, due to having to change equipment which was supposedly operating and in good condition. The outgoing check is usually made by another man and, if the operation is satisfactory, he is held accountable for the operation of that equipment until it arrives at the next terminal.

Any service man performing work on the plane as it passes through an intermediate station is held accountable for operation of any of the equipment that he must readjust. When failure of equipment occurs it is necessary for some person to determine whether the failure could have been prevented before departure. If so, the service man responsible is informed of this occurrence and this serves the double purpose of educating the man in his service work and making him more careful in the future in his inspections.

The airlines have an expert corps of men doing this work and it is seldom that equipment fails in service. On one airline having about thirty-five planes, there have been three or four different months in which no radio failure occurred which interrupted transmission from the plane or reception on either the range or company frequencies. This is the result of careful training and expert service personnel.

The remainder of the form is divided into several parts. Near the top is a section, (A), devoted to inspection at each arrival or departure of the plane from one of its terminal stations. If the equipment is numbered (as are the major units), the number of the piece of equipment is recorded if it is in good operating condition. If it is not operating, a note is made at the bottom of the form with the number of the new unit added. In other spaces will be found the service man's number, showing that he has inspected and approved such parts.

The next major group, operation B, is done at about every forty hours and includes a further check into the operation of the units. The A and B sections are both filled out. The last group (C) is completed at about each 1000 hours and is a major inspection of all the units. Thus the inspection work is divided according to time so that inspections are made of each part on definite occasions. Instructions accompany each form so that when a part is checked off this automatically means that a certain service has been made on the unit.

A space for the comments made by the pilot is provided below these groups. This is copied from the log book on the plane. Verbal remarks are sometimes made when no trouble has been encountered but some phenomenon has been noted by the pilot which is of interest.

As soon as one of the forms has been made out in triplicate the forms are distributed in the following manner. One copy is sent to the major overhaul station so they may have a record of the operation of the equipment. The second copy is sent to the communications department for their analysis of troubles encountered and so that modifications of the equipment may be made to insure the best operation. The third form is kept at the station as their record of service.

Two more forms, Figs. 211 and 212, are required on which failed and repaired equipment is reported. These forms have the same distribution as the previous one and should be carefully analyzed. The data found on these forms are tabulated in a monthly analysis, Fig. 213. This last form is distributed to the interested departments for further analysis and to serve as a record of performance.

Failure and Delays—Each airline has its own particular method of recording failures of equipment. Each has its own terms which define the type of failure. One method used by a transcontinental line was shown in Fig. 213. The definition of the various terms is as follows:

Radio Failure: (First in importance)

1. Failure of either range or company (two-way) reception, or both.
2. Failure to contact ground stations on two successive schedules.
3. Failure of transmitter.

Equipment Failure: (Second in importance)

Failure of any unit including any accessory and any one receiver. This must not cause loss of reception on either range or company frequencies (two-way). The auxiliary receiver can be used to continue reception if one range receiver has failed.

Radio Delay: (Third in importance)

Any delay caused to airplanes on the ground due to radio service.

Form No. M-62 (Rev. June '35)

TRANSCONTINENTAL & WESTERN AIR, INC.
SYSTEM MAINTENANCE AIRCRAFT

RADIO FAILURE REPORT

STATION__________

PLANE NO.__________

DATE__________ PILOT__________ FLIGHT__________

PILOTS REPORT__________

FOUND__________

CORRECTION__________

SIGNED__________

Original (WHITE) Copy—Retain
BLUE Copy—Dispatch to Maintenance Dept.—Kansas City, Mo.
PINK Copy—Dispatch to Communications Dept.—Kansas City, Mo.

Courtesy TWA

FIG. 211—RADIO FAILURE REPORT

Form No. M-132

TRANSCONTINENTAL & WESTERN AIR, INC.
SYSTEM MAINTENANCE

RADIO REPAIR REPORT

STATION________________

DATE________________

PLANE No.__________ (removal from)

TRANSMITTER TYPE__________ TWA No.__________

RECEIVER TYPE__________ TWA No.__________

CONTROL UNIT TYPE__________ TWA No.__________

POWER UNIT TYPE__________ TWA No.__________

TAIL CONE TUNING UNIT__________ T. U. RECEPTACLE__________

MICROPHONE__________ HEADSET__________

JUNCTION BOX__________ RELAYS__________

JACK BOX__________ RACK__________

MISCELLANEOUS (describe)__________

REASON FOR REPAIR__________

FOUND__________

CORRECTION__________

SIGNED__________

Original (WHITE) Copy—Retain
BLUE Copy—Maintenance Dept.—Kansas City, Mo.
PINK Copy—Communications Dept.—Kansas City, Mo.

Courtesy TWA

FIG. 212—RADIO REPAIR REPORT

Form C-131—1M—6-35 La Rue

TRANSCONTINENTAL & WESTERN AIR, INC.
AIRCRAFT RADIO EQUIPMENT OPERATION

MONTH ____________ YEAR __________

NUMBER OF FAILURES — Part of unit which failed

W4A Transmitters ______ ____________________

W4A Receivers ______ ____________________

17A Receivers ______ ____________________

Control Units ______ ____________________

Power Units ______ ____________________

Other Units ______ ____________________

Total ______

LAST SERVICE BEFORE FAILURE

No. Failures	Station	No. Failures	Station	No. Failures	Station
______	Newark	______	Kansas City	______	______
______	Pittsburgh	______	Albuquerque	______	______
______	Chicago	______	Los Angeles	______	______

NUMBER OF EQUIPMENT CHANGES

W4A Transmitters ______ Reasons ____________________

W4A Receivers ______ Reasons ____________________

Control Units ______ Reasons ____________________

Power Units ______ Reasons ____________________

Trailing Wires ______ Reasons ____________________

17A Receivers ______ Reasons ____________________

Other Units ______ Reasons and unit ____________________

REPAIRS TO EQUIPMENT

W4A Transmitters ____________________

W4A Receivers ____________________

17A Receivers ____________________

Control Units ____________________

Power Units ____________________

Other Units ____________________

REMARKS ____________________

TROUBLES NEEDING CORRECTION ____________________

Communications SPECIFICATIONS ISSUED

No. ______ To Correct ____________________

No. ______ To Correct ____________________

No. ______ To Correct ____________________

Miscellaneous Remarks on the above ____________________

Courtesy TWA

FIG. 213—RADIO OPERATION REPORT

Equipment Troubles—Before discussing trouble shooting, either on the ground or in the air, a review of the types of trouble that will be found at times in various parts of the equipment will be given together with the best methods of inspection of such equipment.

Wiring and Terminal Boards—Wiring that is in conduit cannot be inspected with ease as it requires removal of wires. Wiring in terminal boxes can be inspected and should be carefully examined at frequent intervals. Under no condition should wires in terminal boxes be unnecessarily moved or pulled for inspection. Such a procedure will very quickly result in failure, particularly at soldered terminals. The terminals may be best inspected by tightening with a socket wrench, with the power "off." The terminal can be tightened and it will be found that even the best lockwashered terminals will loosen a small amount in time due to vibration. A certain amount of cold flow will occur also in the covering of lead-dipped lugs to which wires are attached, causing looseness.

All terminal box covers should have a blueprint of the terminal connections, shellacked to the inside for rapid identification of terminals when trouble testing is required. The covers of such boxes should be fastened on with a wing nut or any good patent fastener. Where there is a chance of the box being exposed to water, the lid should be equipped with a tight rubber gasket, fastened to the lid with cement.

Fuses—Fuses in terminal boxes may be inspected visually and may best be tested by operation of the equipment. Fuses should be changed and discarded about every six months to assure that the fuse will not fail due to oxidation and crystallization from vibration and heating. A record of such fuse changes may be made in the box. Fuse clips can be inspected at intervals by removing the fuse and tightening the clip if it is loose.

Plugs and Hoses—Plugs may be inspected by removal of the back of the plug at infrequent intervals. Good plugs will not fail nor will the connection to the back of them if the soldering job was well done and the back is not removed too often. The male part of the connector should be cleaned occasionally. A good method of doing this is to puncture a hole in the end of a rubber eraser, push the eraser onto the male prong and rotate it. This will clean off the dirt and the prongs should be cleaned with carbon tetrachloride after this is done. The female part of the plug usually will not require cleaning, nor is this easy to do in most plugs. In some types of plugs a rag dipped in carbon tetrachloride may be inserted into the female element to clean it.

Mechanical connections of the two halves of the plugs should be kept clean and bright and covered with a very light film of light oil.

Hoses can be inspected visually and the connections at the ends of the hoses must be kept tight and clean. If the hose is exposed to any liquid it should be covered with a rubber sheath. Where hoses may be rubbed by the clothing or chafed mechanically, a suitable cover should be provided. Particular attention should be paid to the condition of hoses with respect to kinks or dents. If the hose had been mechanically dented it is very likely that the wires in the hose will have had the insulation removed from them and be subject to grounding. Hoses in this condition should be replaced and the damaged wires in the hose should be replaced.

On some simple radio installation in flying boats the wiring is exposed for improperly installed wire in conduit will soon fail due to the salty atmosphere.

Switches and Control Equipment—All switches should be operated when inspected, as the switch will often show its condition by appearing to be rough in operation or will be harder than normal to throw. All units which can be inspected from the back should be looked at to see that contact surfaces are clean and good pressure is being made on the contacts.

Autosyns—Autosyn units will usually require a thorough cleaning, particularly if they "stick." The small ball bearings should be washed very carefully in carbon tetrachloride and then oiled with the lightest high quality instrument oil available. The end contact on the rotor should be cleaned in the same manner as any relay contact point and the tension should be positive.

Dirt in autosyns will always lead to jumpy indication and for this reason care should be taken that all parts of the unit are cleaned out with dry compressed air. Any oxidized material should be cleaned from the stator cylinder.

Resistance between the three stator connections should be checked to assure that the delta connected stator is electrically good, for if one wire opens the indication will be very erratic, as well as incorrect.

A master antosyn may be set up in the shop with a locked rotor. All other autosyns connected to it should show a given reading as marked on their scale. If this is not true, the needle on the autosyn under test should be moved to exact agreement.

Relays—Relays should be inspected at frequent intervals to insure that the contacts are clean. A previous section on relays has shown the best methods of cleaning and adjustment. When inspecting relays great precaution should be taken not to touch the relay itself but rather to watch its operation.

Solenoid relays which break high current are usually so sealed that the contacts cannot be inspected unless the relay is removed and opened. For this reason such relays are often mounted on some main quickly removable assembly. They should be taken apart, inspected and cleaned at intervals. Those that show very little pitting may be serviced less frequently than those whose contact surfaces show more wear.

Magnetos—Magnetos must be very carefully serviced in accordance with instructions applying to each make and model of magneto. The gaps in the distributor must be very carefully adjusted for when the gaps become large the radio noise produced by the magneto will increase enormously. An increase of gap spacing of thirty per cent may cause the noise level in the radio equipment to be very high, whereas with the normal spacing the noise will cause no disturbance whatsoever.

The fit of the distributor cover to the magneto frame should be kept as tight as possible and the metal contact surfaces should be kept particularly clean. The shielding on the ground return lead should be kept in good condition and the filter in the ground lead should be carefully inspected where such filters are used. If the filter connections should open it will be impossible to stop the engine unless the gasoline supply to that engine is turned off.

The magneto switch in the cockpit should be cleaned at intervals and kept in smooth operating condition.

The ignition wiring from the magneto to the plugs requires more maintenance than most other parts of the electrical system. The wires are operated at very high temperatures due to their proximity to the engine. On many engines it will be found impossible to use a fabric covered wire for the temperature will be so high that the fabric will char and lead to a breakdown of the insulation. Rubber covered wire with no fabric cover is usually more satisfactory but the formation of ozone, due to corona on the wire, will cause rapid deterioration of the wire. In time holes, which appear as pinpricks, will rupture and cause a short circuit in the system. Some ignition wires have a coating of metallic dust over the outside of the wire to prevent the corona from forming and destroying the wire. Where such wire is used it is very important, at bends in particular, that the surface not be scratched. Some wires are covered with "neoprene" to lengthen their life.

When the engine cools off or when the outside temperature of the air drops there is very likely to be condensation in the shielding of the ignition wires. This condensate may run down the wire to lower plugs and cause a short circuit in the elbow where the plug and wire are joined. A small rubber gasket may be installed at the point where the shield is screwed to the plug, particularly on plugs below the center of the engine where water will run downward to the plug.

Many engines are so timed that the two magnetos do not fire at the same instant., In this case the magneto which fires last will have a larger voltage because of increasing pressure in the cylinder head due to the first explosion. Thus these wires will show corona effects and more trouble will be encountered than on the wires attached to the leading magneto.

Bonding—Most bonding can be inspected visually and the bonding should be removed, cleaned and reattached at the main structural overhaul period after about 5000 hours of flying.

Generators—Generators are usually removed and serviced in the manner explained before and at the time of main engine overhaul. The effects of vibration on the generator will cause more than the usual sparking and in some cases it has been found advisable to go so far as to remove the pigtails from brushes as the extra mass presented by the pigtail caused the brush to vary in tension on the commutator. Brush operation will usually be found to be best at rather higher tensions than employed on machines which are not mounted on main engines.

Dynamotors—Occasional inspection of the condition of the commutators on dynamotors should be made by removing the cover on the end of the machine and visually inspecting the operation of the machine for sparking and visually inspecting the commutators with the machines turned off. When the machines are running, high voltages are present and under no condition should fingers be placed in or near the brushes. The methods of dynamotor overhaul were treated previously.

Microphones—Microphones may be checked by operating them on the interphone system or by modulating the transmitter and listening to the sidetone. The microphone, except for the carbon cup, should be cleaned on major overhaul and the contact pressure of the push button tested.

Microphones should not be hung in the cockpit where water due to windshield leakage may dampen them, especially when windshields are opened for landings in heavy rain.

Microphones are disease carriers, especially for the common cold. The microphone should be well wiped off with a rag soaked in seventy per cent alcohol (stronger or weaker solutions are less effective) at each change of crew. Alcohol will not damage the microphone while some other disinfectants cause failure of the unit.

Cords should be inspected and plugs well cleaned.

Headphones—The operation of headphones may be tested by listening to a particularly loud signal. The caps should be tight and where rubber caps are used they should be cleaned and kept in good condition. The plug should be polished so it will make good contact and frayed cords should be replaced before a break occurs.

Particular care should be taken in soldering new cords to plugs. The wire tinsel used for connections in some types is very hard to solder. The tension should be carried on the braid and not by the wires. Good replacement cords are arranged so that the braid may be attached in each headphone or screwed into the top of the plug to carry the tension.

The cords of both headphone and microphone must be arranged so that there is no opportunity for either themselves or their cords to become entangled in the main airplane controls. Inspection will show whether this is possible. One excellent place to hang the units and to place their jackbox is above and behind the pilots' heads. In this position the cords may be short and the units may be hung out of the way in a place where they will not be subject to water.

Receivers—Receivers may be tested by a listening test on the air which will usually determine whether the receivers are operating properly. In some cases a small signal generator is supplied which may be set at a given level to test operation of the equipment more thoroughly.

The operation of all control equipment should be checked and tunable receivers should be tuned to several stations as a check of the accuracy of the calibration as well as the operation of the equipment.

If noise has been reported the receiver should be checked, either while the main engines are being run upon the ground or by pounding the receiver with the hand while listening to it on a weak station.

Transmitters—A meter kit is supplied to test most transmitters so that the currents, particularly the antenna current, may be examined under conditions of modulation. On aircraft equipped with trailing wires there will be a dummy antenna which may be fastened to the tail receptacle of the plane and the current may be read using a proper dummy antenna.

It should be borne in mind that the plane battery or battery cart will deliver lower than normal voltage to the transmitter when the generator is not operating, with the engines stopped, on the ground. In some cases if a low battery or battery cart is used, the transmitter may appear to be quite below normal and some relays may not work.

When transmitters are set up on the maintenance test bench it is well to test their operation on both reduced and increased battery voltage. Ten and fourteen volts are usually supplied on test benches for this reason.

Test Bench Equipment—In every maintenance shop there are several pieces of equipment which are usually required for rapid servicing of the equipment. The test bench is usually wired in accordance with the wiring in the plane. It should be arranged so that all the removable aircraft radio units can be tested separately or together.

Test Antennas—A dummy loop which is electrically equivalent to that of the airplane installation should be a part of the equipment as well as an antenna which may be attached to receivers for outside listening tests, which occasionally are useful.

Transmitter Test—The transmitter portion of the test bench consists of a power supply that may be an alternating current rectifier or a power unit of the kind used in the plane. Meters are permanently wired into circuits, where possible, so that a test kit is not required. An adjustment of voltage of both the primary supply and the high voltage supply should be made available.

If the transmitter is to be operated normally at 1000 volts, the highest voltage available on the test bench should be 1500 volts which will serve to test equipment for short periods. Any part of the apparatus which is about to break down will usually show up on the high voltage test thus anticipating failures that might occur in the air. Voltage sparkover occurs very much more easily at high altitudes and the raising of the voltage on the ground test will simulate this condition as all voltages will be higher than normal with the 1500 volts applied to 1000 volt transmitters. A dummy antenna, which exactly duplicates the conditions found in flight, should be part of the test bench and this will allow adjustment of the transmitter so that it may be installed in the plane with no further adjustment.

Receiver Test—A signal generator should be supplied for the testing of receivers and sensitivity limits should be set which the aircraft receiver must meet to be satisfactory. Dummy antennas should be provided so that the receivers may be adjusted on the bench and so require no readjustment of the antenna stage in the airplane itself. Thus relatively unskilled personnel may install replacement receivers with safety.

The rectifier or dynamotor used to supply current to the receiver should be arranged so that the current and the voltage of the plate supply to the receiver may be read. Limits will be set on this current and by this means faults in tubes or in the receiver may often be spotted.

The receiver under test is attached to the generator through a shielded cord. The ground connection of the cord is attached to either the receiver ground binding post or the chassis, when no post is provided. The shielded wire is connected to the antenna binding post through a suitable dummy antenna. Each type of plane antenna will require a different dummy antenna and a suitable number of dummy antennas should be provided. A small capacitor may be used in almost all instances, although low impedance transmission lines to receivers may require a dummy that consists of a low resistance alone.

The receiver sensitivity control is usually adjusted for maximum sensitivity. The generator is then tuned to the exact frequency of alignment which is either determined by crystal control in the signal generator or by beating the generator against a standard crystal. Except for tunable receivers that are not crystal controlled, it is very important that the generator be tuned to the exact frequency. As a matter of fact even tunable receivers should be tuned to given frequencies and the generator should be set against a crystal standard.

If crystal standards are not used it will be found that the calibration accuracy will get worse as time goes on. Where radio range stations can be heard they may be used as a frequency standard thus saving the cost of the crystals. The receiver should be adjusted with the constantly modulated generator and not with the frequency standard if it happens to be a radio range or broadcast station.

It will be found that signal generators have an output control which governs the amount of signal to the receiver. This should always be adjusted for the minimum signal which will give a readable output from the receiver on an output meter which is always used in adjustments of the receiver. As the receiver approaches alignment the output of the generator must be turned down to keep the receiver output approximately at the same low level. Sometimes the frequency of the generator will be varied by changing its output. When this occurs the frequency of the generator may have to be reset during the alignment process.

A calibration should be provided on the generator control so that the generator input must be below a certain level for a certain output voltage. If this is not true the receiver either is not aligned or there is some difficulty in the receiver, such as a poor tube. Thus all receivers should be at least as sensitive as the standard adopted.

Other adjustments appear on some generators. Some are arranged so that the percentage modulation may be changed. This adjustment may be maintained at thirty per cent for most tests. Modulation frequency may also be adjustable. Marker receiver test generators should be provided with 400, 1300 and 3000 cycles modulation at 100 per cent modulation while the standard receiver adjustment frequency is either 400 or, preferably, 1000 cycles.

Dynamotor Test—A dynamotor test stand should be provided with a power source for "running in" the machines after overhaul. Dummy resistive loads will serve to show the voltage at normal current operation. The same stand may also be arranged with a noise tester to determine the amount of noise from the brushes of the machine. The easiest way to do this is to connect a sensitive rectifier alternating current voltmeter across the direct current supplied by the generator half of the machine. A small condenser is inserted in series with the voltmeter which allows the passage of alternating but not direct current. The voltmeter for this purpose should not be left across the machine and a push button should be provided for test. The meter may be equipped with a sensitive and a non-sensitive scale, each on a separate push button, so that the rough test may be determined first without danger of blowing out the meter in case the noise is very high, as it may be on a very dirty machine.

Magneto, Generator and Starter Tests—A magneto drive unit may be arranged to turn the magnetos at their usual speed and a set of gaps may be provided to allow the operation of the machine to be tested. The same is true for the generator supply and an electric motor should be provided for testing the generator under full load. A jig for testing starters which consists of a coupling for connection of the starters to a brake load may be used.

Generator Control Box Test—Voltage regulators (control boxes) should be tested on a board with a source of variable voltage with an ammeter and a very accurate voltmeter. Arrangements can be made to provide a generator test stand on the same piece of equipment so that a generator may be tested through the control box with an artificial load, which should be variable in steps. Sometimes a spare generator is left on its motor drive assembly so as to act as a source of voltage. The generator speed control may be provided through a frictional transmission or with a variable speed drive motor.

Megger—The megger is a special form of ohmmeter, reading in megohms rather than ohms. It is used for measuring the leakage through condensers only if they have sufficient voltage rating to be used at the megger voltage.

A high voltage megger is a device consisting of a source of about 1000 volts in series with which is a sensitive microammeter (50 microamperes full scale). Sufficient resistance is incorporated in series with the voltage, meter and test prods so that with a dead short circuit at the prods the meter will read full scale.

Spark plugs and ignition shielding may be tested by attaching the two test prods to the unit to be tested. The voltage is turned on and the meter will rise to full scale due to charging of the capacity of the unit under test. The meter will then decrease in deflection (increase in resistance reading) until the needle stops moving. At this point the resistance of the unit may be read on the calibrated scale. A jumping indication shows a defective unit. The needle should move steadily and then stand still.

To adjust the megger, the test prods are shorted together, just as with any ohmmeter adjustment, and the variable series resistor adjusted so the meter reads full scale or zero resistance.

Spark plugs should show more than fifty megohms, either on the engine or on the test bench. If they do not they should be dried out in an oven at about 300 degrees Fahrenheit for an hour or more. If this does not remedy the difficulty the spark plug is showing leakage which is not due to moisture.

Resistance of ignition cable in a shield should show "infinite" resistance, i.e: considerably higher than 500 megohms. The shield is attached to one prod and the wire to the other prod.

Moisture on the surface of plugs and other parts tested may lead to the belief that such apparatus is defective so that all parts should be dry when tested. In warm weather, in particular, handling of a unit before test may deposit enough moisture on the surface to give a deceptive reading.

Meggers can be built at lower voltages for testing of magneto condensers and the like. Condensers of any kind will show their age very definitely on a megger. When resistance starts to drop the unit is nearing the end of its life. For this reason the megger is a very valuable instrument in the electrical and radio shop. Various limits may be set from experience. Electrolytic condensers (which always must be attached with proper polarity) normally show a rather high leakage but, here again, when the leakage starts to increase the unit is reaching the end of its life.

Tube Test—A tube tester which indicates the performance of the tubes and is capable of showing noisy tubes and the like should be used. It is very desirable to test new tubes that are put in service as some poor tubes will be received in every batch of new tubes. When trouble is encountered in a receiver, its tubes may be tested by this device. The tube tester will save its cost many times over in preventing troubles that require time to track down as well as in preventing defective tubes from getting into service.

Miscellaneous Shop Equipment—A number of other smaller pieces of test equipment that are portable will also be found. Several volt-ohmmeters are necessary during the inspection and testing process on most radio and electrical equipment.

A small lathe for turning down armatures is required. The lathe will also be useful for making such small parts as special insulators and fittings that are otherwise unobtainable. A small drill press has infinite number of uses.

Jigs and other arrangements may be made for the rapid disassembly, reassembly and adjustment of spark plugs and for cleaning and testing them under air pressure equivalent to a somewhat higher pressure than found in the cylinder (120 pounds per square inch).

Perhaps one of the most useful things that can be added to a service shop is a board showing the numbers of all planes and which is divided into columns which show special services to be performed and especially the status of certain minor changes in aircraft. All installations on modern transports are standardized so that any change made on one plane of a type must be made on all others like it. This also serves to notify the engineering staff of the airline communication department of the status of airplane changes. Radio service men can place tags opposite the proper plane for each operation completed.

Outside Field Test Cabinets—Spare radio apparatus is often stored at fields other than the main repair base. As a cabinet must be supplied to protect the units it is often desirable to wire the cabinet so that the whole assembly will operate as a ground station. Thus the unit may be used for testing supposedly defective radio apparatus removed from planes before shipping it to the base for repair or it may be used as an auxiliary transmitter at the field in case of a ground station break-down or power failure.

As a rule, the unit will have an antenna attached to it as well as a dummy antenna for absorbing energy while testing transmitters. A receptacle can be provided for attachment of a battery cart to furnish power for operation, as an emergency ground station.

Approved Type Certificate—The Civil Aeronautics Authority requires that transport aircraft radio apparatus must be constructed to meet certain requirements and be able to pass a number of tests. These specified requirements are considered to be very good practice and show the qualities that should be found in any mobile radio equipment. While certain minor revisions will be incorporated from time to time the general specifications are based on experience gained in the last ten years.

The remainder of the material in this section is directly quoted from the August 21, 1937, specification issued by the United States government. The specification has since been issued as C.A.R. 04.503 by the Civil Aeronautics Authority with some minor changes (July 1, 1939).

AIRCRAFT RADIO EQUIPMENT

1) *Provisions for Approval:*
2) Pursuant to the provisions of the Air Commerce Act requiring the Secretary of Commerce to provide by regulations for the examination and rating of and the issuance of certificates to all air lines engaged in interstate or foreign air commerce as to their qualifications for such service, the following specifications are prescribed for approval of radio equipment installed aboard aircraft engaged in scheduled interstate or foreign air commerce.
3) These specifications shall apply to all important items of radio equipment which may be required by regulations to be installed aboard certificated aircraft engaged in scheduled interstate or foreign air commerce.
4) For the purpose of these specifications an item of radio equipment shall be considered important if, by malfunctioning, it can endanger the safety of the aircraft, or the cargo or passengers in the aircraft or persons or property on the ground.
5) Because of the fact that the development of aircraft radio equipment is constantly increasing in scope and variety, there undoubtedly will be important items of radio equipment developed from time to time for which specific provisions are not made in these specifications. In such cases the general procedure for approval will be in accordance with these specifications, and the manufacturer of the radio items in question shall apply to the Bureau of Air Commerce for special rulings particularly applicable to that product.
6) All items of radio equipment falling within the meaning of these specifications, irrespective of their classification and used in certificated aircraft engaged in scheduled interstate or foreign air commerce shall:
    (a) Satisfactorily fulfill the purpose for which they are intended.
    (b) be free from undue hazard, both in themselves and in their method of operation, and
    (c) be constructed of suitable and dependable materials.
7) *Identification Data:*
8) Each type or model of an item of radio equipment for which approval is requested shall be assigned a model name or model number by the manufacturer so that it may be distinguished from all other types of models of item of equipment.
9) Each unit of an approved type or model of radio equipment shall be plainly and suitably marked to indicate that it has been approved by the Bureau of Air Commerce.
10) Each unit of an approved type or model of radio equipment shall bear the following identification data:
    (a) Manufacturer's name and address.
    (b) Model number or model name and weight to the nearest pound and ounce.
    (c) The serial number and date of manufacture of the particular unit.
    (d) Type certificate number.
11) This information shall be displayed in a conspicuous place on the unit and in such a manner that it may not be easily erased, disfigured,

or obscured. Any other information may be added by the manufacturer at his discretion.

12) *Procedure relative to approval:*

13) Manufacturers are directed to advise the Bureau when finished model (laboratory service test sample) is ready for test. The Bureau will assign an Inspector to the manufacturer's plant who will witness the tests outlined in these specifications. If it is found necessary to make changes in design or construction, such changes may be made and the tests continued repeating any previous tests which the Inspector may deem necessary. Upon satisfactory accomplishment of the tests outlined in these specifications the Inspector will immediately advise the Bureau and the manufacturer that the model under test has complied with these specifications. The Inspector will then submit all supporting evidence of tests, model, or type designation, weights, dimensions, etc., and statement of approved changes to be made in production to the Bureau, upon receipt of which a letter bearing the approval number will be issued by the Bureau.

14) Upon completion of the first production unit and prior to disposition of any units or models through trade channels the manufacturer will advise the Bureau that such production units are ready for final inspection at which time an Inspector will be assigned to determine compliance with the specifications of the original test sample and approved changes thereto. Any minor change made in the unit after this inspection must be separately approved by the Bureau. If the change desired is of a minor nature and there is no question of such change or changes affecting the characteristics of the unit, particular attention being given to the factor of reliability, a new letter of approval will not be required. If the change is of a major nature and radically alters the characteristics of the equipment the manufacturer shall be required to assign a new type designation to the unit and if an inspection is deemed necessary by the Inspector such tests as may be required will be conducted by the manufacturer and upon satisfactory accomplishment of same a separate letter of approval will be issued in accordance with the regular procedure governing the issuance of such approval for a new unit or model. Certain basic measurements for purpose of comparison with that obtained from subsequent production units will be made a part of the letter of approval.

15) *Approval of Units:*

16) Before approval each unit of radio equipment shall be subjected to the inspections and tests outlined below and in the order given, and shall have been found to comply with the specifications as given therein. Insofar as tests and ratings are concerned, in all cases not specifically covered, the unit shall, as a minimum requirement, conform with the standards of the Radio Manufacturers Association, or, where these do not apply, with the standards of the American Institute of Electrical Engineers, and deviations from these standards shall be made only where an undeniable gain in reliability and/or performance results therefrom.

17) The manufacturer submitting the unit shall provide the necessary apparatus and facilities for making the required inspection and tests.

18) The term "unit" (of radio equipment) as used herein shall mean any assembly of parts and material which in itself can be (and usually is) used for the ultimate purpose intended.

19) Where the term "operate" is used in conjunction with a time period it is to be interpreted as meaning operation during the period stated according to a pre-determined duty cycle. In general, for receivers and receiving equipment, this duty cycle shall be continuous operation for the period indicated, while for transmitters and transmitting apparatus the duty cycle shall be some combination of intermittent operation, the time periods of which shall be determined by the service for which the equipment is designed.

20) *General Inspection:*

21) An inspection, and, where necessary, measurements shall be made to determine that the unit submitted for approval, complies with the requirements outlined herein. As a matter of guidance the more important factors involved in such inspection are noted below.

22) *Mechanical Construction and Design:*

23) The mechanical construction and design shall be such as to provide the greatest possible convenience and safety to personnel in installing and operating the unit. To this end, adequate provisions shall be made to prevent personnel from coming in contact with dangerous voltages.

24) In the design of the unit, the construction of all chassis, bases, and other structural components shall be such as to impart suitable mechanical strength and rigidity to the assembly.

25) The unit shall be so dimensioned, and the dimensions so limited as to accuracy, that the various parts can be removed and replaced without difficulty insofar as alignment of contours and screw holes is concerned. Where interchangeability is a factor the tolerances allowed shall be decreased accordingly.

26) Component parts shall, insofar as possible, be so located and disposed with respect to each other and to other material that they can be inspected, removed, and replaced without unduly disturbing or damaging other parts. In general, parts should not be stacked one upon the other nor should common mounting screws or devices be used unless such design can be justified by gains in reliability or other equally desirable characteristics. In particular, parts which require periodic cleaning and inspections shall be located accordingly.

27) Fragile parts shall be so mounted that under service conditions destructive strains will not be encountered.

28) Sufficient space shall be provided around heat producing parts to permit adequate cooling and to prevent damage to adjacent parts.

29) Except in special cases no part shall be supported wholly by its pigtails and in any case the supporting leads shall not exceed ½ inch in length.

30) The method of securing parts or structures to each other shall be such that any loosening which occurs can be immediately and effectively corrected. In this connection self-threading screws shall be restricted to those places where the design features are consistent with such use.

31) Wherever machine screws engage threads tapped in sheet material the threaded material shall be of ample thickness to assure the permanence and strength of the threads. This requirement is particularly applicable where the screw in question will be frequently removed and replaced for purpose of routine servicing and inspection.

32) All machine screws and nuts shall be secured by a suitable device or method to prevent loosening due to shock or vibration.

33) All wire shall be of sufficient cross section, of correct temper, and of proper construction to provide ample current carrying capacity and mechanical strength.
34) All wiring shall be suitably supported so as to obviate breakage and minimize changes in frequency or output due to severe vibration and acceleration or shock encountered under flight conditions.
35) Wires shall be secured and terminated in such a manner that no bending moment sufficient to fatigue the metal of the conductor to the point of failure will be developed under conditions encountered in service.
36) The insulation on wires shall be such as to provide high dielectric strength and high leakage resistance under any possible combinations of climatic conditions encountered in service. The insulation shall be fabricated of material which is not susceptible of igniting into flames in the event of abnormal currents which would heat the conductor to incandescence. In general, insulation must be of the slow burning type and should not be impregnated with highly inflammable substance.
37) Adequate protection against abrasion of all wires and connections shall be made.
38) Where textile-covered wire has been bared at the ends or at intermediate points for the purpose of making connections to terminals or posts, the exposed end of the insulation to a distance of at least ½ inch shall be painted with a suitable air-drying varnish or cement. Care shall be taken to prevent such varnish from running into the areas where good electrical connection is desired.
39) Exceptions to the use of varnish may be made in the following cases:
   (a) Where part of the terminal is wrapped or clamped around the insulation and holds the braid in place.
   (b) Where the outer braid has been thoroughly coated with lacquer or varnish by the manufacturer.
40) All wiring used in aircraft radio equipment shall, insofar as is practicable, be distinctly color-coded in order to facilitate testing and the location of faults.
41) Metallic shielding used over conductors shall be terminated at a suitable distance from the exposed conductor and in such a manner as to assure a satisfactory and permanent termination.
42) Conductors utilizing metallic shielding which is unprotected by an outer insulation sleeve shall be so secured as to prevent the shielding from coming into contact with exposed terminals or conductors.
43) All splices shall be properly insulated consistent with the voltage involved.
44) All soldering for electrical connections shall be done with rosin flux only.
45) Any process of surface finishing or structure soldering, welding, etc., which uses material of the slightest degree of corrosiveness shall include means by which all traces of the corrosive materials are removed.
46) Mechanical strength of a soldered connection shall not depend upon the solder but shall have been assured by suitable mechanical fastening of the various members involved.
47) In no case shall electrical connections between metals parts or wire smaller than #14 A.W.G. be made by mechanical clamping only. Such connections shall be soldered or shall be made by means of soldering lugs or terminals.

48) Plugs and receptacles used to connect the circuits of the unit to outside circuits or to connect the circuits of the various parts of the unit to each other shall be suitable from the standpoint of:

(a) Mechanical ruggedness.
(b) Method of locking.
(c) Voltage insulation.
(d) Current-carrying capacity.
(e) Method of connecting terminating conductors.
(f) Method of terminating shield (if used).
(g) Method of mechanically supporting terminating conductors.

49) All movable controls shall be so designed and constructed that accidental displacement shall not result under service conditions.

50) Insofar as possible all controls, switches, levers, and the like which are manipulated during operation of the equipment shall be of such rugged design and construction as to permit their operation, without damage and by unskilled personnel, under the most adverse service conditions.

51) *Electrical Design:*

52) The electrical design and construction of the unit submitted for approval shall be such as to indicate that the performance which will be obtained under service conditions will be satisfactory for the purpose intended.

53) The electrical characteristics—input, output, frequency range, etc., of the unit shall be in accordance with the published ratings.

54) Vacuum tubes used in the unit shall not be operated at voltage or current levels beyond the recommendations of the tube manufacturers.

55) The performance of the unit shall not vary beyond the operating limits upon the replacement of any or all the tubes of the unit by good tubes of the particular type recommended by the manufacturer of the unit. An exception to this requirement may be made, however, where the unit in question is designed for some highly specialized use in which case selected tubes may be used but provided the manufacturer so states (with due emphasis) on some placard or plate permanently attached to the unit and in the instruction bulletins pertaining to the unit.

56) In any receiver intended for the reception of radio range signals where the directive properties of these signals or the field strength intensity are used as an aid to navigation, the (sensitivity) volume control shall function in such a manner that for signal field strengths within the ranges encountered under service conditions, the output signal shall increase with increasing signal input, the control being always adjusted to prevent overloading of the receiver. The circuit shall be such that improper operation of the control shall not result in a reversal of the signal quadrants, provided the output level of the receiver is not greater than 500 milliwatts when the carrier is modulated 30%.

57) The control by means of which the sensitivity and/or volume of a receiver used for the reception of radio range signals is adjusted shall be so designed as to produce smooth operation through its entire range and the design shall be such that the minimum perceptible difference in output can be obtained, within the operating limits of the receiver, for any signal input up to one volt.

58) *Workmanship:*

59) All soldering shall be done in such a manner that good connections, both electrically and mechanically, are obtained. To this end care should be taken to prevent rosin from entering into areas of pressure contact.

60) The unit submitted for approval shall be constructed and assembled in a neat and workmanlike manner and the workmanship throughout shall be of the highest order.

61) All dimensions and tolerances shall be held as close as is consistent with approved shop practice.

62) Extreme care shall be taken, when soldering conductors to terminals, lugs, etc., which are in pressure contact with other conductors, to prevent the rosin flux from entering into the area of contact.

63) *Materials:*

64) All materials entering into the construction of the unit shall be suitable for the particular use intended. As a guide to the selection of suitable materials certain specifications are given below:

65) Materials which will ignite or explode from an electric spark or from the degree of heat encountered under service conditions shall not be used, and, in general, the use of combustion sustaining materials shall be held to a minimum.

66) Insulating materials used for structural parts shall have such characteristics that no permanent deformation which will cause unsatisfactory operation shall result from temperature changes and other climatic conditions.

67) Insulating materials as a whole shall be such as to assure no degradation of electrical properties sufficient to cause unsatisfactory operation under the service conditions which will be encountered.

68) All bolts, studs, screws, and nuts, wherever practicable, shall conform to the National Screw Thread Commission fine or coarse thread series.

69) The use of magnetic materials, except where required for specific magnetic purposes, shall be kept to a minimum.

70) Electrolytic condensers of the liquid or semi-liquid type shall not be used. However, the so-called "dry-type" electrolytic condenser may be used, providing the design of the condenser itself is suitable for the purpose intended and the use of such a condenser in the circuit involved is in accordance with the best engineering and design practice.

71) *Humidity Test:*

72) The radio unit submitted for approval shall be subjected to a humidity test as outlined below:

73) The unit shall remain for a period of 48 hours in an atmosphere of clear vapor maintained at not less than 95% humidity and at a temperature within plus or minus 3° of 50° C. The percentage of humidity and the time element involved may be less than the maximum specified above if in the opinion of the Inspector the design of the unit is such that any inherent defects will be disclosed at a lower percentage of humidity or within a shorter period of time. At the end of the humidity exposure period the unit shall, before the elapse of ten minutes, be operated for a period of two hours at ambient room temperatures and at the voltage specified as normal for the unit. The performance of the unit during this period of operation shall not be below the minimum necessary for the service for which it is intended. Immediately after the end of the period of operation

specified, an inspection of the unit shall disclose no evidence of corrosion or other condition the presence of which, or the continuance of which will lower the performance of the unit below the minimum necessary for the service intended.

74) *Temperature Test:*

75) The unit submitted for approval shall then be subjected to a temperature test as outlined below:

76) The unit shall be placed in an ambient temperature of + 55° C. for a period of time sufficient to allow the material of the unit to assume the same temperature as the surrounding air, after which it shall be operated for a period of one hour at a supply voltage 20% above that specified as normal for the unit in the case of 12 volt or 24 volt DC equipment, and 10% above that specified as normal in the case of 115 volt AC equipment. The performance of the unit during this period of operation shall not be below the minimum necessary for the service intended. The unit shall then be placed in an ambient temperature of — 40° C., for a period of time sufficient to allow the material of the unit to assume the temperature of the surrounding air after which (with the unit still in the cold chamber) it shall be operated at a supply voltage 10% below that specified as normal for the unit for a series of five cycles of operation consisting of 2 minutes of operation followed by a period of non-operation sufficient to allow the temperature of the unit to return to — 40° C. At the end of this series of cycles of operation the unit shall be operated for a period of 15 minutes. The performance of the unit during these periods of operation shall not be below the minimum necessary for the service intended.

77) *Spray Test:*

78) Those units submitted for approval which are usually located in the aircraft where rain and spray is encountered during the normal course of operation of the aircraft, shall be so constructed that they will not be susceptible to being damaged by such rain or spray. In any case where the ability of the unit to meet this requirement can not be determined from a superficial inspection it shall be subjected to a test substantially as outlined below:

79) The unit shall for a period of two hours, and at normal room temperature be sprayed with a stream of commercially pure water of sufficient volume to flush the surface of the unit against which it is directed and of a velocity approximately equal to that of rain drops in a heavy rain. The stream shall, during the two hour period, be directed at the unit from various angles so that actual conditions will, insofar as possible, be simulated. The unit shall be operated during the time it is sprayed. At the end of the spraying period the unit shall remain, without being disturbed, for a period of 48 hours at ambient room temperatures after which it shall be operated for a period of two hours. During the period of operation specified the performance of the unit shall not be below the minimum necessary for the service intended. A visual inspection shall disclose no evidence of conditions the presence of, or the continuance of which, is detrimental to the proper functioning of the unit.

80) *Vibration Test:*

81) The unit submitted for approval shall then be subjected to a vibration test as outlined below:

82) The unit shall for a period of 4 hours be vibrated at any frequency from 30 to 60 cycles per second, and with an amplitude and wave form sufficient to produce a vertical acceleration of 10g. During a portion of this period of vibration the unit shall be operated in order to detect any variation in output, frequency, etc., or any other harmful effect which may be caused by the vibration. During this period of operation the performance of the unit shall not be below the minimum necessary for the service intended. At the termination of the period of vibration, a visual inspection shall be made and shall disclose no condition, produced by the vibration, the presence or the continuance of which is detrimental to the satisfactory performance of the unit. The unit shall then be vibrated over a range of frequencies varying continuously from 25 to 150 c.p.s., and at an amplitude such that the vibrations can readily and easily be felt by placing the hand on the unit. The period of time taken to cover the frequencies from 25 to 150 c.p.s., or vice-versa at a uniform rate of change shall not be less than 125 seconds. During this varying frequency vibration test the component parts and structures of the unit shall be observed for evidence of resonant vibration. Parts which show such tendencies shall be re-designed or remounted to eliminate the characteristics. Where re-designing or remounting is not practicable the manufacturer shall satisfactorily demonstrate that the resonant vibration of the part will in no manner be detrimental to the performance of the unit. The number of such cycles of vibration shall be whatever is necessary to obtain the information desired.

83) *Orientation Test:*

84) The unit submitted for approval shall then be operated for a period of 15 minutes in each of the eight attitudes assumed at the 45° and 90° positions when the unit is rotated through the four planes normal to the mounting plane of the unit as mounted for normal operation. Performance during these periods of operation shall not be below the minimum necessary for the purpose intended.

85) *Drop Test:*

86) The unit submitted for approval shall then be subjected to a drop test as outlined below:

87) At least 100 free drops from heights varying from 6 to 18 inches shall be made. No shock absorbers will be attached to the unit, but commercial sponge rubber of a thickness not to exceed one inch may be interposed as a damping medium between the unit and the substantially solid platform upon which the equipment will come to rest. During this test the equipment will be released in the free drop in such a manner as to insure it coming to rest in the same attitude for which it was designed and to further insure the stress to be equally distributed. During this test it may be required that transmitters be in the operation, standby, or inoperative position. The purpose of this test is to determine the adequacy of the mounting of various component parts, wiring, tube mounts, etc., and any abnormal failures will be deemed sufficient reasons for disapproval of the units. It is also the purpose of this test to determine the suitability of connecting plugs and particular attention will be given to the end that no failures from this source will be evident.

88) *Pressure Test:*

89) The unit submitted for approval shall then be operated in a pressure

chamber under conditions of pressures ranging from 31 to 8.52 inches of mercury. The period of operation at the lower pressures shall be sufficient to disclose any defects which will be aggravated by reduced pressure, and during the entire period the performance of the unit shall not be below the minimum necessary for the service intended. Where it is not practicable to operate the entire unit under the conditions of pressure specified the major assemblies shall be operated at these pressures with voltages and other electrical conditions equal to those which exist when the parts and assemblies are operated within the unit.

Trouble Shooting—Trouble shooting is a skill which is developed after considerable experience. The first requirement in trouble shooting is to stop for a few seconds and think before attacking the problem. The evidence may not be correct and it is well to see that the trouble is as reported. Instances will arise where one man will tell another that a certain receiver fails to function whereas he meant that the transmitter or another receiver would not function.

A few simple tests may be made which will localize the trouble to some unit or a group of units. If electric, units may fail due to the battery voltage being low or absent. It is best to ascertain first that the battery is in proper condition. Relays may fail with very low voltage as do oscillators in transmitters.

Suppose that a ten frequency transmitter has failed. The first thing to do is to be sure that it has failed and then try another frequency. If this will not work then the failure involves something common to all frequencies. Possibly it is a common tube which can be readily inspected. Perhaps the shift mechanism is stuck between two frequencies. Maybe there is no high voltage. A number of such things may be checked at one time while at one location.

The most important thing in trouble shooting is to follow any single clue until a positive answer is determined. Changing ideas of the trouble fault may lead to missing the trouble.

Principally, the fundamentals of the apparatus in use must be very thoroughly understood. After this a plan of attack should be undertaken and not changed until it is exhausted.

The order of trouble shooting might be as follows:

1. See that battery voltage is normal with the fault occurring as reported and check control switch positions.
2. Check fuses by testing.
3. Check tubes visually.
4. Check relays.
5. Plan definite lines of attack following each to completion.

Trouble Charts—It is well to supply charts of trouble which are based on experience with the equipment in use. Fig. 214 shows a chart used by pilots and mechanics on a transcontinental airline.

RADIO TROUBLE SHOOTING DOUGLAS DC-2, DC-3 and SDT
(Summary of Existing Instructions)

FIRST CHECK FUSES VISUALLY, THEN WITH BACK OF FLASHLIGHT, AS A BREAK SOMETIMES CANNOT BE SEEN.

AMPS	FUSE NAME	LOCATION	LOCATION OF SPARE FUSE
40	Main Line	Front Radio Distribution Box	Front Radio Distribution Box
20	Transmitter	DC-2 Tail Rack--DC-3 - SDT Front Radio Dist.	Front Radio Dist. Box
15	Receiver	DC-2 Tail Rack--DC-3 - SDT Front Radio Dist.	Front Radio Dist. Box
15	Relay	DC-2 Tail Rack--DC-3 - SDT Front Radio Dist.	Front Radio Dist. Box
1 (2)	1000 volt	In Power Supply Box (TURN RADIO OFF)	On Power Supply Box
150	Link	In Power Supply Box (TURN RADIO OFF)	On Power Supply Box (Note A)

SYMPTOM (See Note B)	PROBABLE TROUBLE	CORRECTION
All Radio Out	Main line fuse blown	Replace fuse
T-Lights dim or no sidetone	Battery switch off or on cart	Switch to battery
T-Big dynamotor runs - no sidetone	Transmitter fuse blown or either or both 1000v fuses blown	Replace fuses
T-Big dynamotor won't run - no sidetone	Link fuse out (difficult to change)	Replace link
T-R-Sidetone OK - R slightly weak	Trailing wire lost	Release aux. antenna
T-R Humming sidetone - no signals - big dynamotor running	Transmitter stays on	Pull micraphone plug (s)
T-R-Sidetone very mushy or none	R or T not changing frequency	Change manually
R-Receiver out (W4A not 17AX)	Receiver fuse blown	Change fuse
R-Won't dial, shift frequency or tune	Relay fuse blown	Replace fuse
R-Weak reception in ice	Ant. iced (DC3-SDT use aux. instead)	DC-2 Pull cord
R-Won't tune (beacon)	Relay fuse blown	Replace fuse
R-No two-way or range	Electrical system failure	A on emergency battery
R-A-Captain hears signal only	Switch box wrong position of failure	Exchange phones
A-OK on ranges - no two-way	On loop or filter in wrong position	Switch to antenna
R-A-No range signals but voice	Filter switch on VOICE	Switch to Nor. or Range
A-278 kc signals swing or weak	Receiver on loop	Switch to ANT.
T-R-Failure of one frequency	Crystal loose or T or R failure	Listen on region Frequency used if R works, otherwise on alternate (day at night, etc.) AWAIT INSTRUCTION TO TRANSMIT.

(Note A) Fuse seldom fails. To replace, lift small box on power unit after removing two wing nuts. Fuse is flat on base behind 1000v fuse panel.

(Note B) T - W4A Transmitter R - W4A Receiver A - Auxiliary Receiver

Courtesy TWA

FIG. 214—SAMPLE TROUBLE SHOOTING FORM

This chart is particularly valuable where radio faults will occur but once a year or less at a particular station or with a particular pilot. Naturally the chart shown is for a particular equipment complement only.

QUESTIONS ON CHAPTER XIII

1. Wiring on terminal boards is best inspected by:
( ) a. Tightening nuts and visual inspection. ( ) b. Pulling the wires gently at connections. ( ) c. Unscrewing the nut and retightening.

2. Fuses will fail with:
( ) a. Ionization. ( ) b. Rated currents. ( ) c. Age.

3. Relays will give trouble if:
( ) a. Cleaned with carbon tetrachloride. ( ) b. Filed constantly. ( ) c. Contacts rub together instead of touching lightly.

4. Corona does most damage to:
( ) a. Ignition wiring. ( ) b. Spark plug points. ( ) c. Magneto insulation.

5. Condensation will do most damage during engine operation:
( ) a. At spark plug points. ( ) b. Where contacts rub together. ( ) c. Where wires join plugs.

6. Vibration is one of the causes of pitting:
( ) a. Spark plug points. ( ) b. Commutators. ( ) c. Brushes.

7. Wire used in headphone cords is very hard to:
( ) a. Solder. ( ) b. Break. ( ) c. Remove.

8. Transmitting dynamotors should be run in before using as:
( ) a. This assures proper voltage. ( ) b. This shows up noisy bearings. ( ) c. This assures proper brush fit.

9. Receivers should generally be aligned at:
( ) a. Minimum sensitivity. ( ) b. Maximum sensitivity. ( ) c. High output.

10. Receivers are generally aligned:
( ) a. At the high frequency end of a band first. ( ) b. At the low frequency end first. ( ) c. At the center of the band.

11. Generator control box testing requires a very accurate:
( ) a. Ammeter. ( ) b. Ohmmeter. ( ) c. Voltmeter.

12. A megger is a high voltage:
( ) a. Ammeter. ( ) b. Ohmmeter. ( ) c. Voltmeter.

13. A megger will usually give misleading information with:
( ) a. Surface moisture on part tested. ( ) b. Spark plugs. ( ) c. Ignition wiring.

14. A spark plug showing more than about fifty megohms on test is:

( ) a. Defective. ( ) b. Covered with a heavy surface moisture. ( ) c. Satisfactory in insulating qualities.

15. A 400 volt condenser may be tested on a 1000 volt megger:

( ) a. For a short period. ( ) b. Never. ( ) c. Only if an electrolytic.

16. An ohmmeter is principally useful for testing:

( ) a. Dead circuits. ( ) b. Only resistors. ( ) c. Live circuits.

17. A voltmeter is useful for:

( ) a. Dead circuits. ( ) b. Only resistors. ( ) c. Live circuits.

18. Transmitters should not be opened and inspected with:

THE HIGH VOLTAGE ON

CHAPTER XIV

USE OF QUESTIONS

The best way to study this book is to answer the questions on each subject. This process has usually been inconvenient, especially where a large amount of writing has been required. Further, many people will read in an easy chair or on an airplane. Therefore, the questions in this book have been given in a form which require only a light check mark at the proper answer to the question. Some small amount of calculation will be required in answering some of the questions but usually this can be done mentally.

By the use of this more popular method of answering questions, it has been possible to greatly increase the number of questions. It is highly desirable to finish each set of questions before looking up the answers for it is very easy to see the next answer unintentionally. As it is a great nuisance to continually turn the pages to the answers and then back again to the questions, a system has been instituted which will overcome even this. Each set of answers that follows is given a code number by which it is possible to correct any group of questions by one quick reference to the answers. In addition, the answers to some few questions have been explained where necessary.

The use of the code number in answering questions is accomplished in the following manner. The code number for a certain chapter might be 3442. This means that the "C" (third) answer is

Number of Question	*Correct Answer*	*Order of Answer*
1	(c)	3
2	(d)	4
3	(d)	4
4	(b)	2
5	(c)	3 (repeating)
6	(d)	4
7	(d)	4
8	(b)	2
9	(c)	3 (repeating again)
10	(d)	4
etc.	etc.	

correct for the first question, "D" (fourth) for the second and third questions and the "B" (second) answer is correct for the fourth question. The series then repeats itself to the end of the group. Thus the answers to the questions using the code 3442 are on page 365.

The following pages are devoted to the answers to questions with explanation where needed. Questions that can be answered by direct reference to the text are not explained.

If this chapter is being read for the first time the reader may now turn back to his set of chapter questions.

ANSWER TO QUESTIONS

CHAPTER I: Code 432 (or d,c,b) then repeating.

Q. 8: 200 ampere hours in, and 200 ampere hours out—no loss.

Q. 9: Difference in charge rates—10 amperes. In 10 hours, which is 600 minutes, the 100 ampere hour charge will be exhausted.

Q. 10: Time for discharge 4 hours—time for charge about 4 hours as efficiencies are not 100 per cent.

Q. 11: Neither charging nor discharging. A battery, after reaching full charge, will have a terminal voltage of 14.2 volts.

Q. 16: The two 10s in parallel are equivalent to 5. This equivalent 5 in parallel with the 5 ohm resistor gives a total equivalent of 2.5 ohms.

Q. 18: A complete cycle reaches zero twice and the third time it reaches zero is actually the beginning of the next cycle. The answer is twice.

Q. 20: Answer (d) is wrong as 10 cycles is not considered a voice frequency.

Q. 25: Loads are respectively 20, 20, 20 and 40 ampere hours or an equivalent 100 ampere hours or 50 amperes for 2 hours. The final charge of the battery was still 60 ampere hours, or half charged.

CHAPTER II: Code 1234 (or a,b,c,d) then repeating.

Q. 1: The North magnetic pole behaves in effect like the south pole of a magnet, for it attracts the north pole of any magnet.

Q. 7: Steady direct current will not be transformed, so zero current will flow in the secondary.

CHAPTER III: Code 213 (or b,a,c) then repeating.

Q. 18: gives a total Z upon simplification of $-j100$ only, which shows a condenser of pure reactance and no resistance.

Q. 20: The (c) answer is wrong because the inductance and the capacitance equality means little, their *reactances* must be the same and this involves selection of the proper frequency.

CHAPTER IV:	Code 4231	(or d,b,c,a)	then	repeating.
CHAPTER V:	Code 2231	(or b,b,c,a)	then	repeating.
CHAPTER VI:	Code 1324	(or a,c,b,d)	then	repeating.
CHAPTER VII:	Code 1213	(or a,b,a,c)	then	repeating.
CHAPTER VIII:	Code 2431	(or b,d,c,a)	then	repeating.
CHAPTER IX:	Code 3211	(or c,b,a,a)	then	repeating.
CHAPTER X:	Code 3132	(or c,a,c,b)	then	repeating.
CHAPTER XI:	Code 3121	(or c,a,b,a)	then	repeating.
CHAPTER XII:	Code 3131	(or c,a,c,a)	then	repeating.
CHAPTER XIII:	Code 132	(or a,c,b)	then	repeating.

BIBLIOGRAPHY

Brimm, D. J. and Boggess, H. E., *Aircraft Engine Maintenance.* Pitman, New York 1939.

Brimm, D. J. and Boggess, H. E., *Airplane and Engine Maintenance.* Pitman, New York 1936.

Everitt, W. L., *Communication Engineering.* McGraw-Hill, New York 1937.

Johnson, K. S., *Transmission Circuits for Telephonic Communication.* D. Van Nostrand, New York.

Olson, H. F. and Massa, F., *Applied Acoustics.* Blakiston, Philadelphia 1934.

Peters, L. J., *Theory of Thermionic Vacuum Tube Circuits.* McGraw-Hill, New York.

Redpath, P. H. and Coburn, J. M., *Air Line Navigation* (in preparation). Pitman, New York.

Reyner, J. H., *Short Wave Radio.* Pitman, London 1938.

Taylor, G. F., *Aeronautical Meteorology.* Pitman, New York 1938.

Terman, F. E., *Radio Engineering.* McGraw-Hill, New York 1935.

Titterton, G. F., *Aircraft Materials and Processes.* Pitman, New York 1937.

INDEX

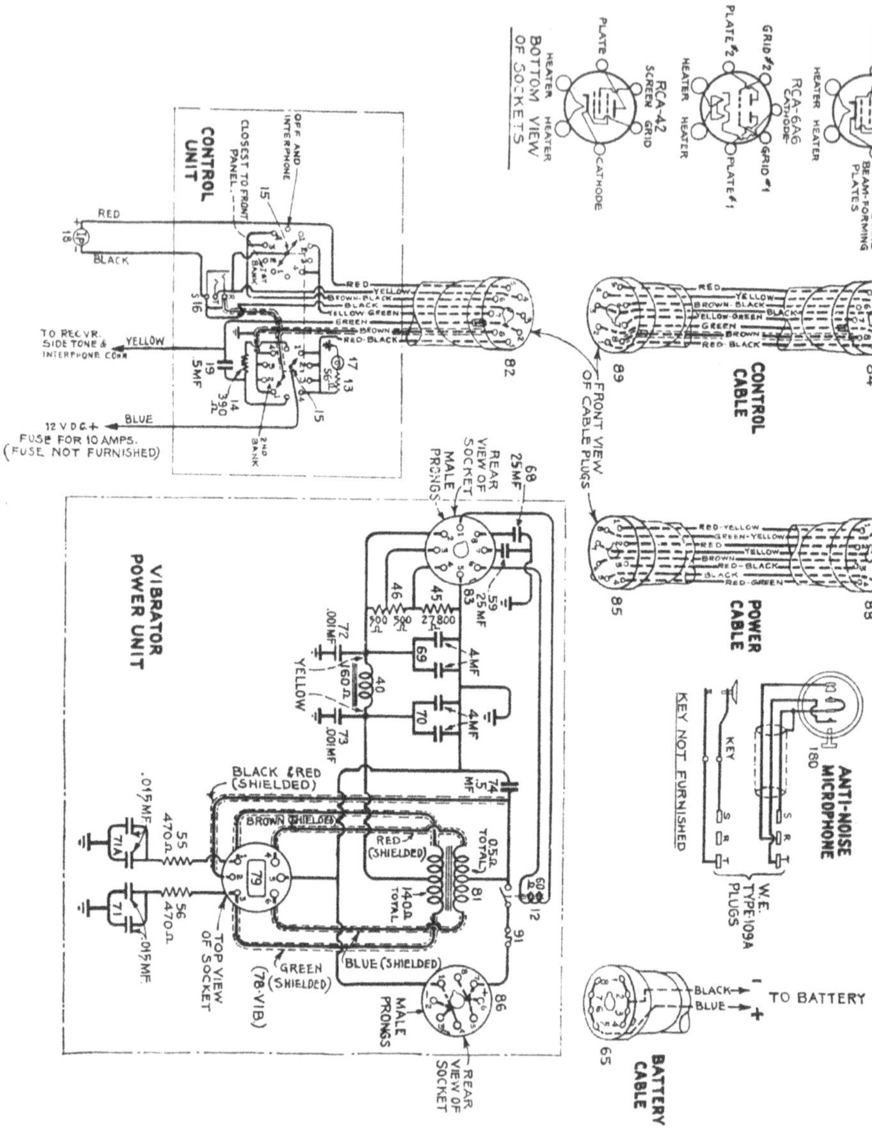

Fig. 68—RCA AVT 7B Transmitter Schematic

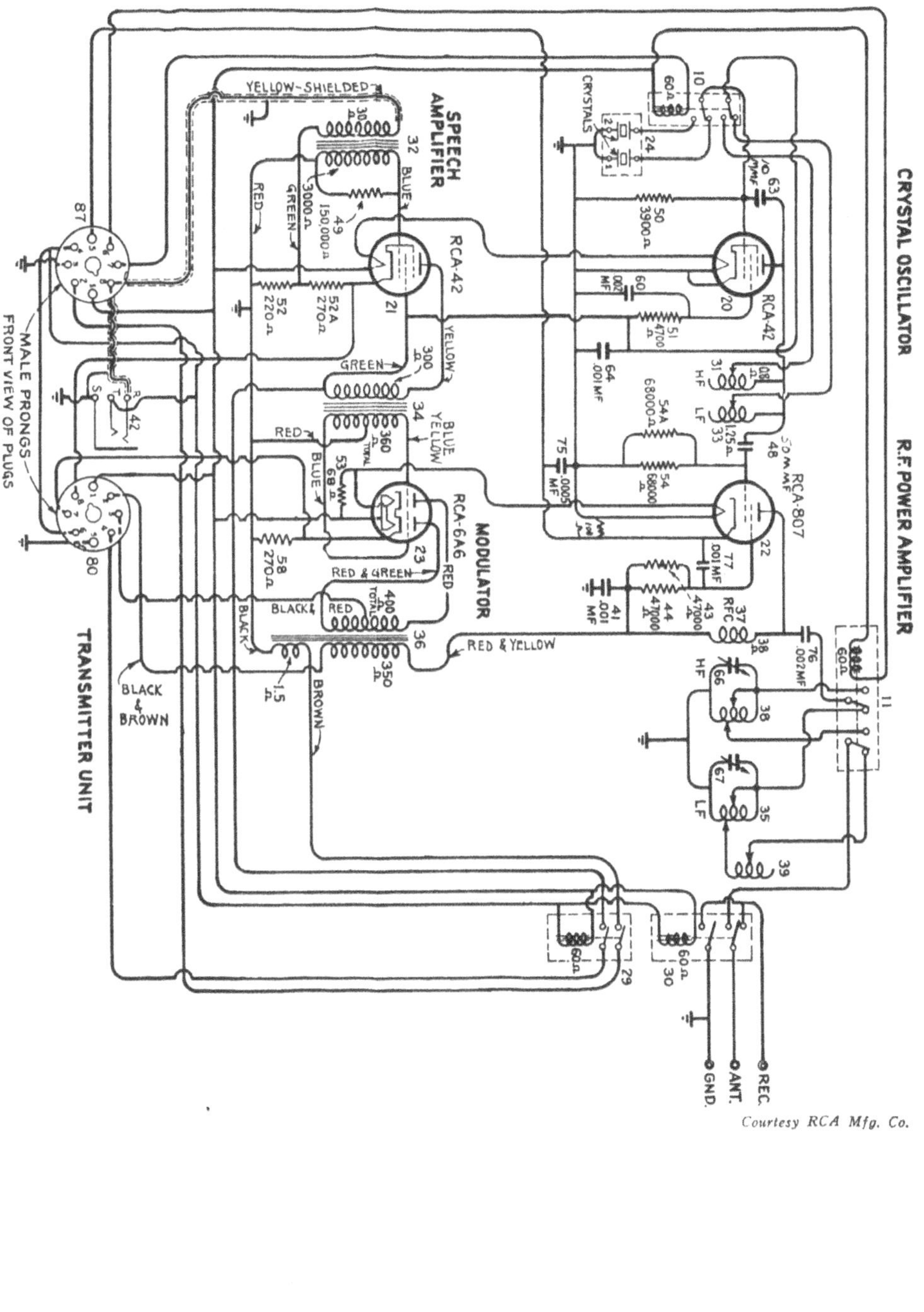

Courtesy RCA Mfg. Co.

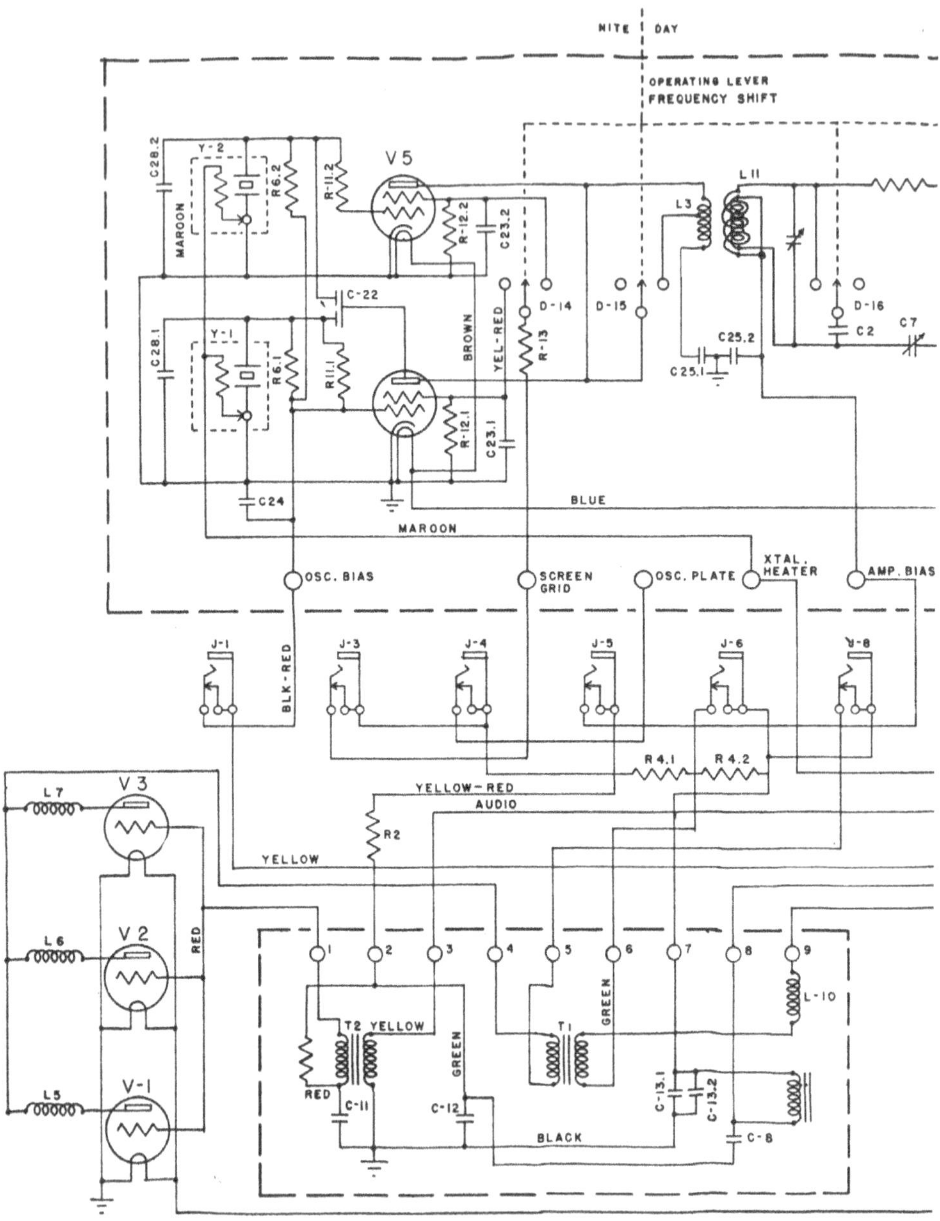

Fig. 73—United Airlines ES 271 Transmitter

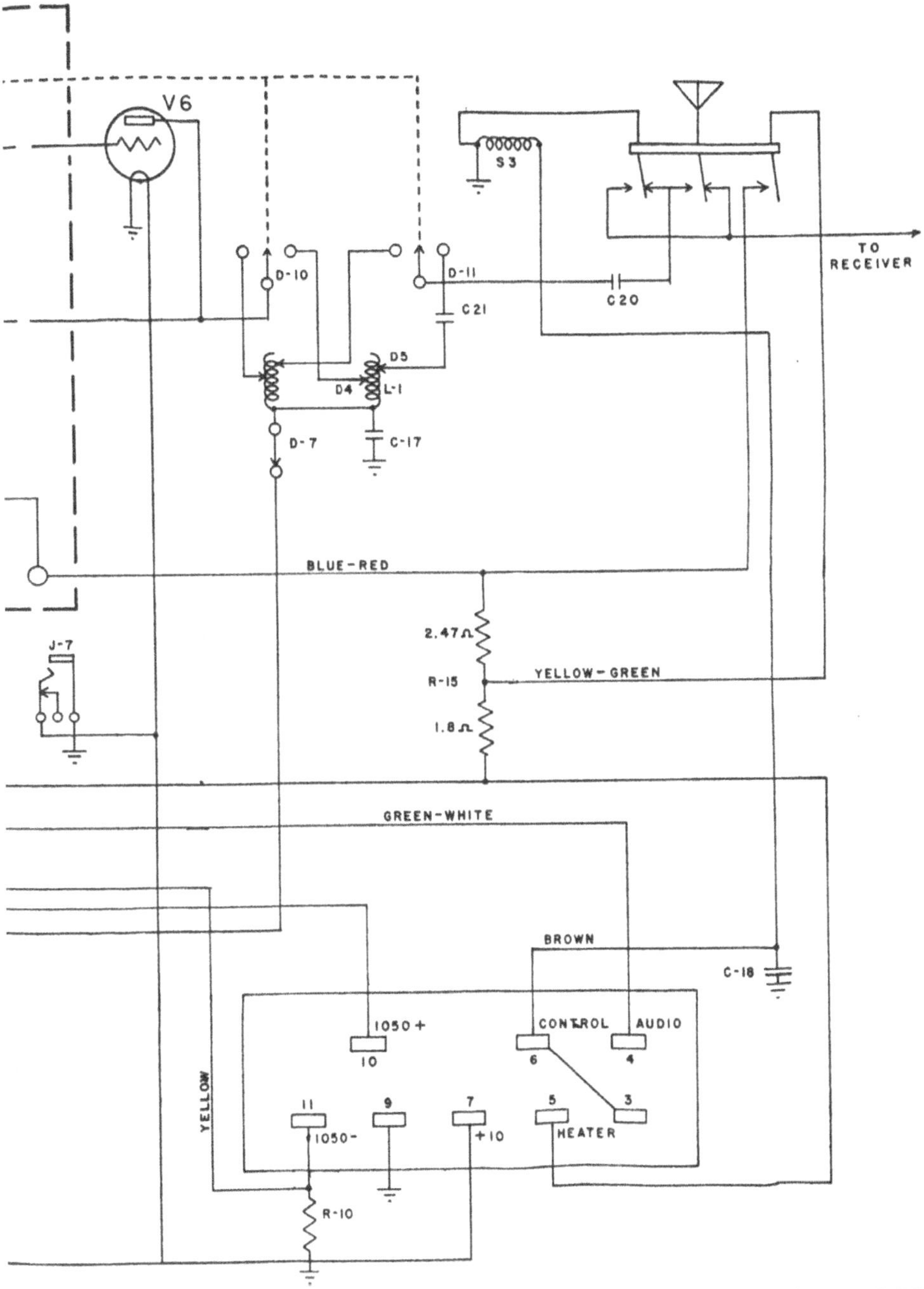

Courtesy UALTC

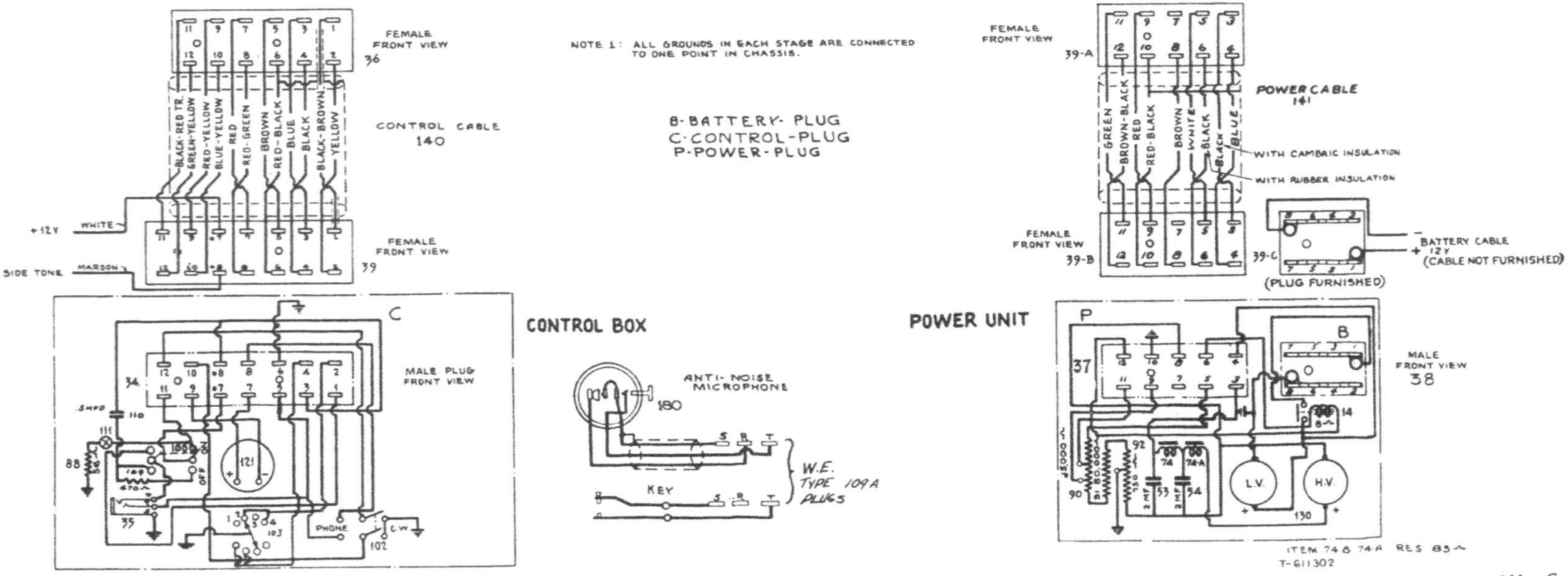

Courtesy RCA Mfg. Co.

FIG. 75—RCA AVT 12B TRANSMITTER SCHEMATIC

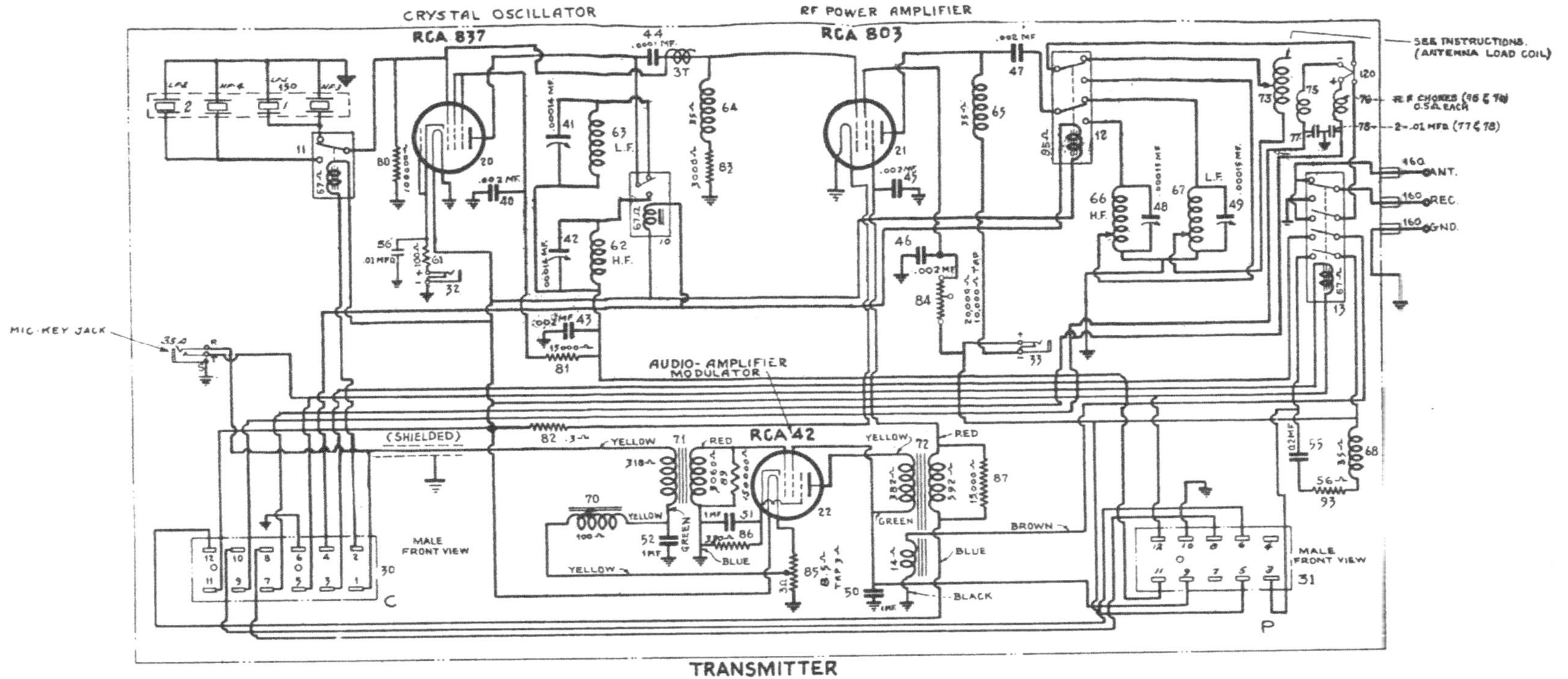

TRANSMITTER

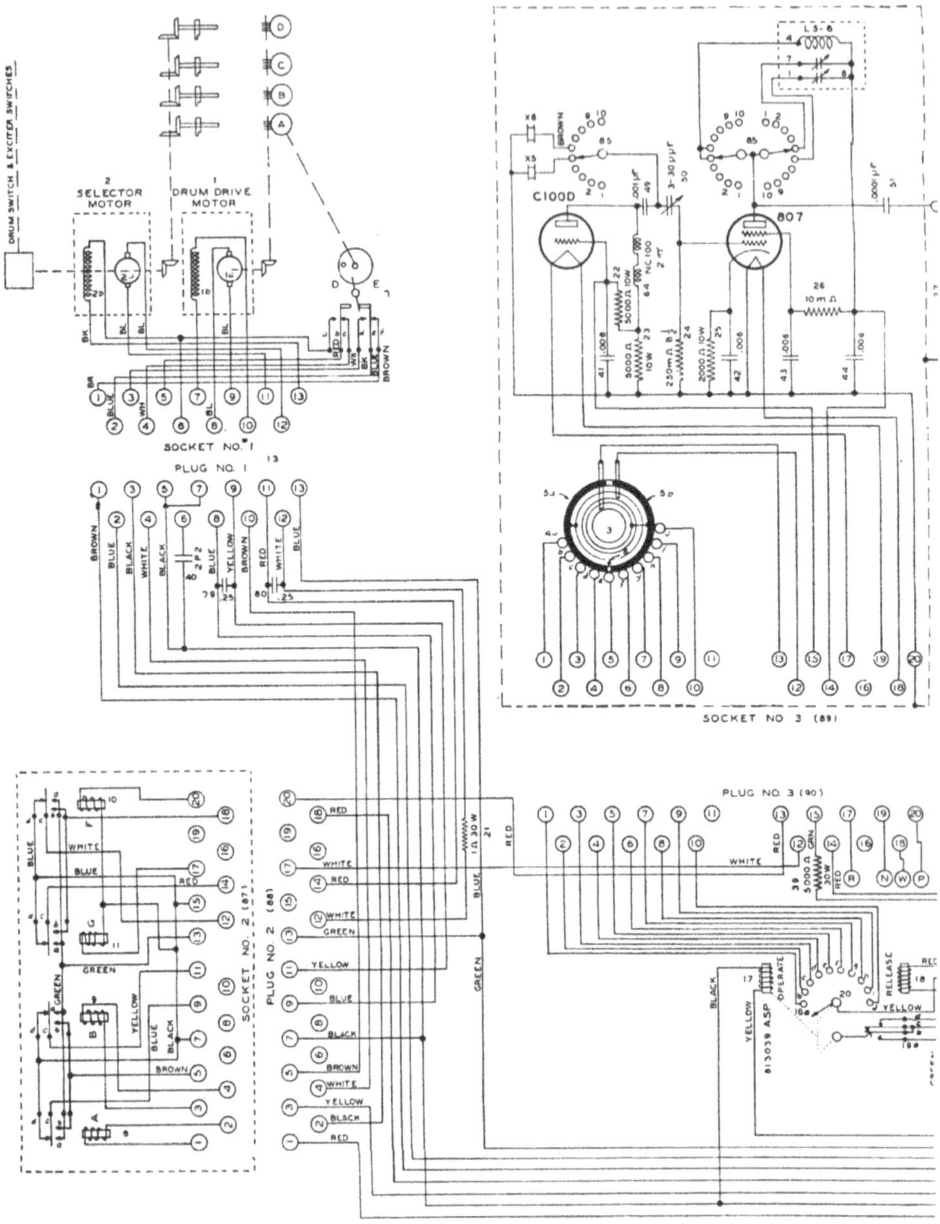

Courtesy Cbs Radio Co.

FIG. 92—COLLINS 17D TRANSMITTER SCHEMATIC

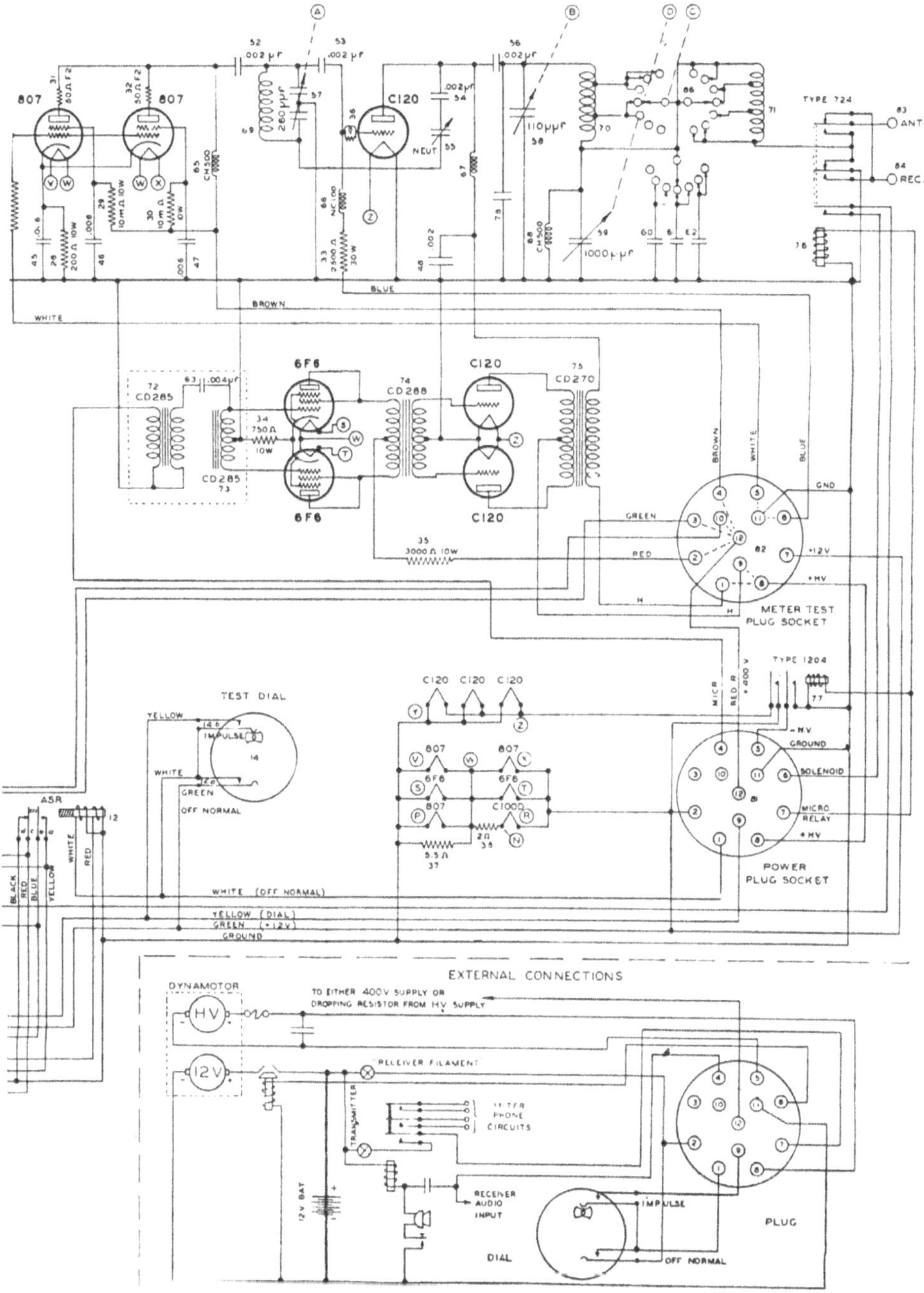

807
807
C120
6F6
6F6
C120
C120
CD285
CD288
CD270
TEST DIAL
IMPULSE
OFF NORMAL
ASR
METER TEST
PLUG SOCKET
POWER
PLUG SOCKET
TYPE 724
ANT.
REC.
TYPE 1204
WHITE (OFF NORMAL)
YELLOW (DIAL)
GREEN (+12V)
GROUND
EXTERNAL CONNECTIONS
DYNAMOTOR
TO EITHER 400V SUPPLY OR
DROPPING RESISTOR FROM HV SUPPLY
RECEIVER FILAMENT
TRANSMITTER
INTER PHONE CIRCUITS
RECEIVER AUDIO INPUT
12 V BAT
DIAL
IMPULSE
OFF NORMAL
PLUG

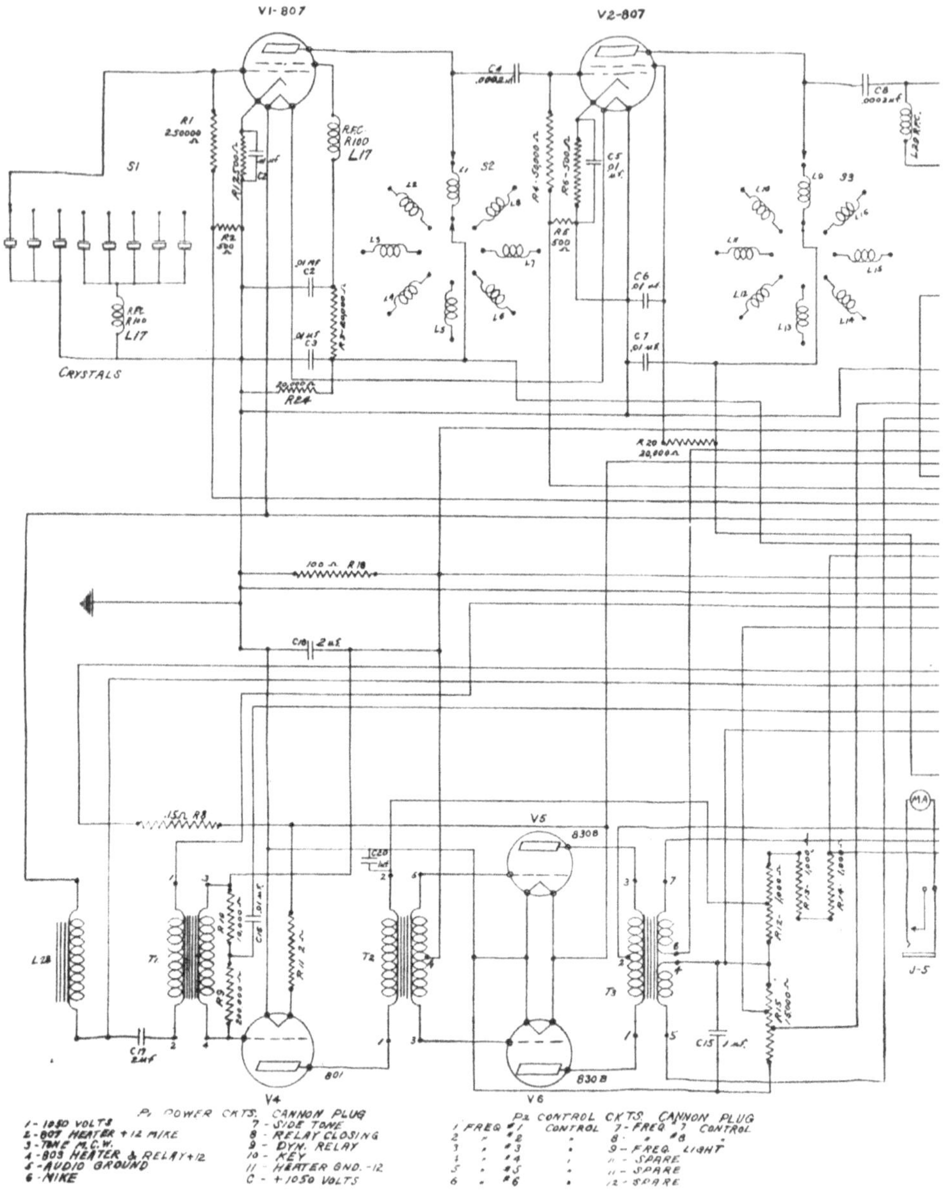

Fig. 96—Bendix TA-2D Transmitter Schematic

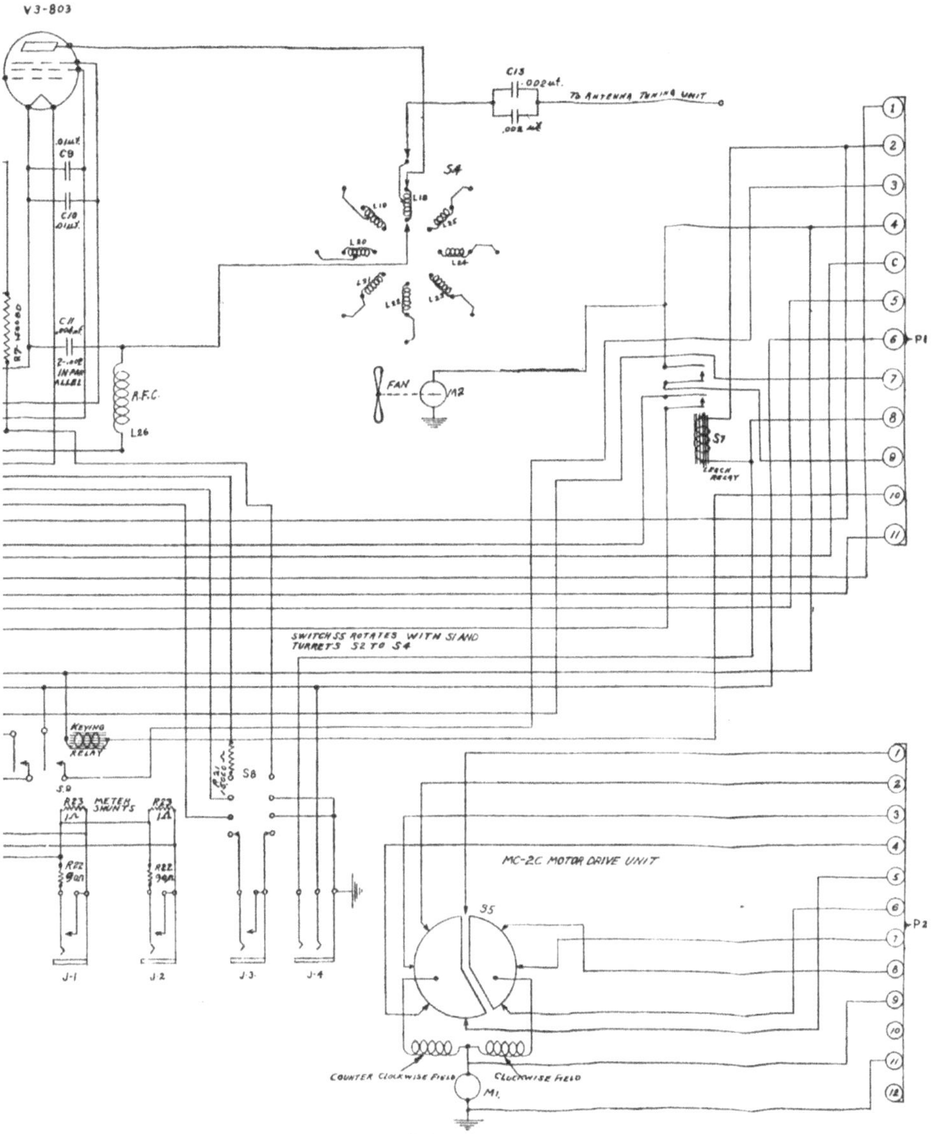

NOTE
APPLYING VOLTAGE TO ANY TERMINAL BETWEEN 1 & 8 INCLUSIVE, ON P2, CAUSES S1 TO S5 INCLUSIVE TO ROTATE TO THE CONTACT WHERE VOLTAGE IS APPLIED, THUS ALLOWING ANY ONE FREQUENCY OR CHANNEL TO BE SELECTED.

Courtesy Bendix Radio Corp.

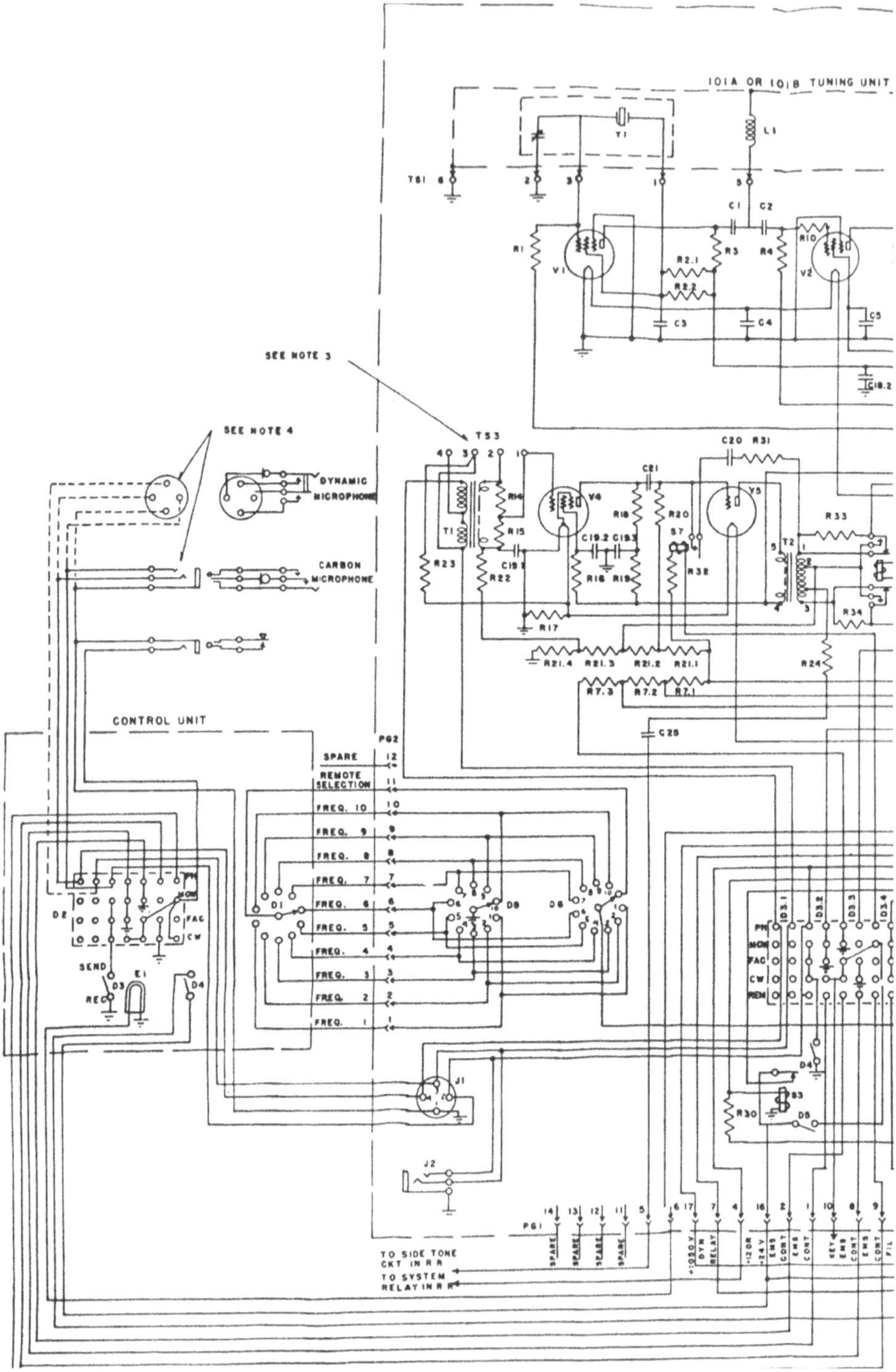

101A OR 101B TUNING UNIT
SEE NOTE 3
SEE NOTE 4
DYNAMIC MICROPHONE
CARBON MICROPHONE
CONTROL UNIT
SPARE
REMOTE SELECTION
FREQ. 10
FREQ. 9
FREQ. 8
FREQ. 7
FREQ. 6
FREQ. 5
FREQ. 4
FREQ. 3
FREQ. 2
FREQ. 1
SEND
REC
TO SIDE TONE CKT IN R R
TO SYSTEM RELAY IN R R

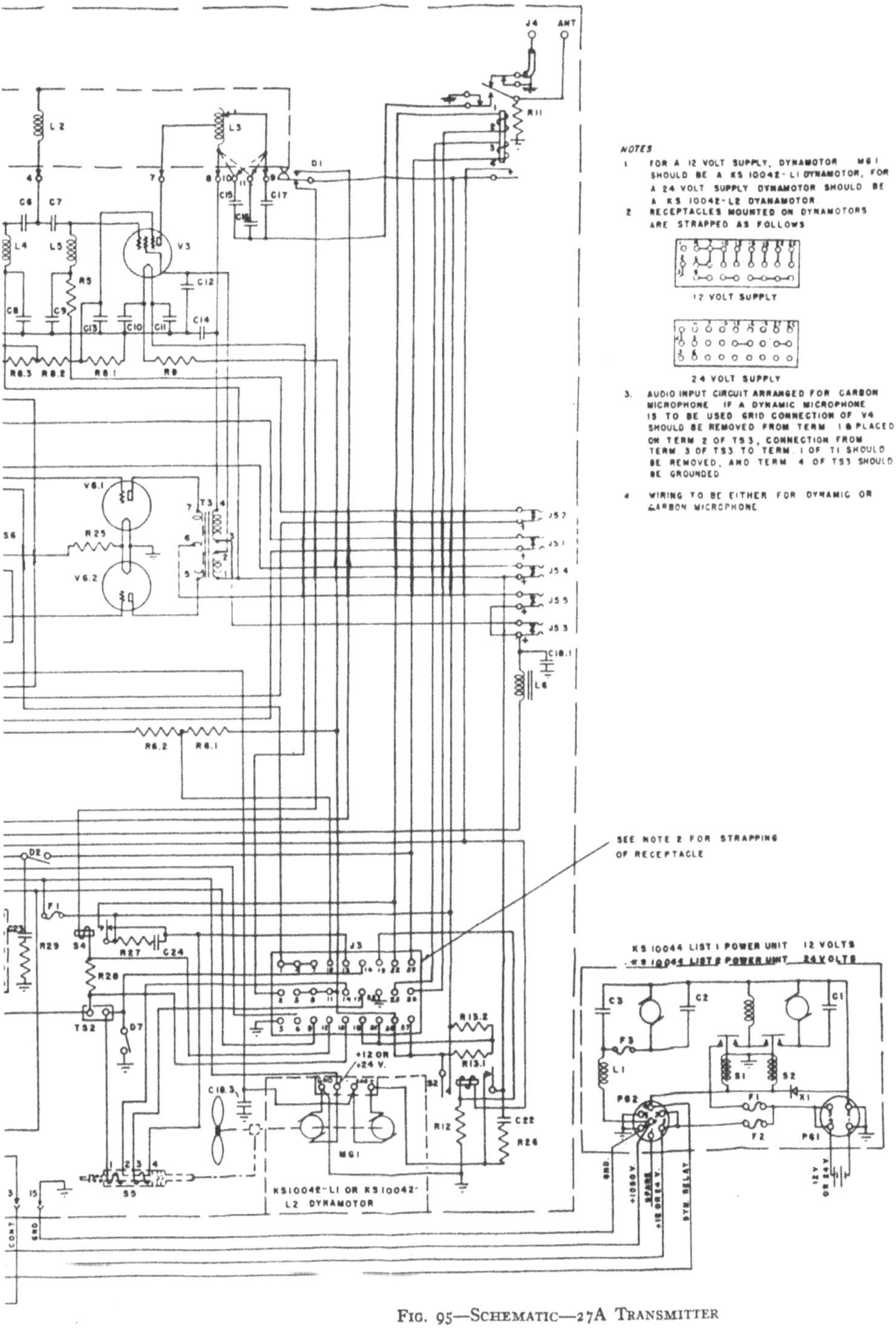

Fig. 95—Schematic—27A Transmitter

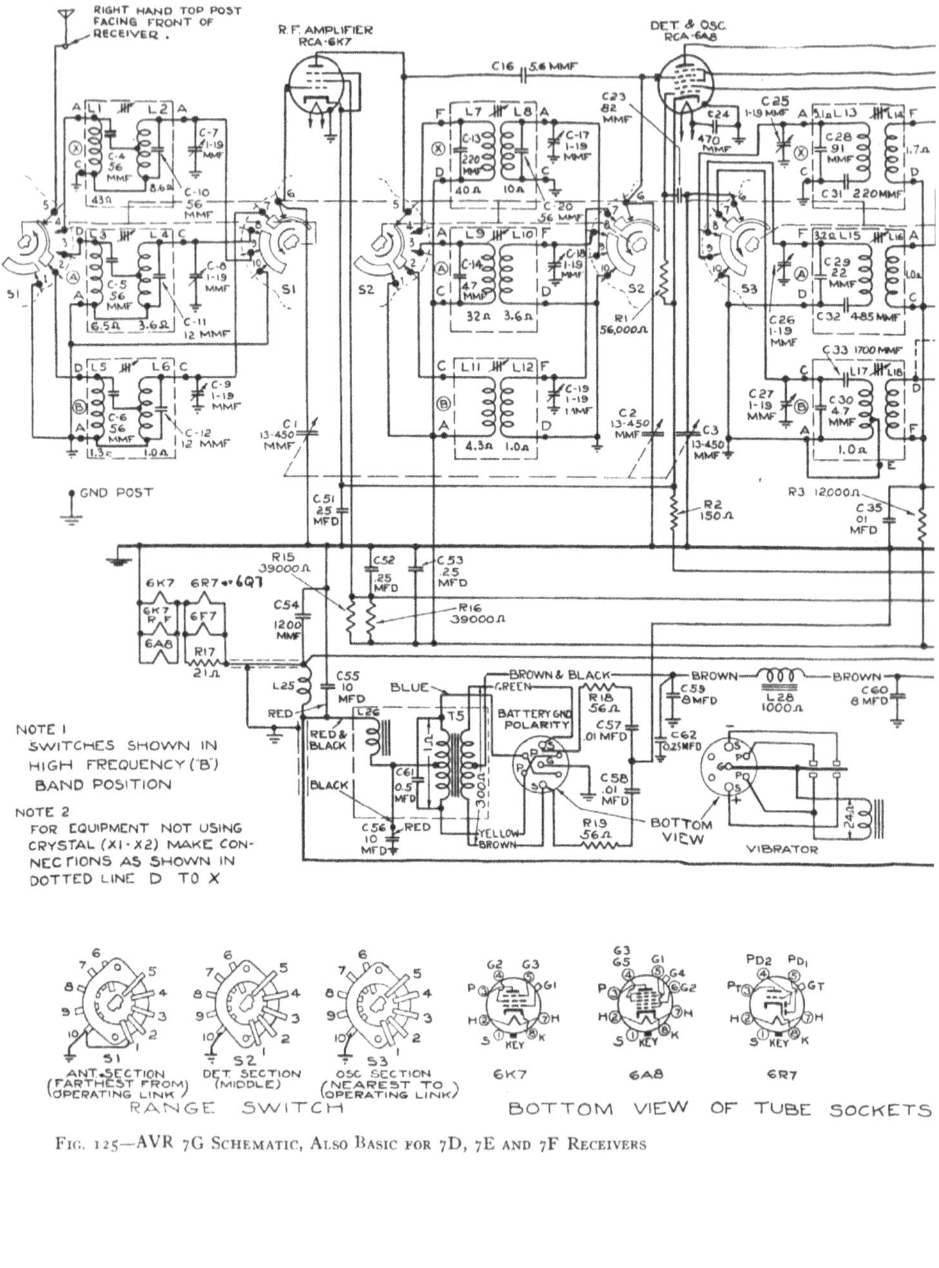

FIG. 125—AVR 7G SCHEMATIC, ALSO BASIC FOR 7D, 7E AND 7F RECEIVERS

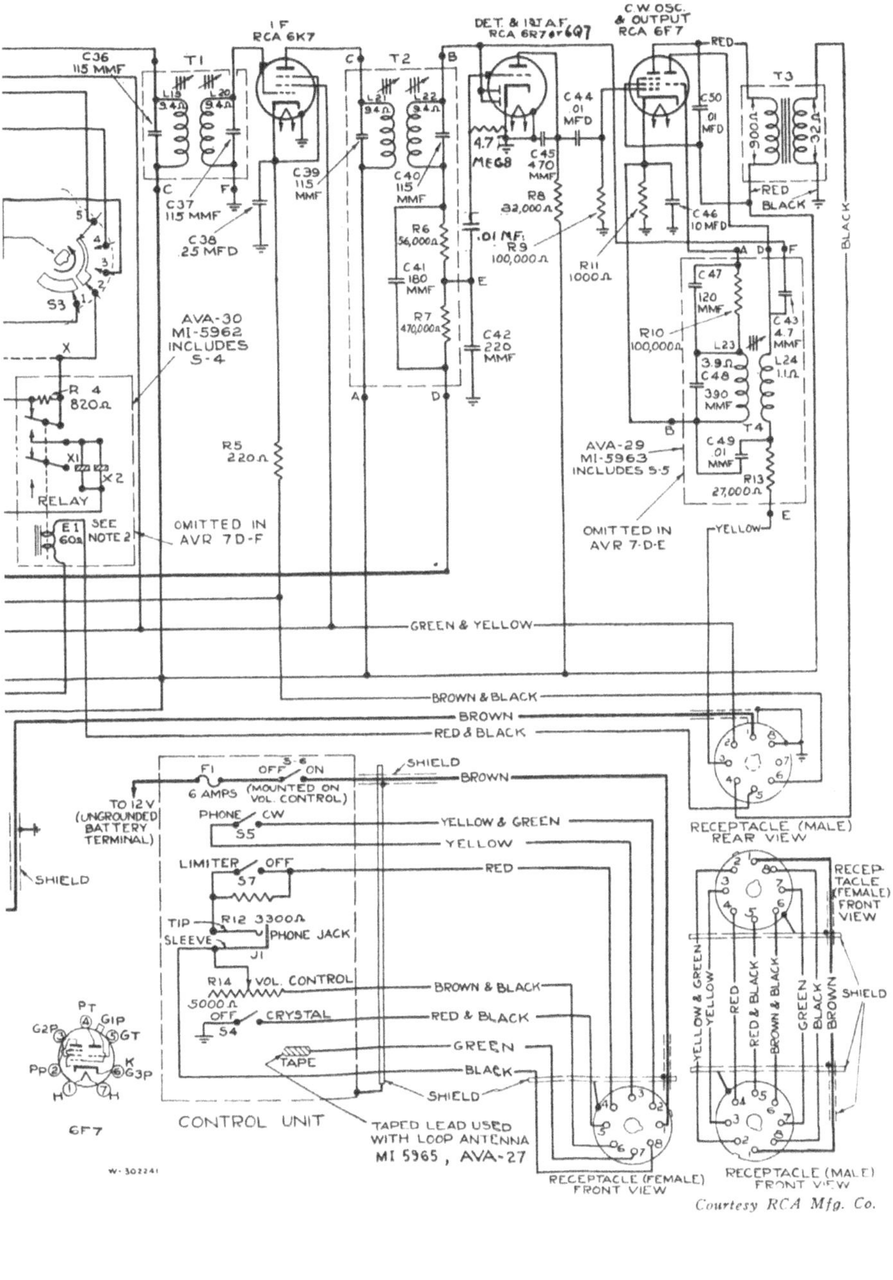

Courtesy RCA Mfg. Co.

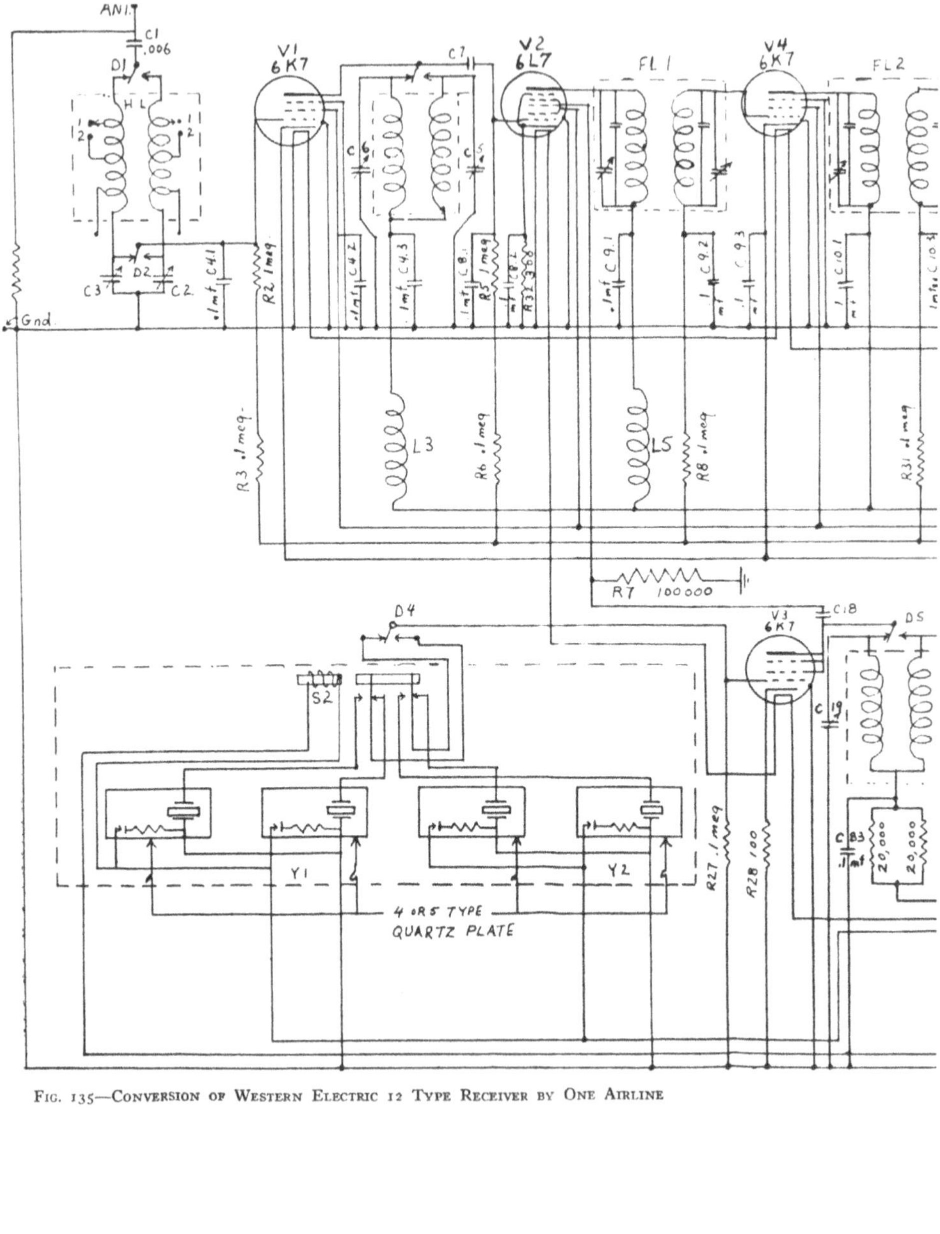

Fig. 135—Conversion of Western Electric 12 Type Receiver by One Airline

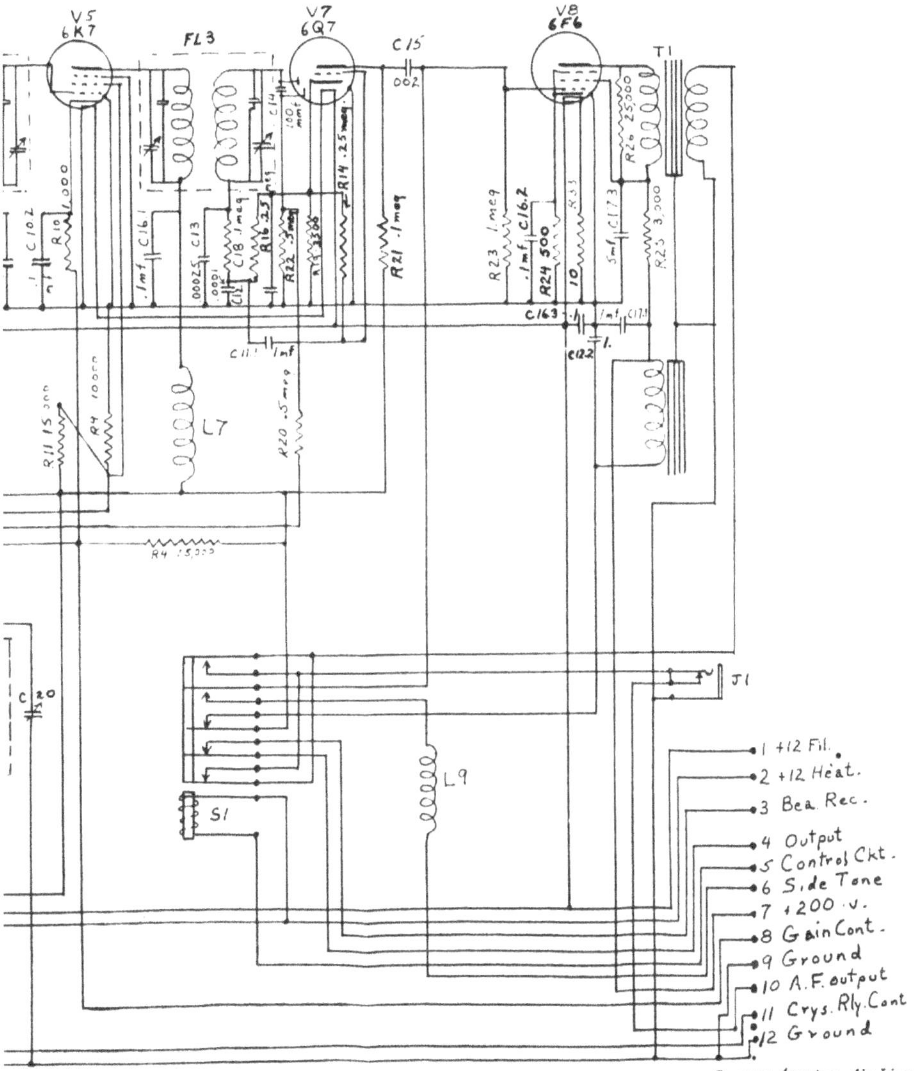

Courtesy American Air Lines

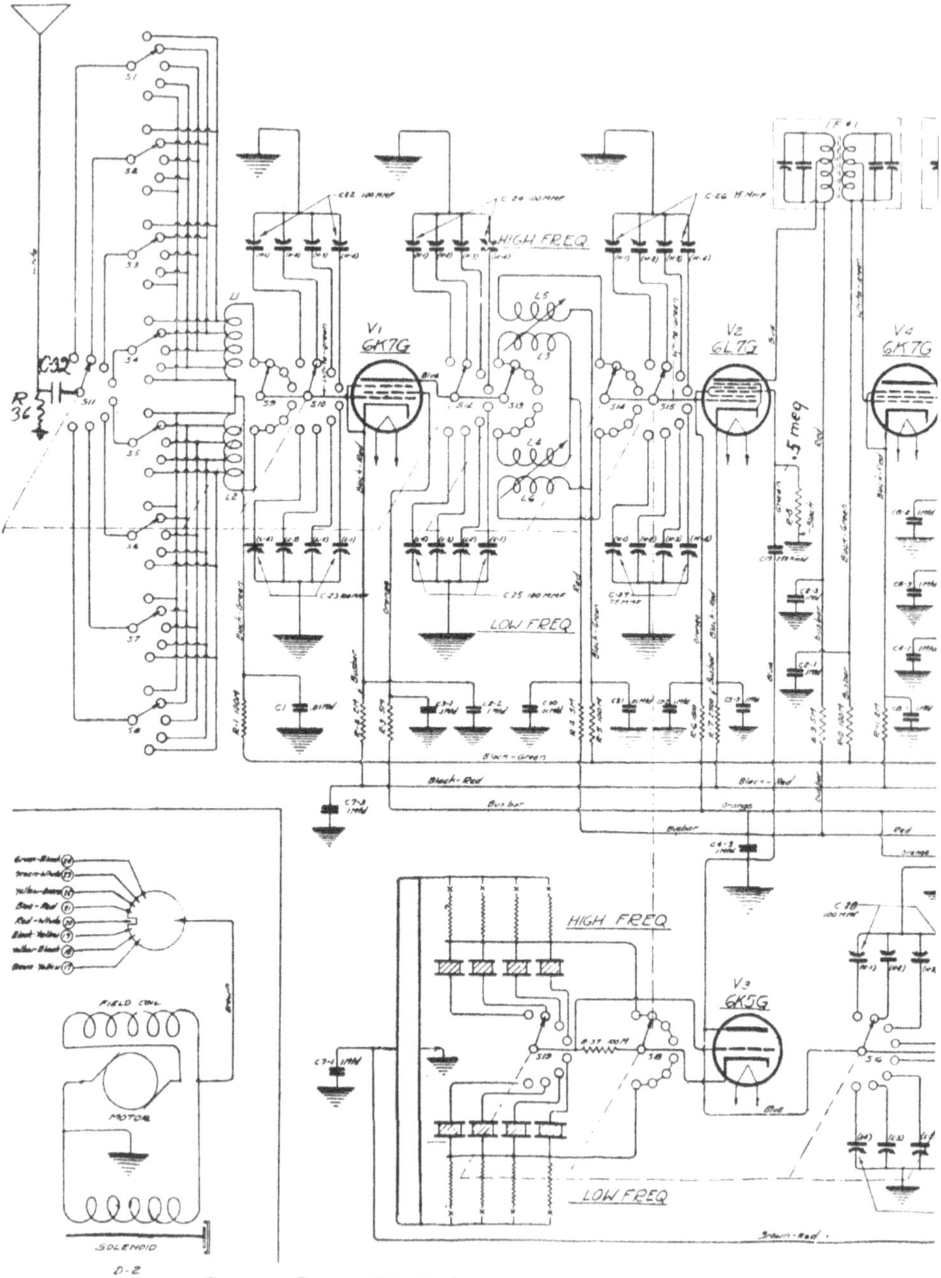

FIG. 145—BENDIX RA-2B RECEIVER SCHEMATIC

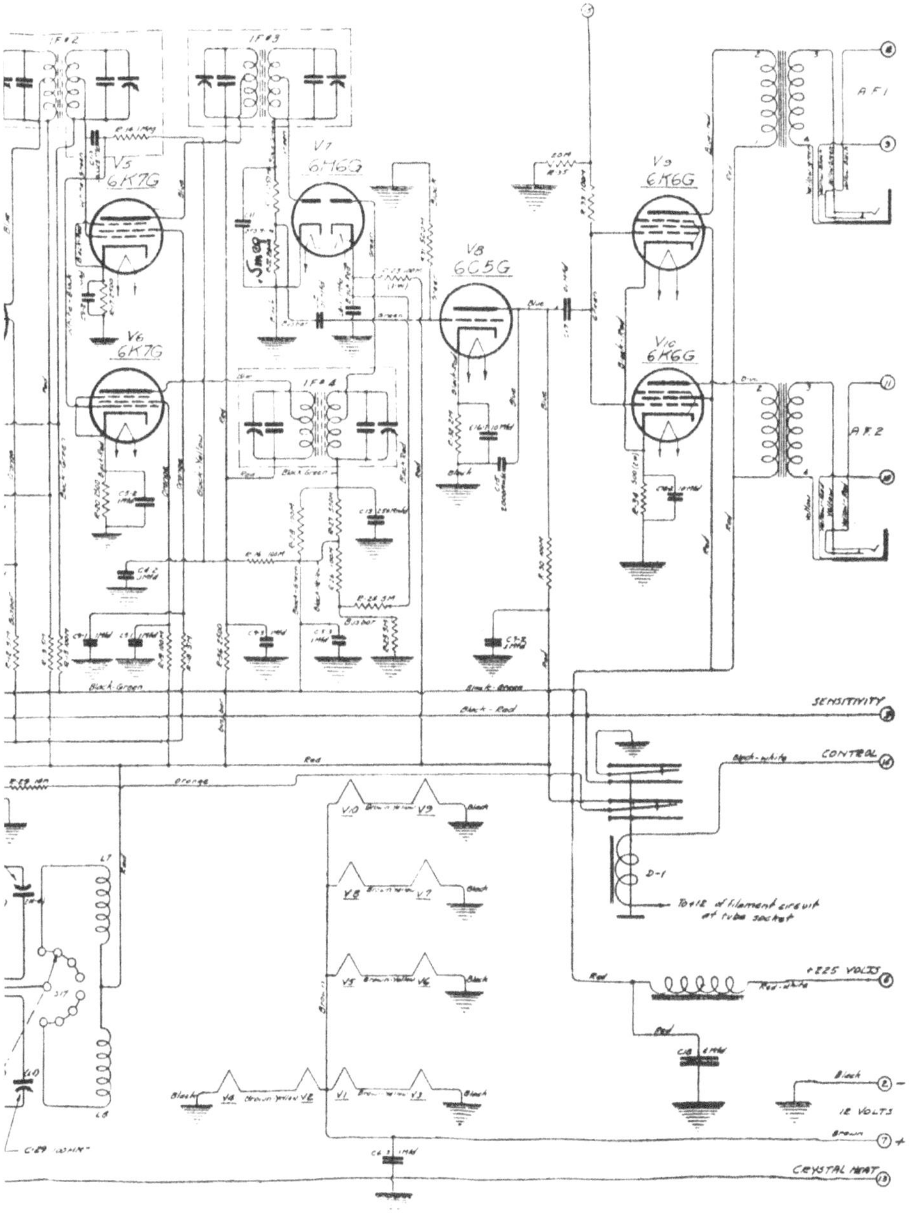

Courtesy Bendix Radio Corp.

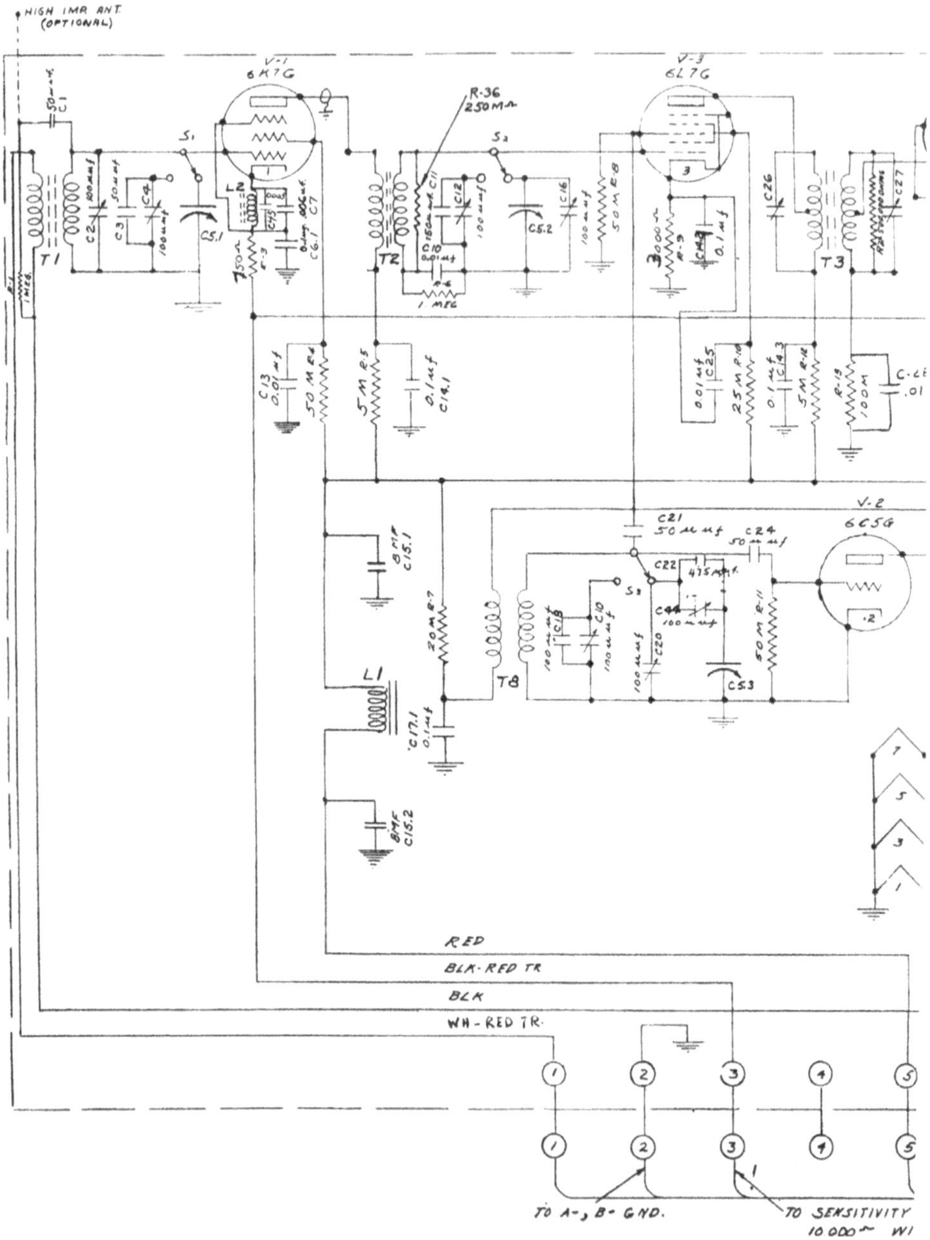

Fig. 148—Schemc RA-4C Receiver

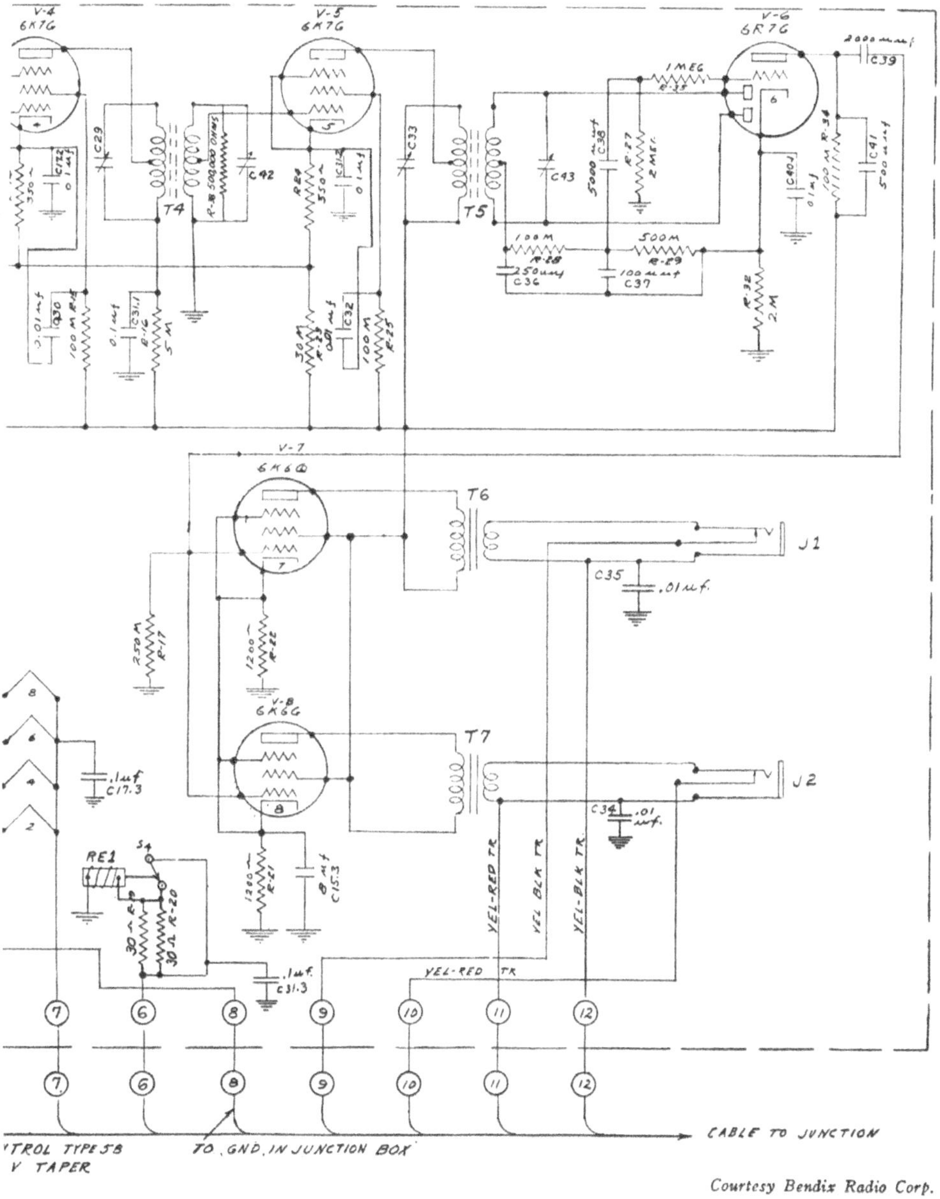

Courtesy Bendix Radio Corp.

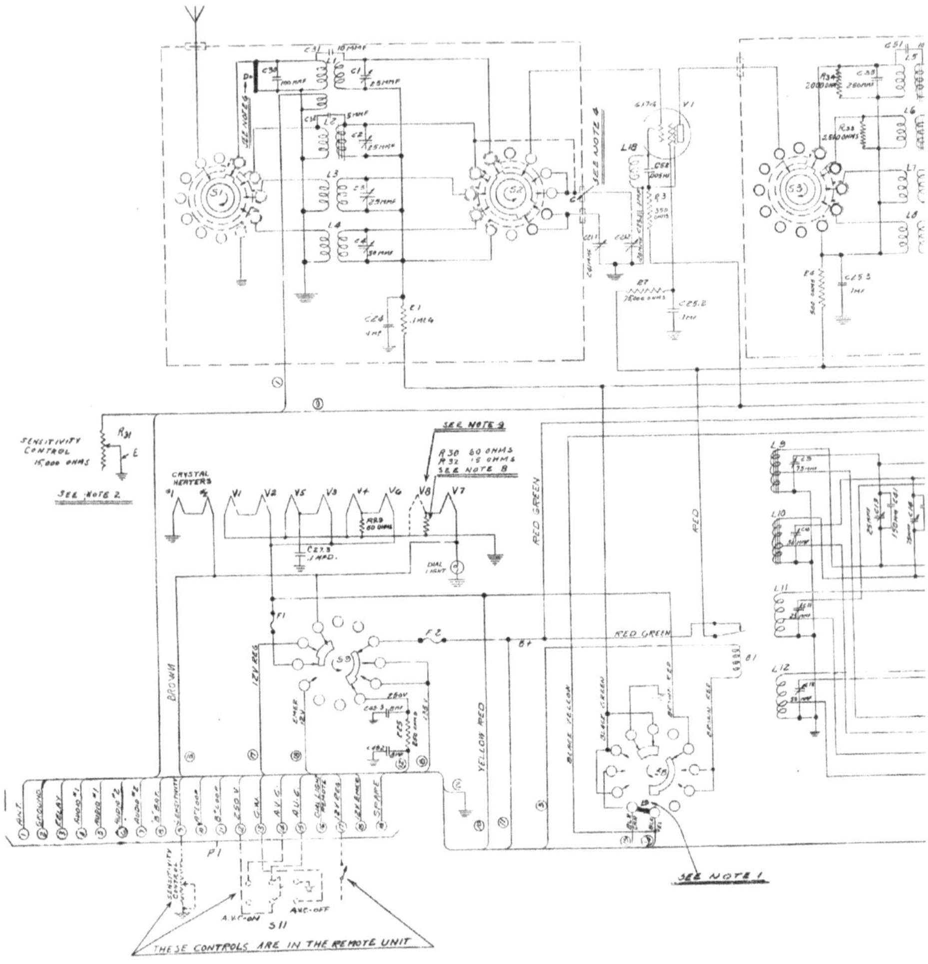

SENSITIVITY CONTROL 15,000 OHMS
R31
SEE NOTE 2
SEE NOTE 9
R30 60 OHMS
R32 15 OHMS
SEE NOTE 8
CRYSTAL HEATERS
V1
V2
V5
V3
V4
V6
V8
V7
C27.3 .1 MFD.
DIAL LIGHT
F1
F2
S9
B+
RED GREEN
RED
BROWN
12V REG
YELLOW RED
SEE NOTE 4
SEE NOTE 1
S1
S2
S3
L1
L2
L3
L4
L5
L6
L7
L8
L9
L10
L11
L12
L18
P1
S11
A.V.C-ON
A.V.C-OFF
THESE CONTROLS ARE IN THE REMOTE UNIT

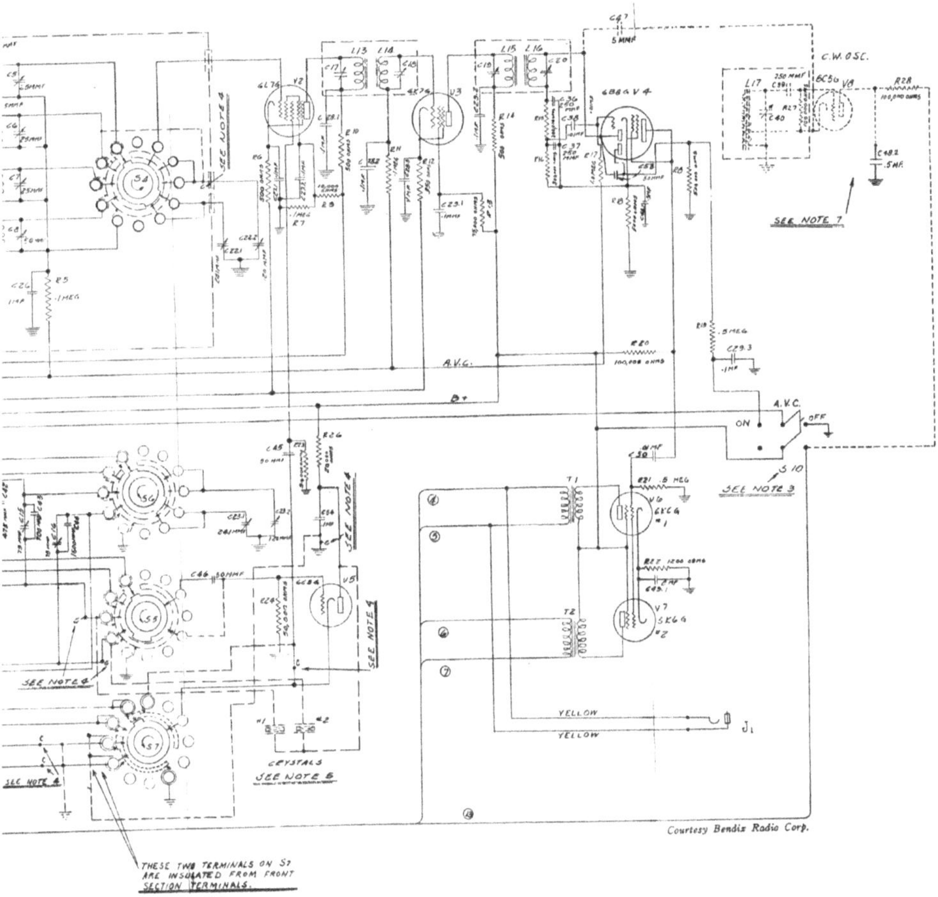

Courtesy Bendix Radio Corp.

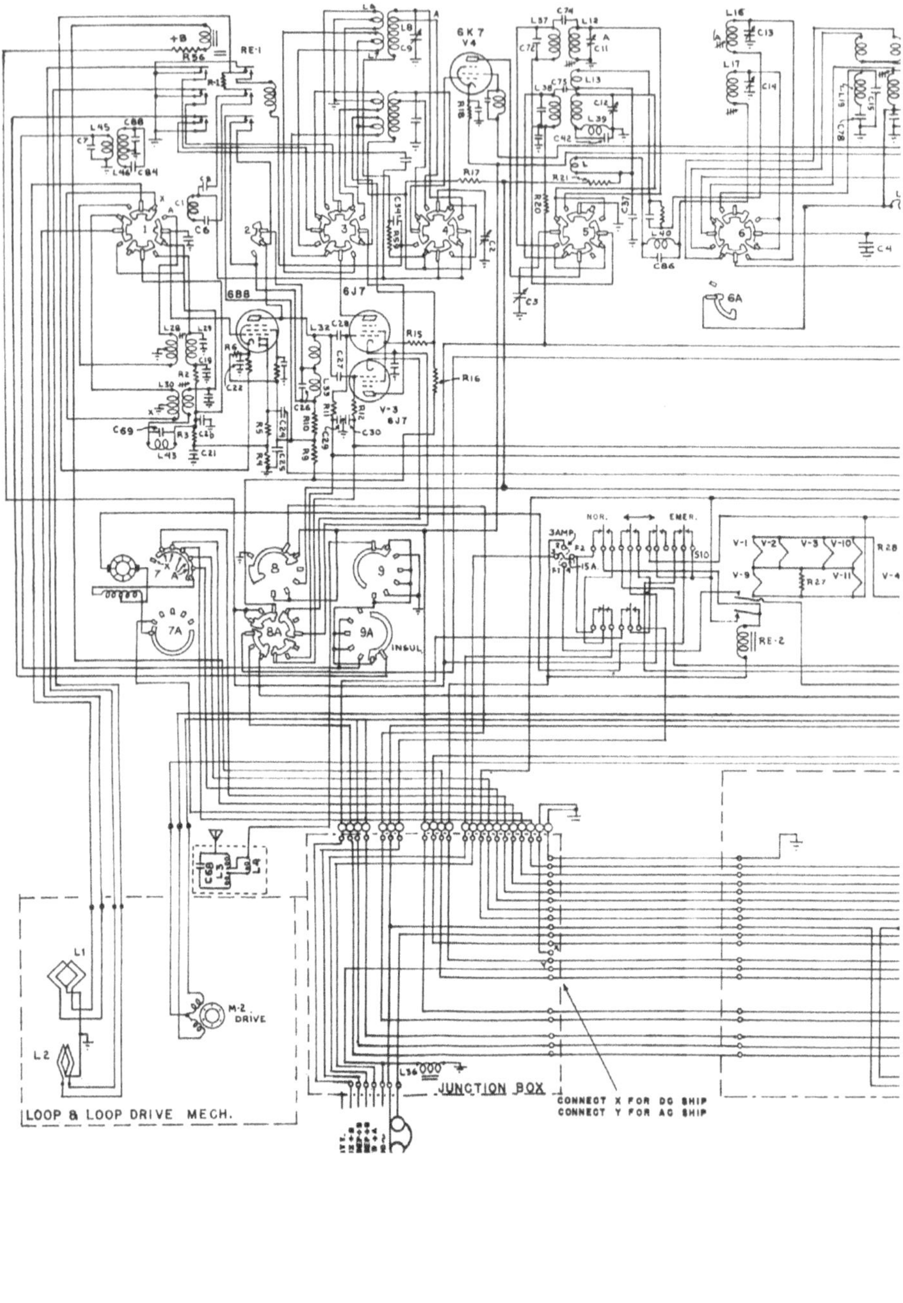

RE-1
R56
+B
R-1
L45
C88
C7
L46
C84
6K7
V4
R18
L8
C9
L37
C74
L12
C72
C11
L13
L38
C75
C12
L39
C42
R21
R20
C37
L40
C86
L16
C13
L17
C14
L19
C78
C15
C4
R17
C34
R59
C2
6A
C3
6B8
6J7
L32
C28
R15
C27
R16
L28
L29
R6
C19
C22
R2
L30
L33
C26
C24
R5
R10
R11
R12
V-3
6J7
C30
C69
C20
R3
L43
C21
R4
C25
R9
C29
NOR.
EMER.
3AMP
15A.
S10
V-1
V-2
V-3
V-10
R28
V-9
R27
V-11
V-4
RE-2
7A
8A
9A
INSUL.
C68
L3
L4
L1
L2
M-2
DRIVE
L36
JUNCTION BOX
CONNECT X FOR DC SHIP
CONNECT Y FOR AC SHIP
LOOP & LOOP DRIVE MECH.

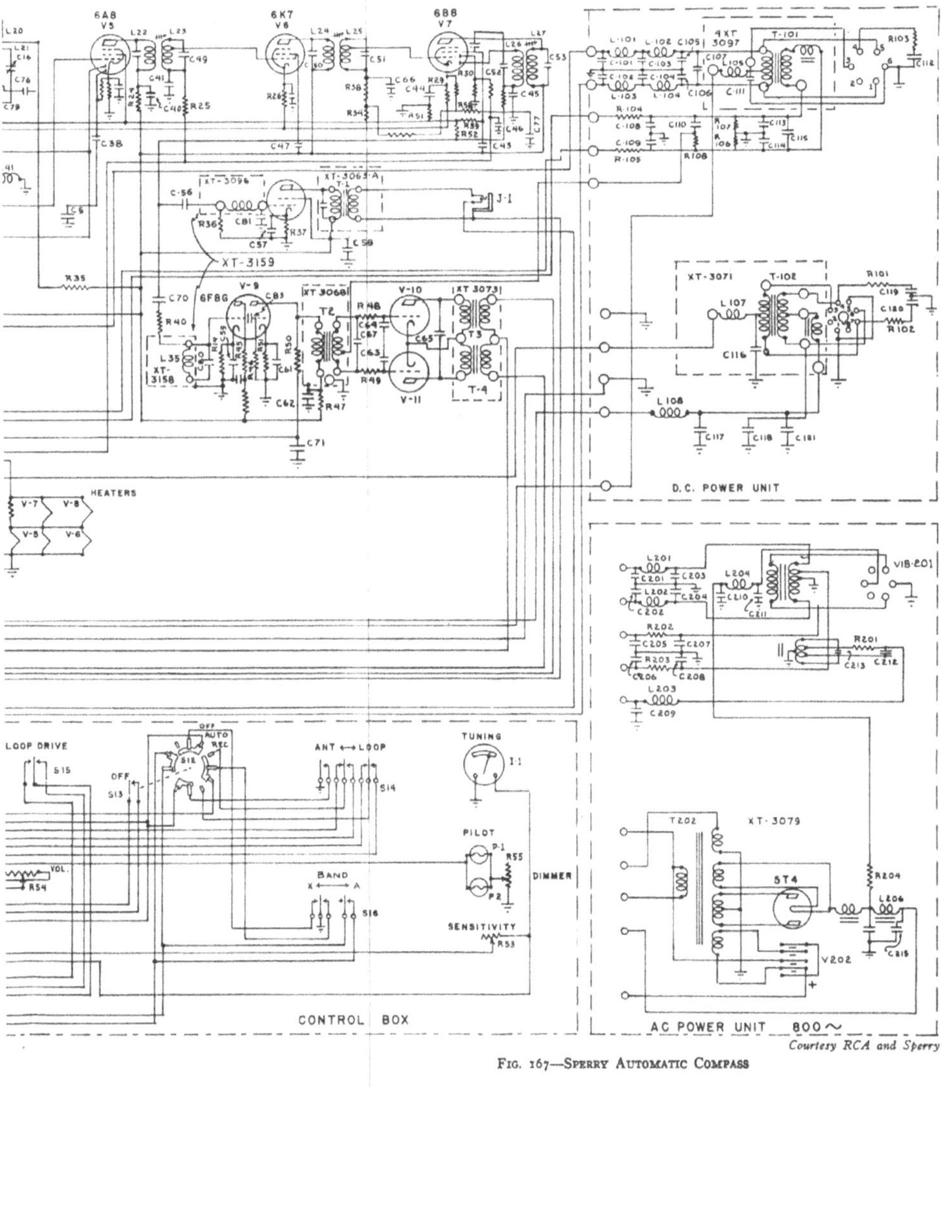

Courtesy RCA and Sperry

Fig. 167—Sperry Automatic Compass

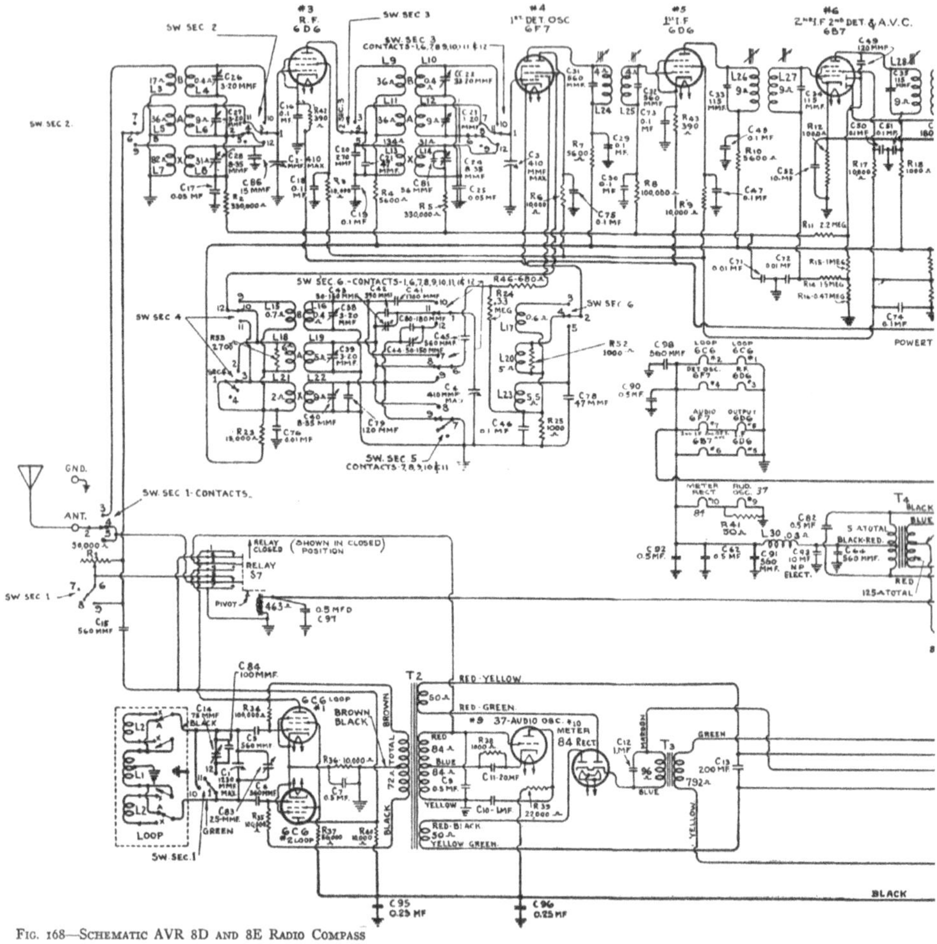

FIG. 168—SCHEMATIC AVR 8D AND 8E RADIO COMPASS

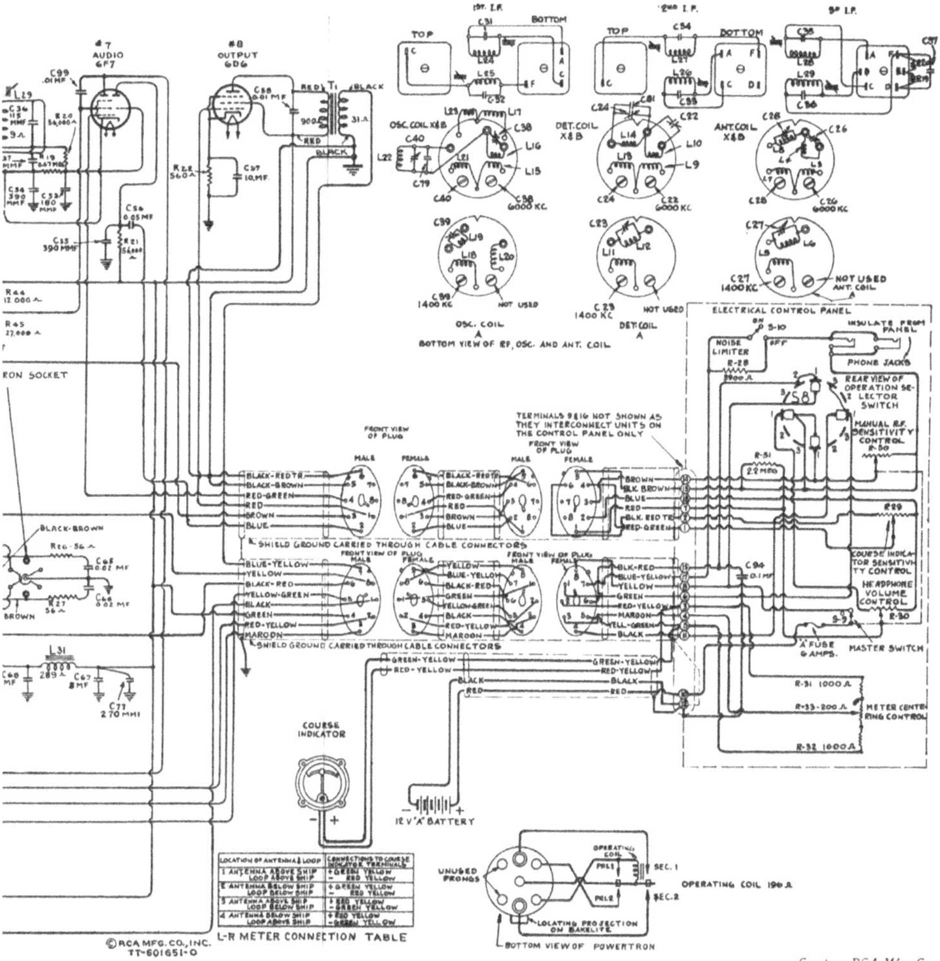

Courtesy RCA Mfg. Co.

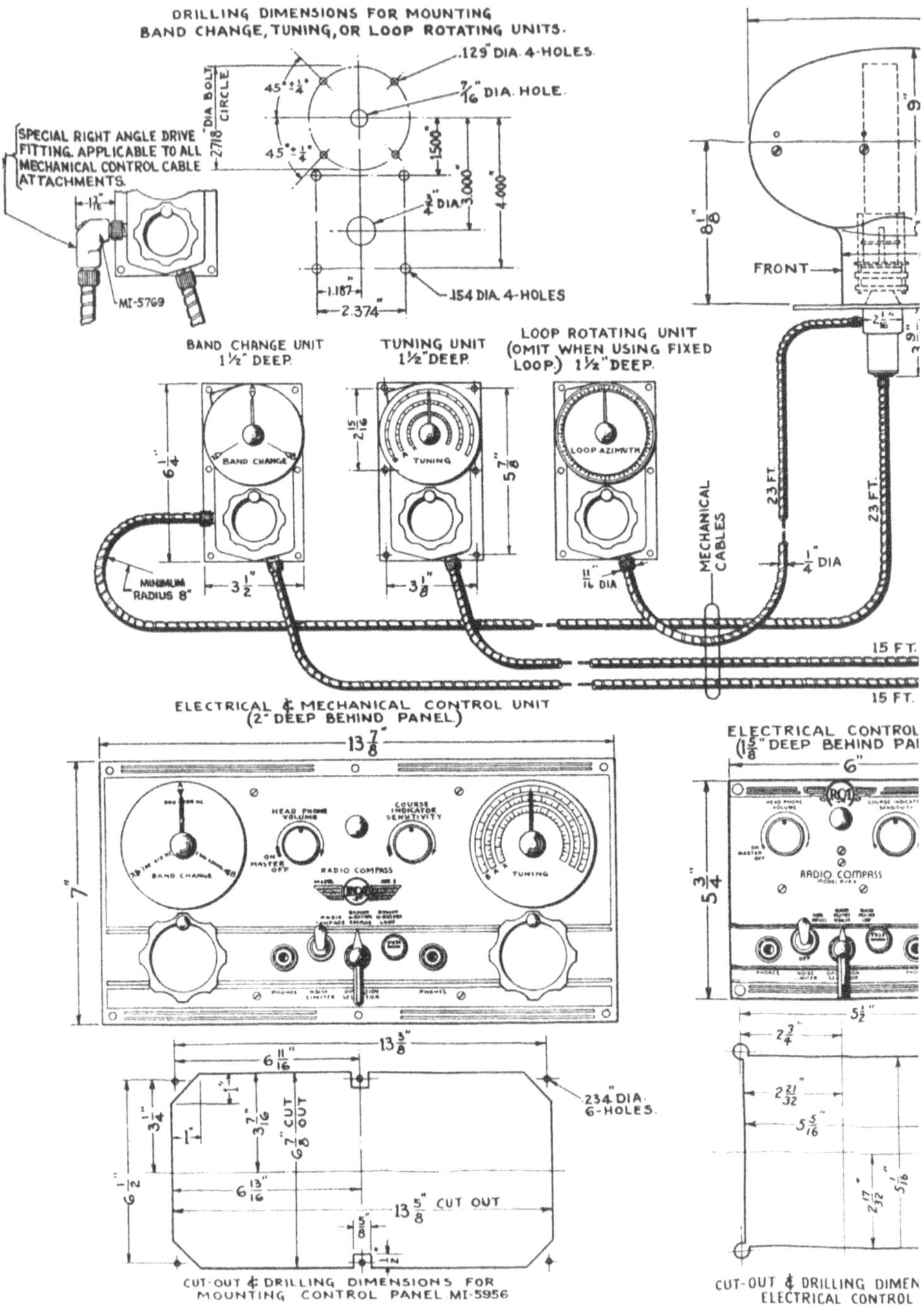
DRILLING DIMENSIONS FOR MOUNTING
BAND CHANGE, TUNING, OR LOOP ROTATING UNITS.
.129" DIA. 4-HOLES.
7/16" DIA. HOLE.
2.718" DIA BOLT CIRCLE
45° ± 1/4°
45° ± 1/4°
1.500"
3.000"
4.000"
1 3/4" DIA.
1.187"
2.374"
.154 DIA. 4-HOLES
SPECIAL RIGHT ANGLE DRIVE FITTING. APPLICABLE TO ALL MECHANICAL CONTROL CABLE ATTACHMENTS.
1 7/16"
MI-5769
9"
8 1/8"
FRONT
2 1/16"
BAND CHANGE UNIT
1½" DEEP.
TUNING UNIT
1½" DEEP.
LOOP ROTATING UNIT
(OMIT WHEN USING FIXED LOOP.) 1½" DEEP.
6 1/4"
BAND CHANGE
2 15/16
TUNING
5 7/8"
LOOP AZIMUTH
MINIMUM RADIUS 8"
3 1/2"
3 1/8"
11/16 DIA.
MECHANICAL CABLES
23 FT.
23 FT.
1/4" DIA
15 FT.
15 FT.
ELECTRICAL & MECHANICAL CONTROL UNIT
(2" DEEP BEHIND PANEL.)
13 7/8"
7"
HEAD PHONE VOLUME
COURSE INDICATOR SENSITIVITY
ON MASTER OFF
RADIO COMPASS
BAND CHANGE
TUNING
PHONES
PHONES
ELECTRICAL CONTROL
(1 5/8" DEEP BEHIND PA
6"
5 3/4"
RADIO COMPASS
5 1/2"
13 3/8"
6 11/16"
.234 DIA. 6-HOLES.
3 1/4"
1"
3 7/16"
6 7/8" CUT OUT
6 1/2"
6 13/16"
13 5/8" CUT OUT
5/8"
2 3/4"
2 21/32"
5 5/16"
2 17/32"
5 1/16"
CUT-OUT & DRILLING DIMENSIONS FOR MOUNTING CONTROL PANEL MI-5956
CUT-OUT & DRILLING DIMEN
ELECTRICAL CONTROL

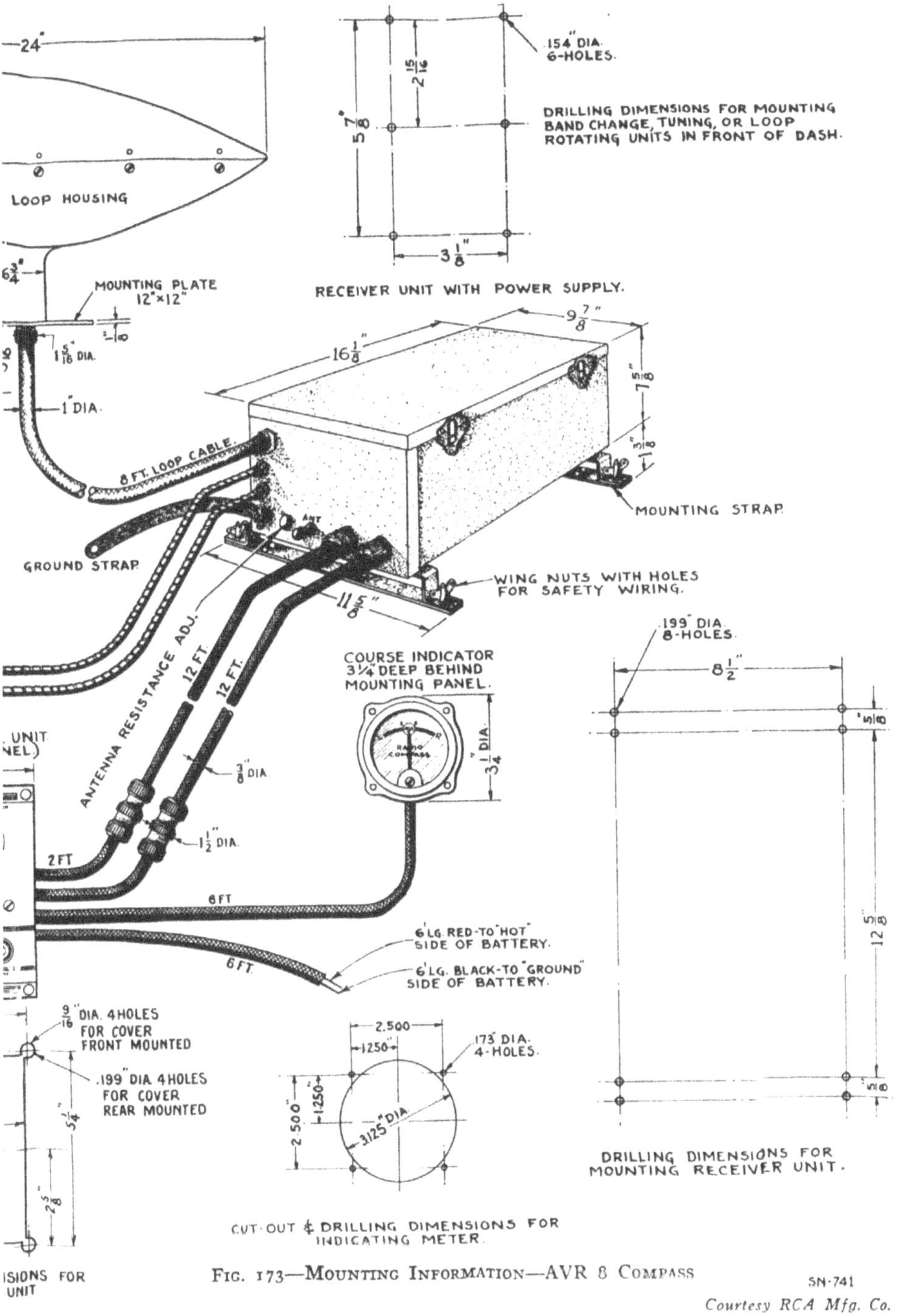

FIG. 173—MOUNTING INFORMATION—AVR 8 COMPASS

Courtesy RCA Mfg. Co.

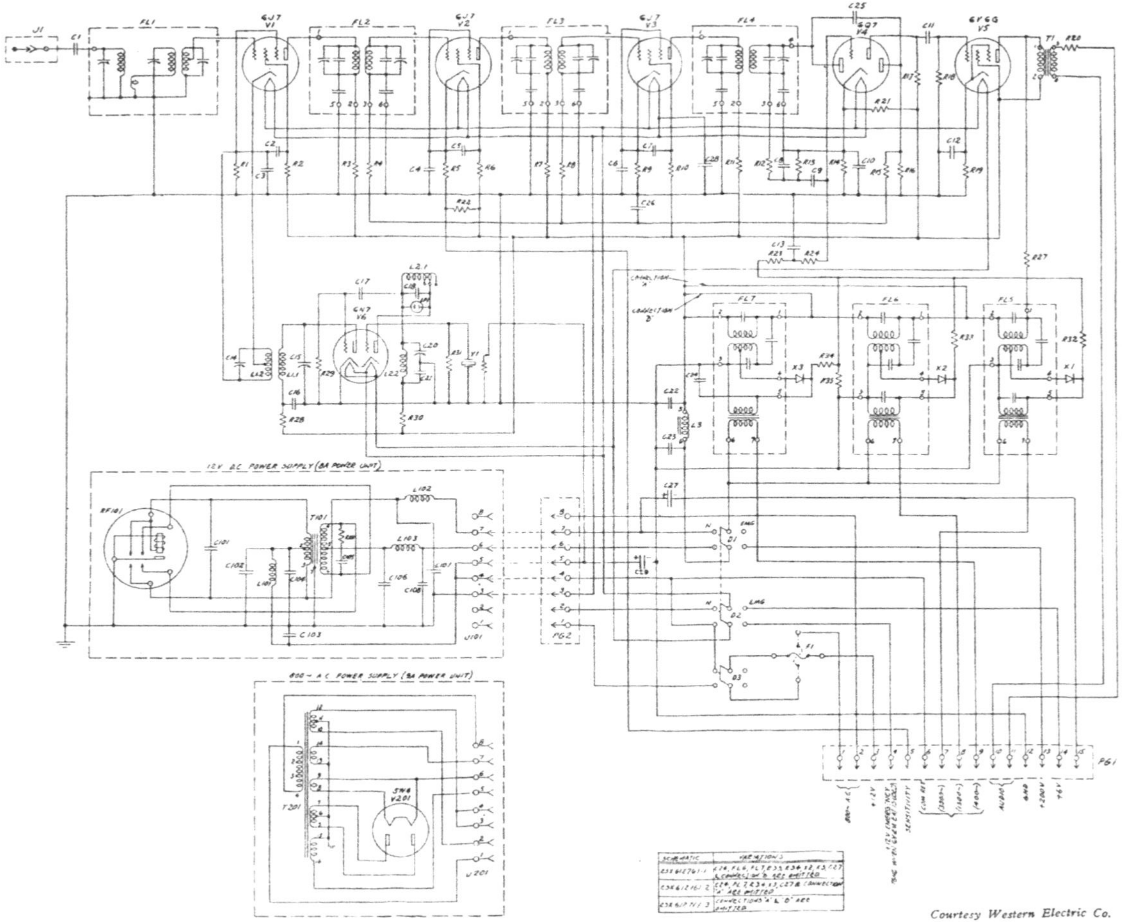

Courtesy Western Electric Co.

Fig. 174—Western Electric 27A Marker Receiver

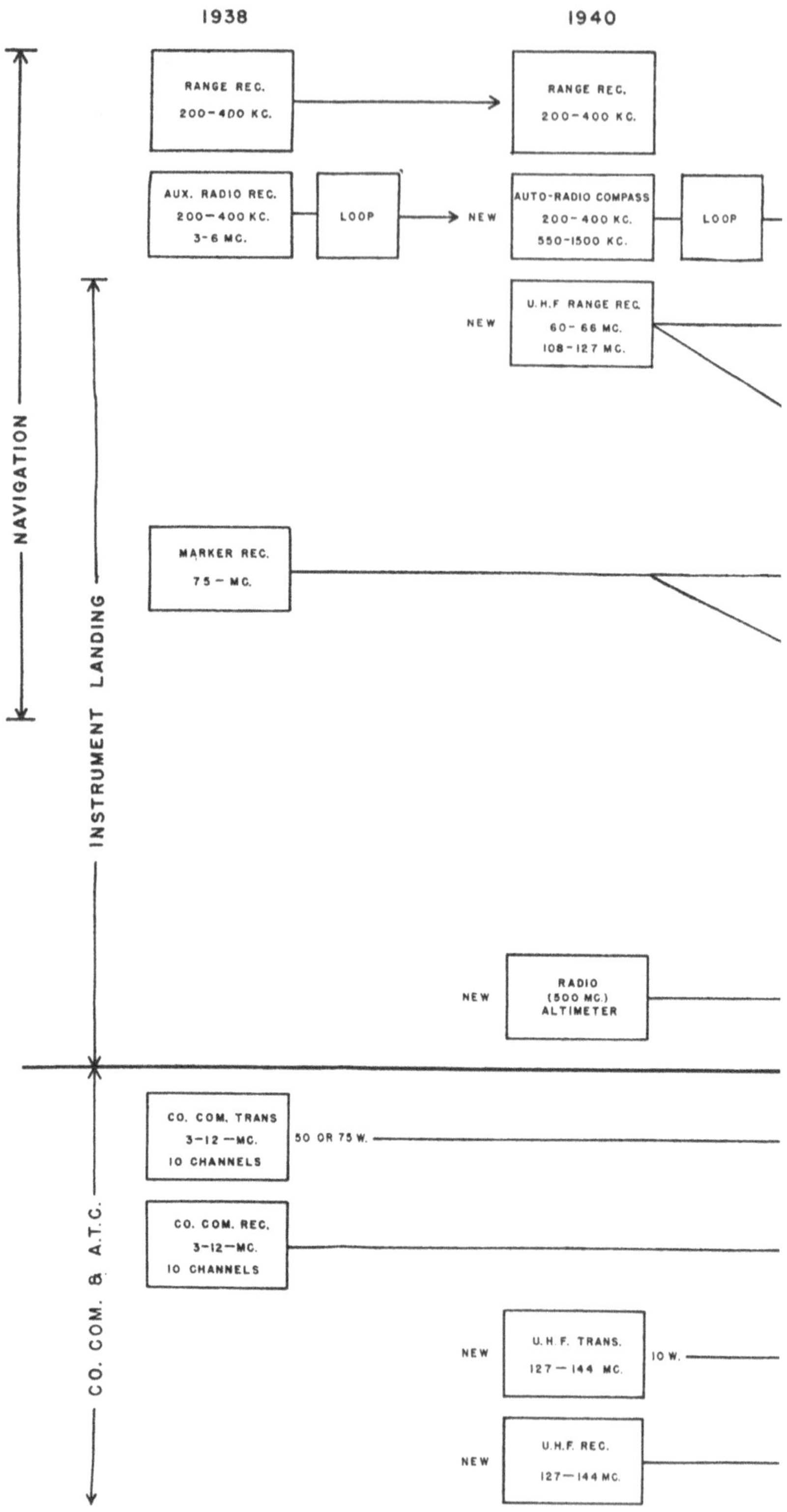

FIG. 209—AIR TRANSPORT RADIO EQUIPMENT—ONE POSSIBILITY IN

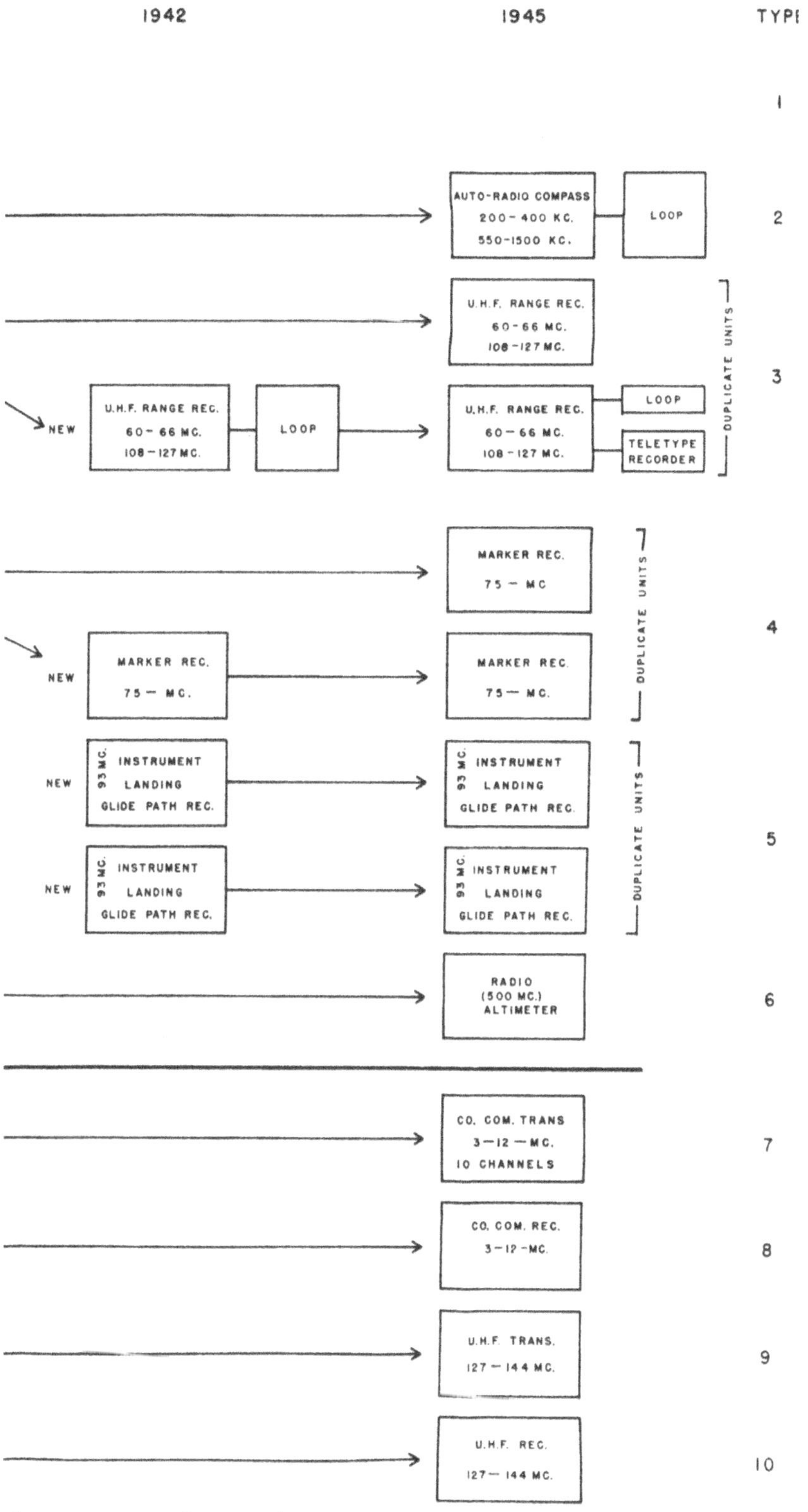

TRANSITION FROM LOW AND MEDIUM TO ULTRA HIGH FREQUENCIES

Made in United States
Cleveland, OH
04 November 2026

47642853R00230